PHYSICS RESEARCH AND TECHNOLOGY

OPTICAL AMPLIFIERS

PHYSICS RESEARCH AND TECHNOLOGY

Additional books in this series can be found on Nova's website
under the Series tab.

Additional E-books in this series can be found on Nova's website
under the E-books tab.

PHYSICS RESEARCH AND TECHNOLOGY

OPTICAL AMPLIFIERS

GALINA NEMOVA
EDITOR

Nova Science Publishers, Inc.
New York

For permission to use material from this book please contact us:
Telephone 631-231-7269; Fax 631-231-8175
Web Site: http://www.novapublishers.com

Library of Congress Cataloging-in-Publication Data
Optical amplifiers / editor, Galina Nemova.
p. cm.
Includes bibliographical references and index.
ISBN 978-1-61209-835-7 (softcover)
1. Optical amplifiers. I. Nemova, Galina.
TK8360.L48O68 2011
621.36'94--dc22

2011005835

Published by Nova Science Publishers, Inc. † New York

CONTENTS

PREFACE

An optical amplifier is a device which receives some input optical signal and generates an output optical signal with higher power without the conversion to an electrical signal during the intermediate stages. The amplification process takes place in a so-called gain medium. Most optical amplifiers are laser amplifiers. Their gain media are either insulators doped with some laser-active ions or semiconductors. The amplification process in such amplifiers is based on stimulated emission. Moreover, various types of optical nonlinearities can be used for optical amplification in addition to stimulated emission. This book presents leading edge research on optical amplifiers from researchers spanning the globe.

Chapter 1 - Programmable logic unit (PLU) is more attractive than different separate logical circuits because, any user can reprogram such device as per his requirement. PLU consist of a logical AND array followed by a logical OR array. This array can be easily designed using all-optical interferometric switches. In very recent years, the discovery of ultra high-speed all-optical *s*emiconductor *o*ptical *a*mplifier (SOA) based switches has brought the revolution in all-optical information processing systems. SOA is similar to a semiconductor laser. Its gain is higher ~ 20 to 30 dB compared to that of laser. The optical gain is caused by the recombination of electron holes. Many all-optical switches based on cross phase modulation (XPM) such as *T*erahertz *O*ptical *A*symmetric *D*emultiplexer (TOAD), *M*ach-Zehnder *i*nterferometer (MZI), *U*ltrafast *N*onlinear *I*nterferometer (UNI) etc have already been established. In an interferometric switch, there is an incoming signal (IS) which can be directed according to the *c*ontrol *s*ignal (CS). If CS is 'ON' then IS transmitted through the transmitted port or bar-port (\\), otherwise it is directed to the reflected port or cross-port ($\times$). Hence CS may be used as programming input of all-optical PLU. 16-Boolean logical operations from this unit can be designed, only by changing programming inputs.

Chapter 2 - Passive optical network (PON) is considered as cost-effective optical access network architecture, and it has already been deployed all over the world. As the bandwidth demand continues to grow, optically amplified long-reach (LR) PON and wavelength division multiplexed (WDM) PON are currently under intensive studies for the next generation access networks providing higher capacity. Semiconductor optical amplifier (SOA) is a promising photonic device for these next generation access networks as a gain block to increase the power budget or as a colorless modulator in the WDM-PON. In this chapter, the author describes the basic principle and the characteristics of the SOA, and presents some integrated devices using SOAs for the next generation access networks. The author also demonstrates the applications of reflective SOA (RSOA) as the colorless optical networking unit (ONU) for the next generation signal remodulation WDM-PON and signal remodulation LR-PON.

Chapter 3 - The semiconductor optical amplifier (SOA) is a versatile device that has found application at the core of all–optical signal processing sub–systems. These sophisticated structures are expected to replace current electronic data processing elements in next–generation reconfigurable photonic telecommunication networks. Besides, there has also been a renewed interest to use SOAs to cost–effectively extend the power budget margins of current metro and access optical networks. However, in contrast to Er–doped fiber amplifiers, whose excited level has a long lifetime of about 10 ms, SOAs exhibit carrier lifetimes in the order of picoseconds and a nonlinear response that is orders of magnitude higher as compared to optical fibers. Therefore, these minute active devices normally become affected by signals modulated at Gb/s rates. As a result of this characteristic, a clear understanding of the nonlinear dynamical behavior of the light–matter interactions that take place within active semiconductor waveguides is necessary. This helps in the design and engineering of optical systems where SOAs play a significant role. In this contribution the author first reviews some of the most relevant nonlinear dynamical phenomena found in SOAs. Then, a discussion is presented on how the finite response time associated to these phenomena affects some applications employed by the optical telecommunications industry, such as in–line amplification and wavelength conversion. Finally, using a well–tested photonic circuits and systems simulator, novel solutions to enhance the performance of the aforementioned SOA–based applications are demonstrated. Narrow–band optical filtering and the combination of several devices in ingenious architectures are among them. The former approach is mainly analyzed through the use of simulated response curves, while the latter is evaluated via eye pattern analysis. In particular, a significant reduction of data pattern–dependent amplitude jitter is numerically demonstrated using an original method. The novel approach consists in combining the effect of interferometer (in this case a Mach-Zehnder interferometer) unbalance and the injection of a modulated holding beam that constitutes the binary complement of the processed signal. This chapter, therefore, not only represents an aid for the neophyte scientist or engineer to become involved in the nonlinear dynamical behavior of semiconductor optical amplifiers, but it provides with new insights in SOA–induced patterning mitigation to the well-informed reader.

Chapter 4 - Extensive research has been done on erbium-doped waveguide and optical fiber amplifiers in materials including silica, alumina, telluride glass, phosphate glass, lithium niobate, silicon and others. Comparison shows that these materials demonstrate different qualities that have a significant impact on the overall performance and applicability of an optical amplifier. Some materials have wider emission bandwidths than others, which can be used to amplify more wavelength channels in wavelength division multiplexing (WDM) systems. Others allow higher erbium concentrations before detrimental effects such as concentration quenching and cluster formation occur, which can translate to equal gain in a more compact device. Extremely low waveguide loss is also possible with some materials and results in improved amplifier efficiency. In the choice of glass host, many researchers have focused on only high silica glass due to its proven reliability and compatibility with conventional fiber-optic components. In this chapter, a Zircornia-based erbium-doped fiber (Zr-EDF) is comprehensively reviewed as an alternative medium for wideband optical amplification with a compact design. With a combination of both Zr and Al, a high erbium doping concentration of more than 3000 ppm in the glass host has been achieved without any phase separations of rare earths. The Zr-EDF is fabricated using in a ternary glass host, zirconia-yttria–aluminum codoped silica fiber through solution doping technique along with modified chemical vapor deposition (MCVD). A Zirconia-based erbium-doped fiber amplifier (Zr-EDFA) is capable to provide a wide-band amplification as well as flat-gain operation in both the C- and L-band regions using only a single gain medium. For instance, at a high input signal of 0dBm, a flat gain at average value of 13 dB is obtained with a gain variation of less than 2 dB within the wavelength region of 1530–1575 nm and using 2 m of Zr-EDF and 120 mW pump power. It was found that a Zr-EDFA can achieve even better flat-gain value and bandwidth as well as lower noise figure than the conventional Bismuth-based erbium-doped fiber amplifier. Using an advanced double-pass configuration, a flat gain of around 16 dB can be achieved with gain variation of approximately 2.5 dB throughout the wavelength range from 1530 nm to 1590 nm using 2 m of the Zr-EDF.

Chapter 5 - Wideband optical fiber amplifiers are key components in optical fiber communication networks. Among various technologies of light amplification, rare earth doped fiber amplifiers (REDFAs) and fiber Raman amplifiers (FRAs) are of prime importance in wideband optical fiber links. Among REDFAs, silica-based erbium doped fiber amplifier (EDFA) is the most common one which is used for supporting dense wavelength division multiplexing (DWDM) systems. EDFAs work in the C-band (1530-1565 nm). Optical fibers based on different materials and various doping components of rare earth ions (RE^{3+}-ions) can provide gain in different ranges from O to U-band. In this chapter, some recent reports on the S (1460-1530 nm), C and L-band (1565-1625 nm) REDFAs are reviewed briefly. Also, the inhomogeneity and high concentration effects in EDFAs are considered. Almost every long-distance optical fiber transmission system uses FRAs. Gain characteristics of these amplifiers such as gain amplitude, bandwidth, and gain ripple are strongly dependent to the relative state of polarizations (SOPs) of pump and signal as well as the pumping configuration. These effects are discussed in this chapter. Recently, semiconductor nano-particles (quantum dots) have attracted attention to construct new gain media for optical fiber amplifiers. High gain and extra large bandwidth are the most important expected advantageous of this type of amplifiers. Authors have assigned the final section of this chapter to quantum dot doped fiber amplifiers (QDFAs)

Chapter 6 - Erbium-doped fiber amplifiers, fiber Raman amplifiers, and hybrid erbium-doped fiber amplifiers/fiber Raman amplifiers are the most important optical components which are used nowadays in the conventional and wavelength division multiplexing systems and networks. In this chapter, these three kinds of optical fiber amplifiers are introduced and studied theoretically. The author presents different configurations for these amplifiers, investigate the simulation methods of them and finally after their simulation, derives the optimum parameters for each one of the configurations. After invention of the fast computational machines or computers, simulation and optimization have become an essential part of the engineering design of devices and systems. The author points out that In past decades we have witnessed the striking developments in the simulation software and optimization methods, which provide us with accurate estimations and predictions that to industrialize efficient devices. The three above mentioned amplifiers have many parameters which will affect their performances. To obtain amplifiers with better characteristics, optimization processes on them are quite necessary. After introducing genetic algorithm, it has become one of the most popular evolutionary population-based optimization processes which have been used in many problems. Newer optimization methods, such as particle swarm optimization, have also proved their efficiencies. Particle swarm optimization is a fast and straightforward optimization method which is used in this chapter to optimized erbium-doped fiber amplifier, fiber Raman amplifier, and also hybrid erbium-doped fiber amplifier/fiber Raman amplifier. Its performance is compared to genetic algorithm during the chapter.

Chapter 7 - Pulsed optical amplification, although has shown its profound application in generation of single mode, high power lasers through master oscillator-power amplifiers in experiments related to laser fusion facilities, it has also shown its ability to be used in other scientific and industrial activities. In this review details of experimental and theoretical investigations for gain measurements using oscillator-amplifier (OSC-AMP) N_2-laser systems will be presented. The OSC-AMP laser systems proved their abilities for gain measurements and also for further applications on understanding the gain behavior when the interest is oriented toward the study of effects of laser geometrical configurations, current density, input voltage, etc. on the gain-parameter, where these areas of research have not been explored thoroughly yet. The work is related almost completely to the author's publications in the recent and past years, although many related topics including Nd: YAG and CVL lasers will be given for clarifying the subject. Simplicity of N_2-laser designs, ease of operation, flexibility for geometrical modifications, and wide ranges of collection of the published theoretical and experimental related works make the N_2-laser as a good candidate for studying the laser behavior. Here, after reviewing the previous research works on the gain measurement, a model based on introducing an equivalent OSC system to apply for a real OSC-AMP system is presented, where without ignoring some terms in the relevant rate equations; it was possible to predict gain-values for N_2-lasers. Although the introduced model is simple, but it shows that it can be used to explain the experimental measurements for N_2-lasers. The model was successfully applied for CVL lasers and the approach may find its suitability for other types of laser systems.

Chapter 8 - A ubiquitous model of bulk amplifying laser media is provided by two-dimensional complex Ginzburg-Landau (CGL) equations with the cubic-quintic (CQ) nonlinearity. Solutions to these equations of fundamental physical significance are solitons and solitary vortices. While the zero- vorticity solitons may easily be stable in this setting, a severe problem which impedes the stabi- lization of vortices and, thus, the observation of such modes in the experiment, is the lack of an effective diffusion term in the physically relevant CGL equations (light does not undergo diffusion in lasing cavities). The formal diffusion term is necessary for the stability of vortex solitons. Recently, it has been reported that the addition of a 2D periodic potential, which may be induced by a transverse grating in the laser cavity, readily stabilizes compound (four-peak) vortices. Nevertheless, the most fundamental axisymmetric "crater-shaped" vortices (CSVs), alias vortex rings, have not been found before in a stable form. In this chapter, the author reports families of stable compact CSVs with vorticity $S = 1$ in the CGL model with the CQ nonlinearity and an external poten- tial of two different types: an axisymmetric harmonic-oscillator trap, and the periodic potential corresponding to a grating (in the latter case, the CSV must be essentially squeezed into a single cell of the grating). This chapter identifies stability regions for the CSVs, as well as for the fundamental solitons ($S = 0$). If CSVs are unstable in the harmonic-oscillator potential, the vortex ring breaks up into robust dipoles. All the vortices with $S = 2$ are found to be unstable, splitting into *tripoles*. Stability regions for the dipoles and tripoles are reported too. The periodic potential does not stabilize CSVs with $S > 1$ either; instead, in this case unstable vortices with $S = 2$ evolve evolve into families of stable square-shaped quadrupoles.

Chapter 9 - Fiber optical parametric amplifiers (OPAs) are based on a highly-efficient four-wave mixing process. Their capability to give very high gain and large bandwidths has made them an attractive candidate for numerous applications. One of them is continuous-wave (CW) fiber optical parametric oscillators (OPOs) using them as a gain medium. In this chapter a novel architecture for CW fiber OPOs is presented, which has allowed us to significantly extend the performance of these devices. To do so we used: (i) a short highly-nonlinear fiber (HNLF) as the parametric gain medium; (ii) a narrowband tunable filter in the fiber feedback loop; (iii) a high coupling fraction from the HNLF into the feedback loop (up to 3 dB). With these new features, we have been able to obtain excellent performance, even with a reduced pump power. With only about 2 W of pump power, we have obtained the following record performance: (i) tuning range of 254 nm; (ii) output power in excess of 1 W at some wavelengths; (iii) external conversion efficiency in excess of 60% at some wavelengths; (iv) line width as low as 8 GHz.

This approach can be used for providing narrowband tunable high-power CW radiation over hundreds of nanometers. Such sources could find applications in remote sensing, optical communication, nonlinear optics, etc.

In: Optical Amplifiers
Editor: Galina Nemova

ISBN 978-1-61209-835-7

Chapter 1

SEMICONDUCTOR OPTICAL AMPLIFIER AND PROGRAMMABLE LOGIC UNIT

Tanay Chattopadhyay *

Mechanical Operation (Stage-II), Kolaghat Thermal Power Station,
WBPDCL, Mecheda, Purba Medinipur,
KTPP sub post office, West Bengal, India

ABSTRACT

Programmable logic unit (PLU) is more attractive than different separate logical circuits because, any user can reprogram such device as per his requirement. PLU consist of a logical AND array followed by a logical OR array. This array can be easily designed using all-optical interferometric switches. In very recent years, the discovery of ultra high-speed all-optical *s*emiconductor *o*ptical *a*mplifier (SOA) based switches has brought the revolution in all-optical information processing systems. SOA is similar to a semiconductor laser. Its gain is higher ~ 20 to 30 dB compared to that of laser. The optical gain is caused by the recombination of electron holes. Many all-optical switches based on cross phase modulation (XPM) such as *T*erahertz *O*ptical *A*symmetric *D*emultiplexer (TOAD), *M*ach-Zehnder *i*nterferometer (MZI), *U*ltrafast *N*onlinear *I*nterferometer (UNI) etc have already been established. In an interferometric switch, there is an incoming signal (IS) which can be directed according to the *c*ontrol *s*ignal (CS). If CS is 'ON' then IS transmitted through the transmitted port or bar-port (\\), otherwise it is directed to the reflected port or cross-port (×). Hence CS may be used as programming input of all-optical PLU. We can design 16-Boolean logical operations from this unit, only by changing programming inputs.

Keywords: All-optical logic, semiconductor optical amplifier, cross phase modulation, programmable logic device

* E-mail: tanay2222@rediffmail.com

1. Introduction

Present day digital communication needs tremendous operational speed (in terahertz range). Electronic systems are incapable of processing a large number of data at high speed (far above gigahertz). The limitations can be overcome if the traditional carrier of information, electrons, is replaced by photon for devices based on switching and logic. All-optical signal processing is attractive in future high speed and high capacity networks in order to avoid in efficient optoelectronic conversion. In very resent years, the discovery of high speed all-optical switches based on semiconductor optical amplifier (SOA) has brought the revolution in all-optical information processing system. The technology of SOA has been evolving rapidly during the resent years and has become mature enough so that it is now key factor in implementation of modern optical communication networks. SOA are commercially available device and have different important properties. Such as, fast and strong nonlinearities, short latency, thermal stability, low power consumption, large dynamic range, short response time, broadband and versatile operation and capability of large scale integration with chip level design.

All-optical logic gates are key elements in all-optical signal processing and capable of performing many networks functions such as, adding, switching, header recognition, data encoding and encryption. The single logic implementation has very finite function in the network nodes. To extend the logic functions and make them smart and flexible, a multifunctional logic unit is very much essential; which can execute any binary logic operation on the same hardware. Hence programmable logic gate is a very sophisticated device because any one can program this device to get any logical operation. Using different interferometric switches programmable logic gates can be easily designed, which is discussed in the next sections.

2. Semiconducter Optical Amplifier

The first report of SOA is more than 40-years ago [1]. At that time, it was known as 'traveling wave laser amplifier' or '*s*emiconductor *l*aser *a*mplifier' (SLA). *S*emiconductor *o*ptical *a*mplifier (SOA) is similar to semiconductor laser except it has no reflecting facets [2]. Also its gain is higher ~ 20 to 30 dB compared to that of laser.

2.1. Design

Today many high performance SOA design have been made in chip level. The amplifier chip is ~ 0.5 to 2 mm long. It has *p*-cladding layer, a *n*-cladding layer and an active region. The active region is sandwiched between these two cladding layer as shown in the figure.1. The resulting heterostructure is forward-biased through metallic contact. Characteristic geometries of active region are a cross section of ~1 μm (width) × 0.1 μm (thickness) and the length of 0.5 to 2 mm. It is usually made of $In_{1-x}Ga_xAs_yP_{1-y}$ (In: Indium, Ga: Gallium, As: Arsenic, P: Phosphorus). The subscripts x and y indicates the fraction of In atoms that are replaced by Ga and the fraction of P atoms that are replaced by As, respectively. The subtract material is InP [3]. The schematic diagram of SOA is shown in the figure1.

2.2. Operating as an Amplifier

Here electron and hole recombine in the active region to produce photons through both spontaneous and stimulated emission process. Due to stimulated emission optical gain of the input signal take place. It depends on the carrier density provided by the electrical bias. In absence of current (electrons and holes) semiconductor would absorb the incident photon. The SOA gain depends on the frequency (or wavelength) and intensity of the incident signal. From linear system theory, it is well known that an oscillator is an amplifier with positive feedback. The gain is given by [4]:

$$G = \frac{G_s}{1 + FG_s} \qquad (1)$$

where, G_s is the forward path gain and F is the amount of feedback gain. If $|FG_s| < 1$, the oscillations will not appear, and the device will be amplifier. This phenomenon is same for SOA.

2.3. Optical Switching with SOA

The basic equation that describes optical amplifier is

$$P_{out} = P_{in} G \qquad (1)$$

where, P_{in} is the input optical signal power, P_{out} is the output power, G is the affective amplifier gain. The saturation energy and unsaturated single-pass amplifier gain of the SOA can be expressed as [5-6]:

$$E_{sat} = \frac{\hbar}{a_N \Gamma} \qquad (2)$$

$$G_0 = \exp\left[g_0 L - \alpha_D L\right] \qquad (3)$$

$$g_0 = \Gamma a_N N_{tr} \left(\frac{I\tau_e}{qwdLN_{tr}} - 1 \right) \qquad (5)$$

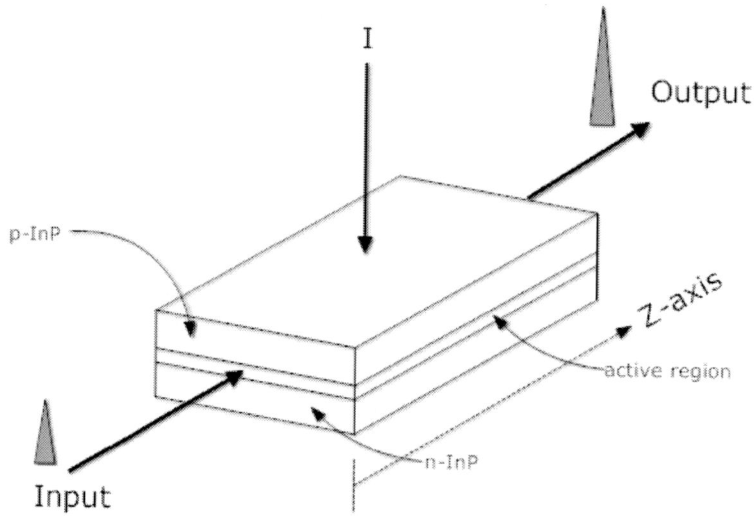

Figure 1. Schematic diagram of semiconductor optical amplifier.

Where, E_{sat} is the saturation energy of the SOA, G_0 is the unsaturated single-pass amplifier gain, Γ= Confinement factor, a_N = differential gain, N_{tr} = Carrier density at transparency, q = charge of a electron, w = width of the active region of SOA, d = depth of the active region of SOA, L = Active length of SOA, α_D = Internal loss of the wave guide, $\omega_0 = c/\lambda_0$ = frequency, c = velocity of light at vacuum, λ_0 = wave length of light, $\hbar$, h = Plank constant. Let us introduce a function $h(t) = \int_0^L g(z,t)dz$ where, $g(z,t)$ is model gain. A pulse when passes through SOA it is amplified and we obtain [7]

$$P_{out}(t) = P_{in}(t) \cdot \exp\left[h(t)\right] \tag{4}$$

$$\varphi_{out}(t) = \varphi_{in}(t) - \frac{1}{2}\alpha h(t) \tag{5}$$

where, $P_{in}(t)$, $\varphi_{in}(t)$ and $P_{out}(t)$, $\varphi_{out}(t)$ are the power and phase of the input and output pulse respectively. α is the line width enhancement factor, which depends on the wavelength of the input light. It is typically in the 3 to 8 dB range for regular heterostructure gain region [8-9]. $h(t)$ can be calculated from the following differential equation [7]:

$$\frac{dh(t)}{dt} = \frac{g_0 L - h(t)}{\tau_e} - \frac{P_{in}(t)}{E_{sat}}\left\{\exp\left[h(t)\right] - 1\right\} \tag{6}$$

Solving this equation we obtain,

$$h(t) = -\ln\left[1-\left(1-\frac{1}{G_0}\right)\exp\left(-\frac{E_{cp}(t)}{E_{sat}}\right)\right] \tag{7}$$

When a high intense pulse (which is called pump beam) is injected into the SOA then, significant depletion of the carrier occur due to stimulated emission. As a result the gain of SOA saturates at time t_s and changes its index of refraction. The gain of the SOA decreases rapidly as [5]:

$$G(t) = \exp\left[h(t)\right] \tag{8}$$

$E_{cp}(t) = \int_{-\infty}^{t} P_{cp}(t')\, dt'$ is the energy fraction contained in the leading edge of the pulse until the moment $t' \le t$. By definition $E_{cp}\left(t \to \infty\right) = E_c$ = total energy of the pump pulse. Here we consider Gaussian pulse $P_{cp}(t) = \frac{E_c}{\sigma\sqrt{\pi}}\exp\left(-\frac{t^2}{\sigma^2}\right)$ as incoming signal. σ is related to full width at half maximum (FWHM) by $T_{FWHM} \cong 1.665\sigma$. Then we can write

$$E_{cp}(t) = \frac{E_c}{2}\left[1+erf\left(\frac{t}{\sigma}\right)\right] \tag{9}$$

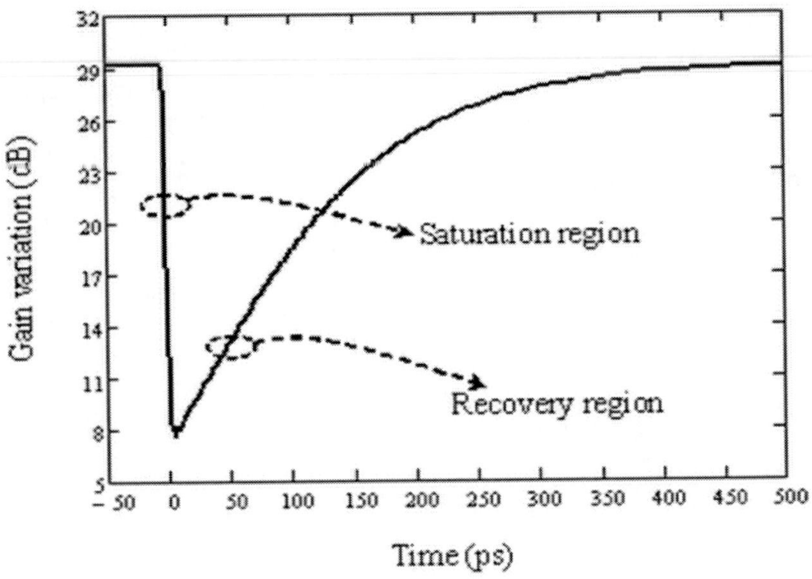

Figure 2. The variation of Gain of SOA with time when a pump pulse is applied to it.

where *erf*(.) is the error function. The SOA saturation time $t_s \approx T_{FWHM}$, then 99% of the pulse transmits through SOA.

As a while the gain recovers due to injection of carriers and can be obtained from the gain recovery formula [10],

$$G(t) = G_0 \left[\frac{G(t_s)}{G_0} \right]^{\exp\left[-(t-t_s)/\tau_e\right]} \quad ; \quad t \geq t_s \tag{10}$$

where, τ_e is the gain recovery time. The characteristic parameters of InGaAsP / InP SOA is shown in the Table 1.

Using these parameters the gain variation of SOA with time is plotted against time (in ps), shown in the Figure 2. SOA is biased with bias current 'I'; the gain variation (G_0) with 'I' using equation (4) and (5) is plotted in Fig-3 for different sets of carrier lifetime (τ_e) of the SOA. From this graph we see that 'I' should be 400 mA for G_0 = 29.146 dB. If another light signal (which is called probe signal) is injected into the SOA simultaneously with pump signal experiences the change of carrier density as a change of gain ΔG and the phase $\Delta\varphi$.

Table 1. Important parameters of InGaAsP/InP SOA

Parameters	Symbol	Value
Injection current of SOA	I	400 mA
Confinement factor	Γ	0.48
differential gain	a_N	3.3×10^{-20} m^2
Line-width enhancement factor of SOA	α	8
Carrier density at transparency	N_{tr}	1.0×10^{24} m^{-3}
width of the active region of SOA	w	1.5 μm
depth of the active region of SOA	d	250 nm
Active length of SOA	L	150 μm
Internal loss of the wave guide	α_D	2700 m^{-1}
wave length of light	λ_0	1550 nm
Gain recovery time	τ_e	90 ps
Unsaturated single-pass amplifier gain	G_0	29.146 dB
Pump pulse energy	E_c	50 fJ
Full width at half maximum of control pulse	σ	2 ps
Facet reflectivity		0.01 %
Polarization dependency		Low
Noise figure		~10 dB

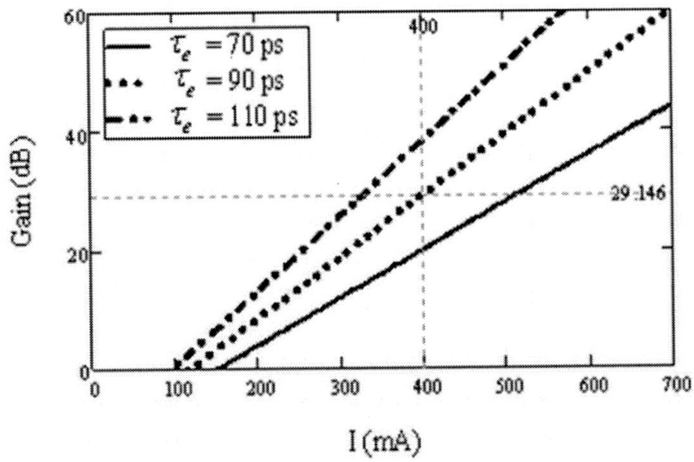

Figure 3. The variation of G_0 in dB with basing current (I) in mA for different τ_e.

The gain and phase change result in the following nonlinear effect.

- The gain change ΔG experienced by the optical probe signal is called '*cross gain modulation*' (XGM) [11].
- The phase change $\Delta\varphi$ experienced by the optical probe signal is called '*cross phase modulation*' (XPM) [12].
- When pump and probe pulse of frequency ω_1 and ω_2 have comparable intensity then, two additional frequencies $2\omega_1 - \omega_2$ and $2\omega_2 - \omega_1$ generates. This is called '*four wave mixing*' (FWM) [13].

In the next section different optical switches based on XPM is discussed. Because, the programmable logic unit (PLU) can be easily designed using it.

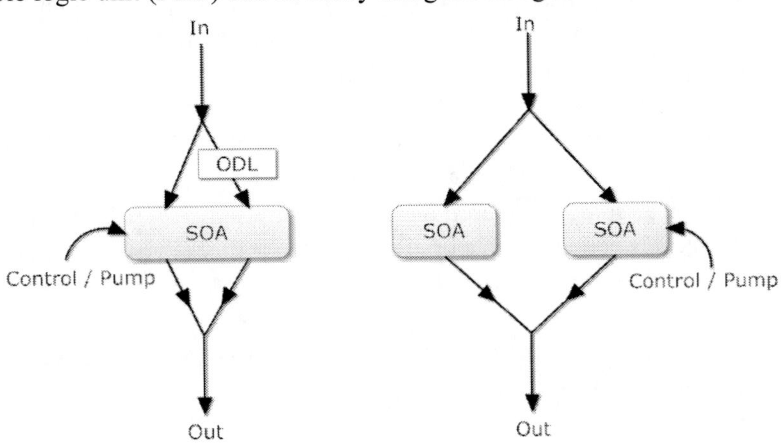

Figure 4. Schematic diagram for understanding cross phase modulation (XPM) using SOA based interferometric switch. ODL: optical delay line.

3. SOA Based Interferometric Switch

SOA makes a revolution in designing high speed (>100 Gb/s) interferometric switches [14]. The size of the switches also reduces to chip level by using it.[1] XPM can be utilized to good effect in interferometric switch. In such devices the light to be switched is split into two optical paths. Then, the two parts are passed through one SOA (at different time) or two SOA (placed in two paths) and a relative phase shift is introduced by optical switching signal (pump beam) entering one of the SOAs', which saturates the gain of the SOA. When the light is recombined, constructive or destructive interference will occurs depending on the phase difference between two paths [16]. The schematic diagram of the basic XPM based interferometric switch is shown in the figure-4. Some of the well known high speed switches are *t*erahertz *o*ptical *a*symmetric *d*emultiplexer (TOAD), *s*emiconductor *o*ptical *a*mplifier on the *M*ach-Zehnder *i*nterferometer (SOA-MZI), *u*ltrafast *n*onlinear *i*nterferometer (UNI).

3.1. Different Switches

3.1.1. Terahertz Optical Asymmetric Demultiplexer (TOAD)

The Figure 5 is a Sagnac interferometer that uses an SOA offset from the midpoint of the loop and is known as a terahertz optical asymmetric demultiplexer (TOAD). It can operate at frequencies in terahertz range [17-24]. There are two couplers; 1) the control coupler provides an input path for the control pulses to enter the fiber loop in order to saturate the SOA, and 2) the input coupler (50:50) where the incoming pulse signal train entering the loop splits equally into clockwise (CW) and counter clockwise (CCW) pulses. CW and CCW pulses arrive at the SOA at slightly different times as determined by the offset Δx of the SOA from the midpoint of the loop. Another strong light pulse is also injected to the loop. It is called control signal (CS). When *CS=ON*, then SOA changes its index of refraction. As a result, the two counter-propagation data signal will experience a differential gain saturation profiles. Therefore cross phase modulation (XPM) take place when they recombine at the input coupler. Then relative phase difference between CW and CCW pulse become ~ and the data will exit from the transmitted port / T-port (port-1 according to the Figure 4). In the absence of a control signal (*CS=OFF*), the incoming signal enters the fiber loop, pass through the SOA at different times as they counter-propagate around the loop, and experience the nearly same unsaturated amplifier gain of SOA, recombine at the input coupler. Then, relative phase difference between CW and CCW is zero (0). Then no data is found at the T-port. Then data is reflected back toward the source and isolated by optical circulator (CR). The port through which it comes is called reflected port (R-port). A filter (F) may be used at the output to reject the control and pass the incoming pulse. 'F' can be polarization filter of band pass filter.

The output power of transmitted port (T-port) and reflected port (R-port) of a TOAD based switch can be expressed as [19-20, 22-23],

[1] As an example, nonlinear optical loop mirror (NOLM) requires few kilometer optical fibers [15]. But introducing SOA into the loop, the size of this interferometric switch becomes to chip label. The device is called terahertz optical asymmetric demultiplexer (TOAD).

$$P_T(t) = \frac{P_{in}(t)}{4} \cdot \left\{ G_{cw}(t) + G_{ccw}(t) - 2\sqrt{G_{cw}(t) \cdot G_{ccw}(t)} \cdot \cos(\Delta\varphi) \right\} \tag{11}$$

$$P_R(t) = \frac{P_{in}(t)}{4} \cdot \left\{ G_{cw}(t) + G_{ccw}(t) + 2\sqrt{G_{cw}(t) \cdot G_{ccw}(t)} \cdot \cos(\Delta\varphi) \right\} \tag{12}$$

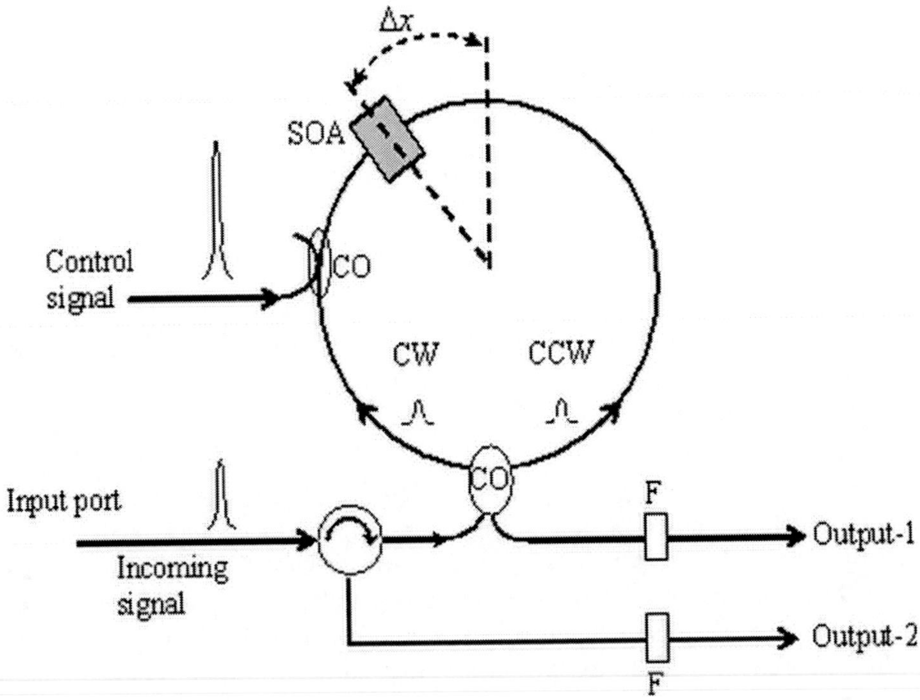

Figure 5. A TOAD based optical switch, where SOA: Semiconductor optical amplifier, CW: Clockwise pulse, CW: Counter clockwise pulse, CO: coupler, F: Filter which blocks control pulse.

where, $G_{cw}(t), G_{ccw}(t)$ are the power gain of CW and CCW pulse, $\Delta\varphi = (\varphi_{cw} - \varphi_{ccw})$ is the phase difference between CW and CCW pulse, can be expressed as $\Delta\varphi = -\frac{\alpha}{2}\ln(G_{cw}/G_{ccw})$. The temporal duration of the switching window (τ_{win}) that depends on the offset position of the SOA in the loop (Δx) is given by $T_{off} = 2\Delta x / c_{fiber}$, where c_{fiber} is the velocity of light inside the optical fiber. More specifically eccentricity of the loop must be less than half the bit period, otherwise the two counter-propagating halves of incoming signal (IS) being processed will not experience the gain dynamics caused by their synchronized control pulses but instead by others resulting in incomplete switching. T_{FWHM} of the control pulse must be as short as possible and ideally less than the switching window so that when CCW pulse is inserted in the SOA the CW pulse already passed through and the SOA gain has started to recover after saturation by the control pulse or else the two components of IS will overlap inside the SOA perceiving its nonlinear properties only partially altered, thus obstructing the creation of the required differential phase shift [24-25].

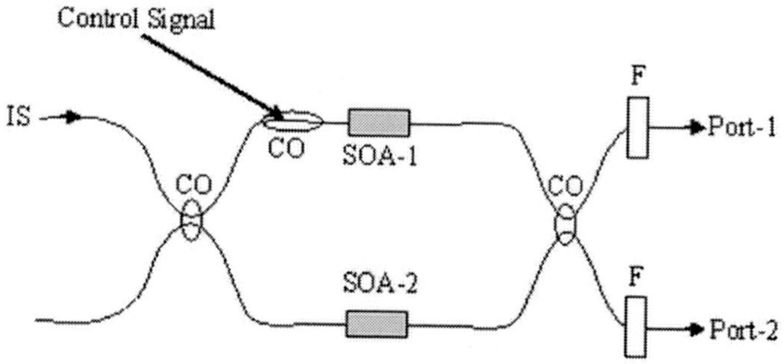

Figure 6. SOA-MZI switch. IS: incoming signal, CS: control signal, CO: coupler, F: Filter which blocks only control signal.

$$\sigma < T < 0.5\xi < \tau_e < 1.5\xi \qquad (13)$$

ξ is bit period. For low switching window eccentricity of the loop (T) should be small. One data when transmit through the switching window, next data cannot pass until the gain recovery of the SOA takes place.

2.1.2. Semiconductor Optical Amplifier on the Mach-Zehnder Interferometer

An MZI switch, as shown in figure-6, is a very powerful technique to realize ultra-fast switching. In this switch a SOA is inserted in each arm of an MZI. The incoming signal (IS) is split equally by this coupler (CO) and propagates simultaneously in the two arms. At the same time, control pulse (CS) is applied to one arm of the MZI, such that one SOA can saturate. In the absence of CS, IS exits from the cross port (lower port / port-2 in the figure). Its power is $P_\times$. However, when both beams are present simultaneously, all one bits are directed towards the bar port (upper port / port-1 in the figure) because of the refractive-index change induced by the CS. Its power is $P_{||}$. The physical mechanism behind the behavior is cross-phase modulation (XPM). Gain saturation induced by the CS beam reduces carrier density inside one SOA, which in turn increases the refractive index only in the arm through which the CS passes. As a result, an additional π phase shift can be introduced on the IP beam because of XPM, and the IP wave is directed towards the bar port (port-1 of the figure) during each one bit. Optical filters (F) are placed in front of the output ports for blocking the CS and pass the IS [26-30].

3.1.3. Ultra-fast Nonlinear Interferometer (UNI)

*U*ltra-fast *n*onlinear *i*nterferometer (UNI) is basically a single arm interferometer [31-37], which is shown in the Figure 7. The main switching principle is based on polarization rotaion of data pulse in the presence of control signal (CS). An input signal pulse is split in two orthogonal polarizations due to a walk-off between the fast and slow axis of a highly birefringent fiber (HBF). The temporal delay (T_{delay}) can be expresed as:

$$T_{delay} = \frac{|n_x - n_y| l}{c} \quad (14)$$

where n_x and n_y are the refractive indices along the HBF axis (fast and slow axis). l is length of PMF and c is speed of light in vacuum. Two signal components travel through the SOA in which a control pulse (marked in black colour in the figure), coupled in via a 3-dB coupler, is temporally coincident with one of the signal pulse components.

The control imparts a differential phase and amplitude modulation between both of signal components in the SOA due to the ultrafast refractive index and gain nonlinearities such as two-photon absorption, carrier heating and other instantaneous virtual electronic processes. When these components pass through the SOA, only the data pulse whose components are separated by the control pulse will experience a differential phase change.

After SOA a second polarization controller (PC) is used to reverse the state of polarization. In another long length of HBF, two components are coupled, so that the separation in time is cancelled. Finally two recombined components subsequently interfere in a PBS set at 45° with respect to the orthogonal signal polarizations. The control pulse is rejected by a filter 'F'. When $CS = ON$ the phase difference occurs and data transmits through port-1. When $CS = OFF$ the no phase difference occurs and the data transmits through port-2.

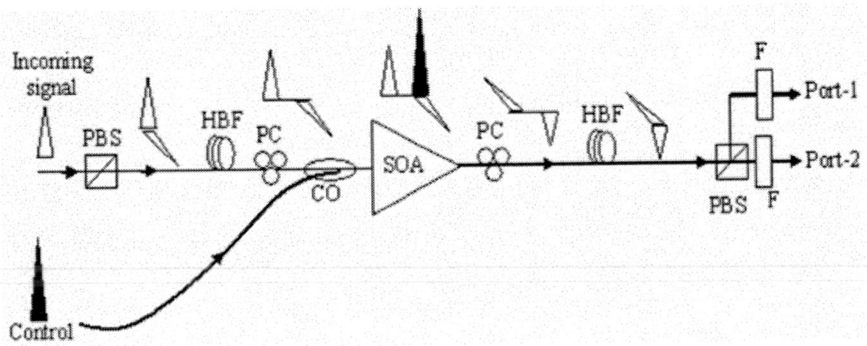

Figure 7. Schmetic diagram of ultra-fast nonlinear interferometer (UNI). F: filter which blocks control signal, CO: coupler, PBS: polarizing beam splitter, PC: polarization controler.

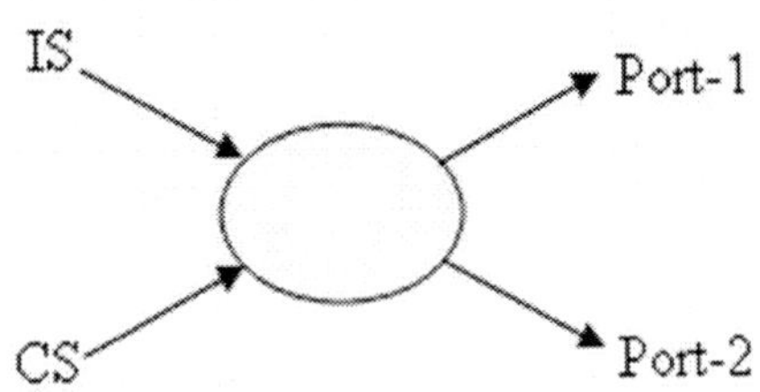

IS	CS	Port-1	Port-2
0	0	0	0
0	1	0	0
1	0	0	1
1	1	1	0

(a) (b)

Figure 8. (a) Block diagram of interferometric switch, (b) the corresponding Truth table.

Table 2. Characteristics of the different SOA based interferometric switches

Device	Switching time	Repetition rate	Control pulse energy (pJ)	Noise figure (dB)	Chip label integration	Advantages	Disad vantages
TOAD	<1 ps	100+ GHz	0.25	6	YES	Thermally stable in cascaded configu-rations	Degradation on the contrast ratio and power penalty
SOA-MZI	<1 ps	100+ GHz	0.25	6	YES	Regeneration capabilities, compact, stable	Limited bit rate operation
UNI	<1 ps	100+ GHz	0.25	6	NO	Cascadability	Polarization sensitive and limited bit rate operation

Table 3. 16-Boolean logical functions (F1 to F16)

Inputs	A	0	0	1	1	
	B	0	1	0	1	
Outputs	F1	0	0	0	0	False (0)
	F2	0	0	0	1	AB
	F3	0	0	1	0	$A\overline{B}$
	F4	0	0	1	1	A
	F5	0	1	0	0	$\overline{A}B$
	F6	0	1	0	1	B
	F7	0	1	1	0	$(A\oplus B)$
	F8	0	1	1	1	$(A+B)$
	F9	1	0	0	0	$\overline{A}\overline{B}$
	F10	1	0	0	1	$(A\odot$
	F11	1	0	1	0	$\overline{B}$
	F12	1	0	1	1	$(A+\overline{B})$
	F13	1	1	0	0	$\overline{A}$
	F14	1	1	0	1	$(\overline{A}+B)$
	F15	1	1	1	0	$(\overline{A}+\overline{B})$
	F16	1	1	1	1	True (1)

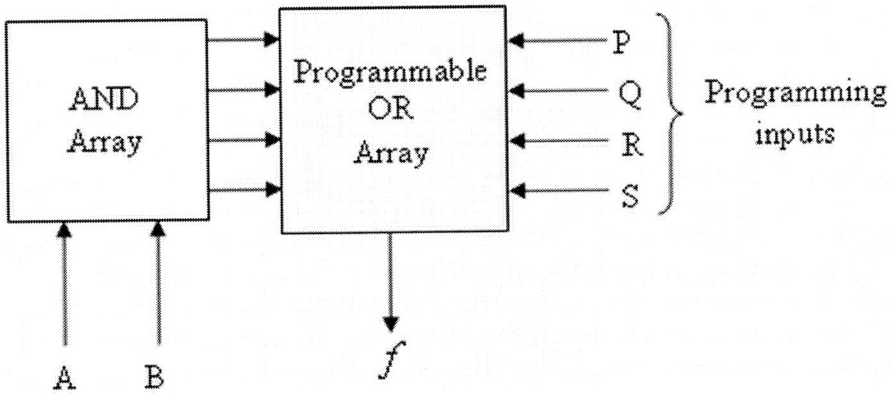

Figure 9. Block diagram of *p*rogrammable *l*ogic *u*nit (PLU).

Table 4. 16-Boolean logical operations according to the programming input (*P, Q, R, S*) of the Figure 3

P	Q	R	S	Output
0	0	0	0	0 (False)
0	0	0	1	$\bar{A}\bar{B}$
0	0	1	0	$\bar{A}B$
0	0	1	1	$(\bar{A}B+\bar{A}\bar{B})=\bar{A}$
0	1	0	0	$A\bar{B}$
0	1	0	1	$(A\bar{B}+\bar{A}\bar{B})=\bar{B}$
0	1	1	0	$(A\bar{B}+\bar{A}B)=(A\oplus B)$
0	1	1	1	$(A\bar{B}+\bar{A}B+\bar{A}\bar{B})=(\bar{A}+\bar{B})$
1	0	0	0	AB
1	0	0	1	$(AB+\bar{A}\bar{B})=(A\odot$
1	0	1	0	$(AB+\bar{A}B)=B$
1	0	1	1	$(AB+\bar{A}B+\bar{A}\bar{B})=(\bar{A}+B)$
1	1	0	0	$(AB+A\bar{B})=A$
1	1	0	1	$(AB+A\bar{B}+\bar{A}\bar{B})=(A+\bar{B})$
1	1	1	0	$(AB+A\bar{B}+\bar{A}B)=(A+B)$
1	1	1	1	$(AB+\bar{A}B+A\bar{B}+\bar{A}\bar{B})=1$ (True)

3.2. Basic Properties

All the above motioned interferometric switches have different advantages and disadvantages. Table-2 summarizes the main characteristics of the different interferometric switches:

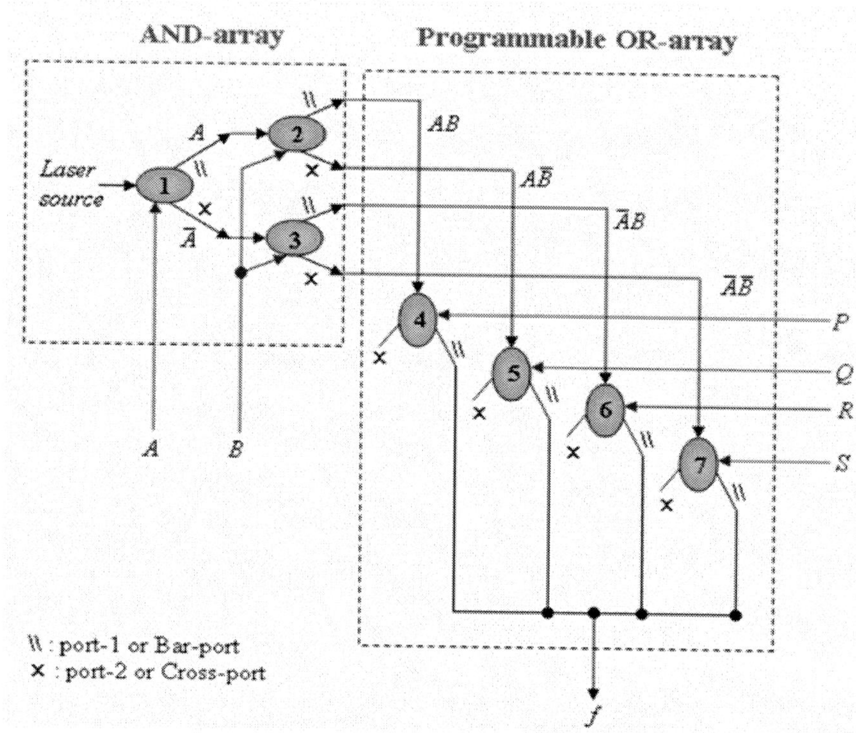

Figure 10. All-optical circuit of *p*rogrammable *l*ogic *u*nit (PLU).

The basic switching principles of different interferometric switches are same. It is clear that in the absence of control signal (CS), the incoming signal (IS) exits through the port-1. In this case no light is present in the port-2. But in the presence of control signal, the incoming signal exits through port-2. In this case no light is present in the port-1. In the absence of incoming signal, port-1 and port-2 receive no light as the filter (F) blocks the control signal. The block diagram of SOA based interferometric switch is shown in the figure-8(a) and truth table is also shown in Figure 8(b).

4. PROGRAMMABLE LOGIC UNIT

Exploiting different interferometric switches many all-optical digital circuits have been proposed that performs different logical operations [38-40]. In 'large scale integration' (LSI) and 'very large scale integration' (VLSI) technology there are hundred and thousand of logic gates internally interconnected to operate in a predefined way. Hence, volume of the circuit, power consumption, quantity production cost and switching time increases if we combine different logical circuits to built integrated circuit (IC). To overcome this problem programmable logic device (PLD) is the best choice, which replaces large number of logical circuits with a single circuits. It is 'programmable' because we can program it to perform any logical operation and logical function. Some literatures have been proposed on PLD [41-42]. P.Nisenson et al reported a Pockels effect real time electrooptic image modulator in optical processing system using of the Itek programmable read only memory (PROM) [43]. R.G.A.Craig et al demonstrated the programmable multifunction eight symmetric two input Boolean logic functions [44]. Programmable binary optoelectronic logic gate that can perform optical routing and logic functions have been experimentally demonstrated by B.Lu et al [45]. Szymanski et al demonstrate optoelectronic field programmable logic devices (FPLD) [46]. 2D optical array logic has been reported by Nishimura et al [47]. Woon-Kyung et al experimentally demonstrated optical logic gates (AND, OR and INVERTER) implemented by VCL-DOT (vertical cavity laser-depleted optical thyristor) for an optical programmable gate array [48]. Y.J.Jung et al demonstrated SOA based read only memory [49]. The generalized optical logic element (GOLE) that can perform any of the 16 Boolean logic operation on signals in an optical beam to very fast switching among functions have been represented by H.J.Caulfield et al in their literature [50]. Programmable logic unit (PLU) consists of a logical AND-array followed by a logical OR-array, as shown in figure-9. The AND-array generates product of input variables and OR-array generates sum-of-product (SOP) expressions [41-42, 51]. A, B are two inputs of AND-array Also P, Q, R, S are the four programming inputs of OR array. The final output is f . By choosing the programming inputs we can get any one of 16-different logical functions. Truth table of 16-boolean logical functions is shown in table-3. In this literature all-optical binary logical states is denoted by 'ON' (1) and 'OFF' (0). The all-optical circuit of PLU is shown in the Figure 10.

Here we require seven interferometric switching elements (1 to 7) to design this circuit. Switch-1 to 3 makes the 'AND' array. Switch-4 to 7 makes the programmable 'OR' array. A constant pulsed laser source is applied as the input signal of the interferometric switch-1. *P, Q, R, S* are the four programming inputs of OR array. All bar ports (port-1) of the switch 4 to 7 are combined by beam combiner to get the final output (f). The operations of this circuit are shown in the Table 4.

CONCLUSION

In this chapter, programmable logic unit (PLD) is designed with the help of SOA based interferometric switches. This is actually programmable logic device (PLD), which can successfully used to design *n*-bit binary logic operation [41], arithmetical operations [41, 52-53], comparator [41], multiplication [54], radix conversion [55-57] etc. This can be easily done by adding more all-optical interferometric switches in 'AND-array' and 'OR-array' [41-42]. Hence all-optical very large scale integration (VLSI) chip can be made. In electronics a programmable IC is manufactured by the selective breaking of some of the interconnection while leaving others intact. The 'fuse blowing' process can be done either by the manufacturer in accordance with the customer instructions or by the customer himself. Once fuse is blown, it cannot be reconnected. Hence wrong programming can damage this IC chip. This designed all-optical PLU has no difficulty like electronics. It can be easily programmed by giving different control signal input (*P, Q, R, S* and *T*) according to the user. In electronics erasable programmable logic device (EPLDs) viz. EPROM and EEPROM can be reprogrammed, but the erasing technique is not so easy like this proposed all-optical PLD design.

ACKNOWLEDGMENTS

This work on programmable logic device was joint work with Dr. Jitendra Nath Roy and Dilip Kumar Gayen. Also I want to thanks Prof H. J. Caulfield, Jim Hardy, Prof K.E.Zoiros, Yaser Khorrami and Tamal Sarkar for giving me suggestions and help every time in the research work. I also want to thanks all my colleagues of Kolaghat Thermal Power Station, who encourage me every time for doing this research work.

REFERENCES

[1] J. W. Crowe and W. E. Ahearn, "Semiconductor laser amplifier", *IEEE J. Quantum Electron.* , 2, (1966), 283-289.

[2] M. J. O'Mahony, "Semiconductor laser optical amplifiers for use in future fiber systems", *J. Lightwave Technol.* , 6, (1988), 531-544.

[3] P. R. Prucnal, I. Glesk, P. Toliver, L. Xu, "Optical switching with SOAs", Chap-7, Optical Switching, Springer, Editor and Principle Author: T. S. El-Bawab, 2006.

[4] A. L. R. Cavalcante, N. S. Riberiro, C. M. Gallep and E. Conforti, "Spectral gain parameters extraction with noise measurement in semiconductor optical amplifiers", 2007 SBMO/IEEE MTT-S International Microwave & Optoelectronics Conf. (IMOC 2007), 207-210.

[5] G. P. Agrawal and N. A. Olsson, "Self-phase modulation and spectral broadening of optical pulses in semiconductor laser amplifiers", *IEEE J. Quantum Electron.* , 25, (1989), 2297-2306.

[6] K. E. Zoiros, G. Papadopoulos, T. Houbavlis and G. T. Kanellos, "Theoretical analysis and performance investigation of ultrafast all-optical Boolean XOR gate with semiconductor optical amplifier-assisted Sagnac switch", *Opt. Commun.* , 258, (2006), 114-134.

[7] K. Obermann, S. Kindt, D. Breuer and K. Petermann, "Performance analysis of wavelength converters based on cross-gain modulation in semiconductor-optical amplifiers", *J. Lightwave Technol.* , 16, (1998), 78-85.

[8] C. H. Henry, "Theory of the linewidth of semiconductor lasers", *IEEE J. Quantum Electron.* , QE-18(2), (1982), 259-264.

[9] J. Wang, A. Maitra, C. G. Poulton, W. Freude and J. Leuthold, "Temporal dynamics of the alpha factor in semiconductor optical amplifiers", *J. Lightwave Technol.* , 25(3), (2007), 891-900.

[10] M. Eiselt, W. Pieper and H. G. Weber, "SLALOM: semiconductor laser amplifier in a loop mirror", *J. Lightwave Technol.* , 13(10), (1995), 2099-2112.

[11] K-I. Chung, T-Y. Kim and S-K. Han, "Analysis of all optical cascaded XGM wavelength converter using CW-holding beam", *Optical and Quantum Electronics*, 34, (2002), 937-950.

[12] L. Schares, C. Schubert, C. Schmidt, H. G. Weber, L. Occhi and G. Guekos, "Phase dynamics of semiconductor optical amplifiers at 10-40 GHz", *IEEE Journal of Quantum Electronics*, 39(11), (2003), 1394-1408.

[13] C. Schubert, R. Ludwig and H-G. Weber, "High-speed optical signal processing using semiconductor optical amplifiers", *Journal of optical fiber communication Report*, 2, (2004), 171-208.

[14] K. E. Stubkjaer, "Semiconductor optical amplifier-based all-optical gates for high spped optical processing", *IEEE Journal on selected topics in Quantum Electronics*, 6(6), (2000), 1428-1435.

[15] N. J. Doran and D. Wood, "Nonlinear-optical loop mirror", *Optics Letters*, 13(1), (1988), 56-58.

[16] D. Nesśet, T. Kelly and D. Marcenac, "All-optical wavelength conversion using SOA nonlinearities", *IEEE Communication Magazine*, Dec 1998, 56-61, 0163-6804/98.

[17] J. P. Sokoloff, P. R. Prucnal, I. Glesk and M. Kane, "A terahertz optical asymmetric demultiplexer (TOAD)", *IEEE Photon. Technol. Lett.* , 5(7), 787-790, (1993).

[18] M. Eiselt, W. Pieper and H. G. Weber, "SLALOM: Semiconductor Laser Amplifier in a Loop mirror", *Journal of lightwave Technology,* 13(10), 2099-2112, (1995).

[19] B. C. Wang, V. Baby, W. Tong, L. Xu, M. Friedman, R. J. Runser, I. Glesk and P. R. Prucnal, "A novel fast optical switch based on two cascaded terahertz optical asymmetric demultiplexers (TOAD)", *Optics Express*, 10(1), 15-23, (2002).

[20] Y. K. Huang, I. Glesk, R. Shankar and P. R. Prucnal, "Simultaneous all-optical 3R regeneration scheme with improved scalability using TOAD", *Optics Express*, 14(22), 10339-10344, (2006).

[21] T. Houbavlis and K. E. Zoiros, "Numerical simulation of semiconductor optical amplifier assisted Sagnac gate and investigation of its switching characteristics", *Optical Engineering*, 43(7), 1622-1627, (2004).

[22] K. E. Zoiros, R. Chasioti, C. S. Koukourlis and T. Houbavlis, "On the output characteristics of a semiconductor optical amplifier driven by an ultrafast optical time division multiplexing pulse train", *Optik*, 118(3), 134-146, (2007).

[23] T. Houbavlis and K. E. Zoiros, "SOA-assisted Sagnac switch and investigation of its roadmap from 10 to 40 Ghz", *Optical and Quantum Electronics*, 35(13), 1175-1203, (2003).

[24] K. E. Zoiros, T. Houbavlis and M. Kalyvas, "Ultra-high speed all-optical shift registers and their applications in OTDM networks", *Optical and Quantum Electronics*, 36(11), (2004), 1005-1053

[25] T. Chattopadhyay, "All-optical symmetric ternary logic gate", *Optics and Laser Technology*, 42, (2010), 1014-1021.

[26] R. Hess, M. C. Gross, W. Vogt, E. Gamper, P. A. Besse, M. Düelk, E. Gini, H. Melchior, B. Mikkelsen, M. Vaa, K. S. Jepsen, K. E. Stubkjaer and S. Bouchoule, "All-optical demultiplexing of 80 to 10 Gb/s signal with monolithic integrated high-performance Mach-Zehnder interferometer", *IEEE Photonics Tech. Letters*, 10(1), (1998), 165-167.

[27] J. Leuthold, P. A. Besse, E. Gamper, M. Dülk, S. Fischer, G. Guekos and H. Melchior, "All-optical Mach-Zehnder interferometer wavelength converters and switches with integrated data-and control-signal separation scheme", *Journal of Lightwave Tech.* , 17(6), (1999), 1056-1066.

[28] P. V. Studenkov, M. R. Gokhale, J. Wei, W. Lin, I. Glesk, P. R. Prucnal and S. R. Forrest, "Monolithic integration of an all-optical Mach-Zehnder demultiplexer using an asymmetric twin-waveguide structure", *IEEE Photonics Tech. Letters*, 13(6), (2001), 600-602.

[29] R. Schreieck, M. Kwakernaak, H. Jackel, E. Gamper, E. Gini, W. Vogt and H. Melchior, "Ultrafast switching dynamics of Mach-Zehnder interferometer switches", *IEEE Photonics Tech. Letters*, 13(6), (2001), 603-605.

[30] J. M. Martinez, F. Ramos and J. Marti, ""All-optical packet header processor based on cascaded SOA-MZIs", *Electronics Letters*, 40(14), (2004), doi: 10. 1049/el:20045209.

[31] N. S. Patel, K. A. Rauschenbach and K. L. Hall, "40-Gb/s Demultiplexing Using an Ultrafast Nonlinear Interferometer (UNI)", *Photon. Technol. Lett.* , 8(12), 1695 (1996).

[32] N. S. Patel, K. L. Hall and K. A. Rauschenbach, "40-Gbit / s cascadable all-optical logic with an ultrafast nonlinear interferometer", *Optics Letters*, 21(18), 1466-1468 (1996).

[33] C. Schubert, S. Diez, J. Berger, R. Ludwig, U. Feiste, H. G. Weber, G. Toptchiyski, K. Petermann, and V. Krajinovic, "160-Gb/s All-optical demultiplexing using a gain-transparent ultra-fast-nonlinear interferometer (GT-UNI)", *IEEE Photonics Technology letters*, 13(5), 475-477 (2001).

[34] Z. Wang, J. Yu, J. Han, and E. Yang, "10-Gb/s optical switch using a scheme of ultrafast nonlinear interferometer (UNI)", *Proceedings of Optical transmission, switching, and subsystems*, 5281, 27-31 (2004).

[35] S. J. Savage, B. S. Robinson, and S. A. Hamilton, "All-optical pulse regeneration in an ultrafast nonlinear interferometer with Faraday mirror polarization stabilization", *Optics Letters*, 28(1), 13-15 (2003).

[36] P. Honzatko, A. Kumpera and P. Skoda, "Effects of polarization dependent gain and dynamic birefringence of the SOA on the performance of the ultrafast nonlinear interferometer gate", *Optics Express*, 15(5), (2007), 2541-2547.

[37] T. Houbavlis, K. E. Zoiros, M. Kalyvas, G. Theophilopoulos, C. Bintjas, K. Yiannopoulos, N. Pleros, K. Vlachos, H. Avramopoulos, L. Schares, L. Occhi, G. Guekos, J. R. Taylor, S. Hansmann and W. Miller, "All-optical signal processing and applications within the esprit project DO_ALL", *Journal of Lightwave Technology*, 23(2), (2005), 781-801.

[38] J. Hardy and J. Shamir, "Optics inspired logic architecture", *Optics Express*, 15(1), (2007), 150-165.

[39] M. A. Karim and A. A. S Awwal, *Optical computing: An introduction*, © 1992 John Wiley & Sons, Inc. Indian Edition.

[40] K. E. Zoiros, "Semiconductor optical amplifier-based interferometric switches: overview and characteristic applications of two main types," in *Handbook of Interferometers; Research, Technology and Applications*, D. Halsey and W. Raynor, Eds. , Nova Publishers, New York (2009).

[41] T. Chattopadhyay and J. N. Roy, "Design of SOA-MZI based all-optical programmable logic device (PLD)", *Optics Communication*, 283 (2010), 2506-2517.

[42] T. Chattopadhyay, (2010, July 18). All-optical programmable VLSI chip: an idea. *SciTopics* (Elsevier), Retrieved July 29, 2010. http://www. scitopics. com/All_optical_programmable_VLSI_chip_an_idea. html

[43] P. Nisenson and S. Iwasa, "Real time optical processing with $Bi_{12}SiO_{20}$, PROM", *Applied Optics*, 11(12), (1972), 2760-2767.

[44] R. G. A. Craig, G. S. Buller, F. A. P. Tooley, S. D. Smith, A. C. Walker and B. S. Wherrelt, "All-optical programmable logic gate", *Applied Optics*, 29(14), 2148-2152.

[45] B. Lu, P. Zhou, Y. Lu, J. Cheng, R. E. Leibenguth, A. C. Adams, J. L. Zilko, K. L. Lear, J. C. Zolper, S. A. Chalmers and G. A. Vawter, "Binary optical switch and programmable optical logic gate based on integration of GaAs/AlGaAs surface-emitting lasers and heterojunction phototransistors", *IEEE Photonics Tech. Letters*, 6(3), (1994), 398-401.

[46] T. H. Szymanski, M. Saint-Laurent, V. Tyan, A. Au, and B. Supmonchai, "Field-programmable logic devices with optical input-output", *Applied Optics*, 39(5), (2000), 721-732.

[47] N.Nishimura, Y.Awatsuji and T.Kubota, "Analysis and evaluation of logical instructions called in parallel digital operations based on optical array logic", *Applied Optics*, 42(14), (2003), 2532-2545.

[48] W.Choi and Y.Choi, "Design of monolithically integrated vertical cavity laser with depleted optical thyristor for optical programmable gate array", *Optics Express*, 16(22), (2008), 18264-18274.

[49] Y.J.Jung, N.Park, Y.M.Jhon and S.Lee, "Design of all-optical read only memory", *Applied Optics*, 48(31), (2009), G21-G27.

[50] H.J.Cauilfield, R.A.Soref, L.Qian, A.Zavalin and J.hardy, "Generalized optical logic elements - GOLEs", *Optics Communication*, 271, (2007), 365-376.

[51] R.P.Jain, *Modern digital Electronics*, © 1984 Tata McGrawHill publishing company limited, 3rd edition (2004), Chap-12.

[52] J.N.Roy and D.K.Gayen, "Integrated all-optical logic and arithmetic operations with the help of a TOAD-based interferometer device—alternative approach", *Applied Optics*, 46(22), (2007), 5304-5310.

[53] D.K.Gayen and J.N.Roy, "All-optical arithmetic unit with the help of terahertz-optical-asymmetric-demultiplexer-based tree architecture", *Applied Optics*, 47(7), (2008), 933-943.

[54] J.N. Roy, T.Chattopadhyay, M.K.Das and T.Sarkar, "All-optical multiplication using SOA-MZI based programmable logic device (PLD)", *2010 International Conference on Communication, Computers and Devices,* (ICCCD-2010), Kharagpur, INDIA, December 10-12. (Accepted).

[55] J.N.Roy, G.K.Maity, D.K.Gayen and T.Chattopadhyay, "Terahertz Optical Asymmetric Demultiplexer based tree-net architecture for all optical conversion scheme from binary to its other 2^n radix based form", *Chin.Opt.Lett.* (OSA) 6(7), 536-540, (2008).

[56] T.Chattopadhyay and J.N.Roy, "All-optical conversion of binary number to quaternary signed digit (QSD) number", *Proceedings of international conference on Trends in optics and photonics (IconTOP 2009)*, March 1-4, 2009, Kolkata, India, 130-137.

[57] T.Chattopadhyay, M.K.Das, J. N. Roy, A.K.Chakraborty and D. K. Gayen, "Interferometric switch based all optical scheme for Conversion of Binary number to its Quaternary Signed Digit form", *IET Circuits, Devices and system*, (special issue on 'Optical Computing Circuits, Devices and Systems'), (Accepted).

In: Optical Amplifiers
Editor: Galina Nemova

ISBN 978-1-61209-835-7

Chapter 2

SEMICONDUCTOR OPTICAL AMPLIFIERS FOR THE NEXT GENERATION BROADBAND ACCESS NETWORKS

Chi-Wai Chow*

Department of Photonics, Institute of Electro-Optical Engineering, National Chiao Tung University, Hsinchu, Taiwan

ABSTRACT

Passive optical network (PON) is considered as cost-effective optical access network architecture, and it has already been deployed all over the world. As the bandwidth demand continues to grow, optically amplified long-reach (LR) PON and wavelength division multiplexed (WDM) PON are currently under intensive studies for the next generation access networks providing higher capacity. Semiconductor optical amplifier (SOA) is a promising photonic device for these next generation access networks as a gain block to increase the power budget or as a colorless modulator in the WDM-PON. In this chapter, we will describe the basic principle and the characteristics of the SOA. After this, we will present some integrated devices using SOAs for the next generation access networks. Then we will demonstrate the applications of reflective SOA (RSOA) as the colorless optical networking unit (ONU) for the next generation signal remodulation WDM-PON and signal remodulation LR-PON.

Keywords: Broadband access networks, passive optical network (PON), fiber-to-the-home (FTTH), wavelength division multiplexed (WDM), semiconductor optical amplifier (SOA)

---

* E-mail: cwchow@faculty.nctu.edu.tw

1. INTRODUCTION

The increase in popularity of the broadband services, such as high definition television, Internet video, music downloading, and online gaming will certainly push the end-users bandwidth demand even higher in the near future. It is well known that, nowadays, there are various technologies for broadband access networks, providing high-speed broadband services to end-users. It is generally agreed that fiber-to-the-home (FTTH) based on passive optical network (PON) is a cost-effective and promising architecture for offering broadband access services [1-2]. 10 Gb/s Ethernet PON (EPON) has now been standardized [3], and the standard supports two configurations. One is symmetric, operating at 10 Gb/s data rate in both upstream and downstream directions. Another is asymmetric, operating at 10 Gb/s in the downstream direction and 1 Gb/s in the upstream direction. Different end-user optical networking units (ONUs) are connected to the optical line terminal (OLT) in the central office (CO) using time-division-multiplexing (TDM).

As the bandwidth demand continues to grow, researchers are planning the next generation PONs. They are going to further increase the capacity and the reach of the present PONs. Hence, high data rate PONs (40 Gb/s – 100 Gb/s) [4-13], optically amplified long-reach PONs (LR-PONs) [14-23], and wavelength division multiplexed (WDM) PONs [24-35] are currently under intensive studies (Figure 1). Under the present TDM-based PON, increasing the data rate of the PON means that the data rate per ONU can be increased, or the split-ratio of the PON can be increased while maintaining the data rate per ONU. LR-PON can merge the metro and access networks into a single system; hence it can significantly reduce the operational cost of the network. WDM-PON is an attractive approach to increase the capacity of the PON by using the wavelength domain. In the WDM-PON, a dedicated wavelength is assigned to each ONU; hence bandwidth sharing in the TDM-PON can be avoided. It can also overcome the contention limitation that will happen in the TDM-PON.

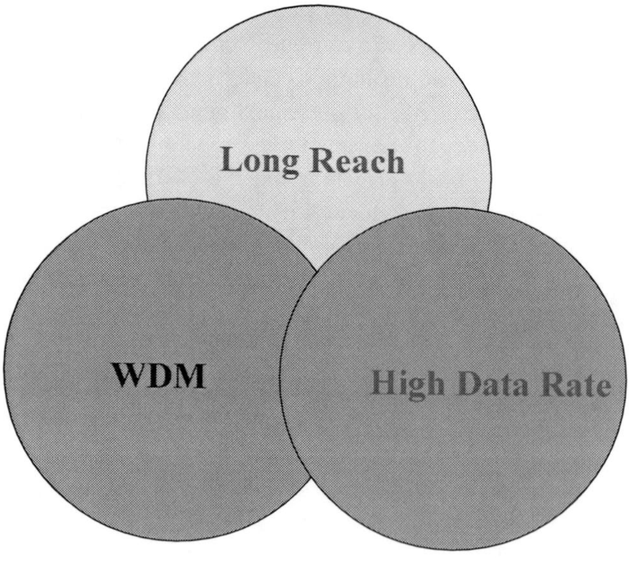

Figure 1. Requirements of future passive optical networks.

Scaling up to the 10 Gb/s/wavelength or 40 Gb/s/wavelength PON is very challenging, since much higher optical bandwidth and optical signal to noise ratio (OSNR) are required. And the receiver sensitivity will decrease. This means that optical in-line amplifiers or optical pre-amplifiers are necessary to compensate the higher OSNR requirement. Besides, when upgrading the present PONs to LR-PONs, optical amplifiers are required to increase the power budget for compensating the power losses due to the long fiber length and the fiber splitter. Besides acting as the gain block for the PONs, it is desirable that the optical amplifiers can be modulated and acting as wavelength independent ("colorless") modulators for the WDM-PONs.

In general, optical amplifiers can be divided into fiber amplifiers and semiconductor amplifiers. Although the fiber amplifiers have dominated the conventional system applications, such as to compensate fiber transmission loss, semiconductor optical amplifiers (SOAs) have aroused attention recently owing to the compact size and relatively low power consumption. Due to the advances in semiconductor fabrication techniques, the performance of SOAs has been improved. SOAs show promising applicability in many areas in optical communication systems, such as optical switches, modulators, optical regenerators, wavelength converters, etc… SOA can also be modulated at a relative high speed (> 2.5 Gb/s) as optical modulator. Because of this, in this chapter, we focus on using SOAs for the next generation broadband access networks.

This chapter is organized as follow: In Section 2, we will describe the basic principle of SOA. In Section 3, we will discuss the characteristics of SOA. Then, in Section 4, we will present some integrated devices using SOAs for the next generation access networks. In Section 5, we will propose and demonstrate using SOA as the colorless ONU for the next generation WDM-PON. Finally, a conclusion will be given in Section 6.

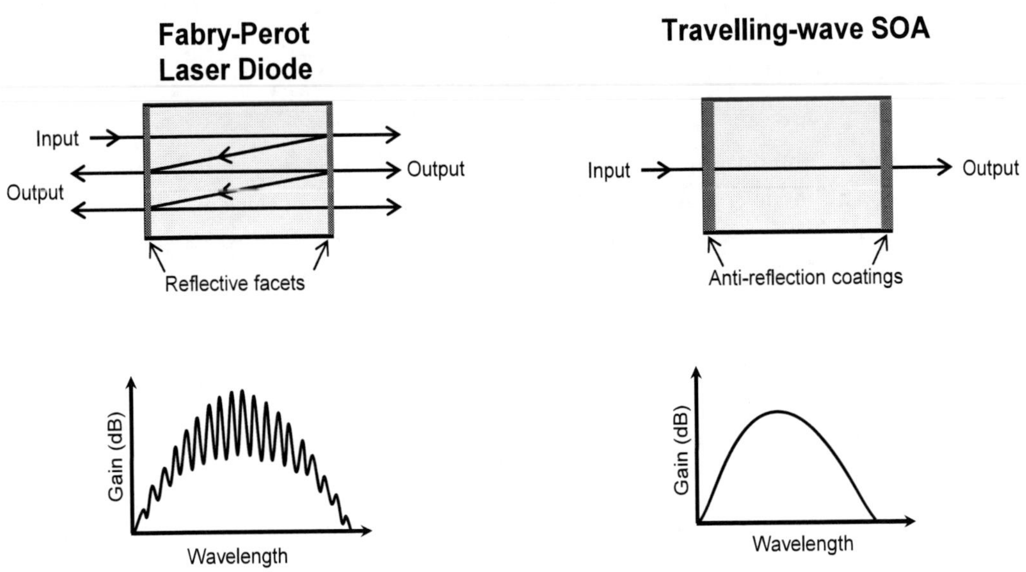

Figure 2. Structures and gain profiles of Fabry-Perot laser diode and SOA.

2. Basic Principle of SOA

A SOA is an optoelectronic device with the primary purpose of providing amplification for optical signal. The first studies of SOAs were around the time of the invention of the first semiconductor laser in 1970's [36-37]. In 1980's, studies of using InP/InGaAsP SOA at 1.3 μm and 1.5 μm wavelength ranges began in order to match with the second and the third low-loss transmission windows of standard single mode fiber (SSMF) [38].

As shown in Figure 2, SOA has a similar structure as a semiconductor laser diode, but having anti-reflection (AR) coatings on both facets with reflectivities $< 10^{-5}$. In a SOA, the carriers are injected from an external current source. These carriers occupy energy states in the conduction band, leaving holes in the valence band. If a photon of suitable energy is launched into the SOA, it will cause stimulated recombination of the carrier and the hole. This will create a new photon having identical properties as the incoming photon. Hence, the process can provide optical amplification to the incoming signal.

The carriers in the conduction band will spontaneously recombine with the hole, hence emitting photons with random phases, directions and a wide range of frequencies. The spontaneous emission is called amplified spontaneous emission (ASE) noise in the SOA.

SOA is usually a double-heterostructure, in which the active area is sandwiched between the p- and n-type areas. As the p- and n-type areas have a higher bandgap than the active area, carriers can be injected into the active area from the applied current. This provides the carrier confinement. Besides, in the double-heterostructure, the active area has a higher refractive index than the p- and n-type areas, this can help optical confinement when light is travelling in the device. Hence the efficiency of the SOA can be further improved.

The small signal gain of a SOA is given by [39]:

$$G(v) = \frac{(1-R_1)(1-R_2)G_S}{(1-\sqrt{R_1R_2}G_S)^2 + 4\sqrt{R_1R_2}G_S\sin^2[\pi(v-v_0)/\Delta v]} \tag{1}$$

where Δv is the longitudinal mode spacing, v is the optical frequency of the input signal, v_0 is the cavity resonant mode frequency and G_s is the single pass amplifier gain, R_1 and R_2 are the facet reflectivities of the SOA. An uncoated SOA has the facet reflectivities about 0.32.

We can define the gain ripple G_r of the amplifier as the ratio of resonant and non-resonant gains.

$$G_r = \frac{1+\sqrt{R_1R_2}G_S}{1-\sqrt{R_1R_2}G_S} \tag{2}$$

From (2), we can find out the relation between the geometric mean facet reflectivity $R_g = \sqrt{R_1R_2}$ and G_r is:

$$R_g = \frac{1}{G_s}\left[\frac{G_r - 1}{G_r + 1}\right] \tag{3}$$

Figire 3 shows the plot of (3) at different gain ripples. We can see that when the gain ripples of the SOA is 1 dB, in order to have a single pass amplifier gain of 30 dB, the geometric mean facet reflectivity should be < 10^{-5}. The facet reflectivities can be reduced by using AR coatings or specialized SOA structures.

High gain and high saturation power could lead to undesired lasing of the SOA, hence, in order to prevent this phenomenon, reducing the facet reflectivity is required. Besides applying AR coating and slanting the active region of the SOA away from the facet plane can effectively reduce the facet reflectivity, as shown in Figure 4 [40]. The reflectivity decreases as the slant angle increases, however, the coupling efficiency between the SOA and the optical fiber decreases due to the far-field asymmetry. Also at large slant angle, the AR coating become polarization sensitive. Optimal slant angle is about 10 degree. Analyses of the slant active area [41, 42] have also been performed. As the slant active area exhibit large far-field divergence, tapering the active area can reduce the far field divergence, hence increasing the output mode size and improving the fiber coupling efficiency. As shown in Figure 4, window regions at both ends of the active area are included to reduce the effective facet reflectivity [43, 44]. This structure is an area between the active area and the facet, having energy bandgap greater than the photon energy. Hence, the stimulated absorption can be avoided. Besides, the reflected signal at the SOA facet will diffract and diverge, so only a small portion of the reflected signal will be coupled back into the active area.

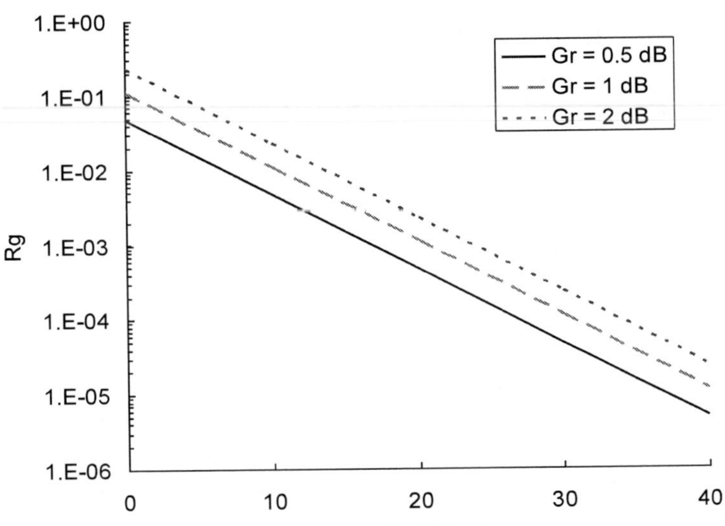

Figure 3. Geometric mean facet reflectivity against single pass amplifier gain at different gain ripples, calculated using (3).

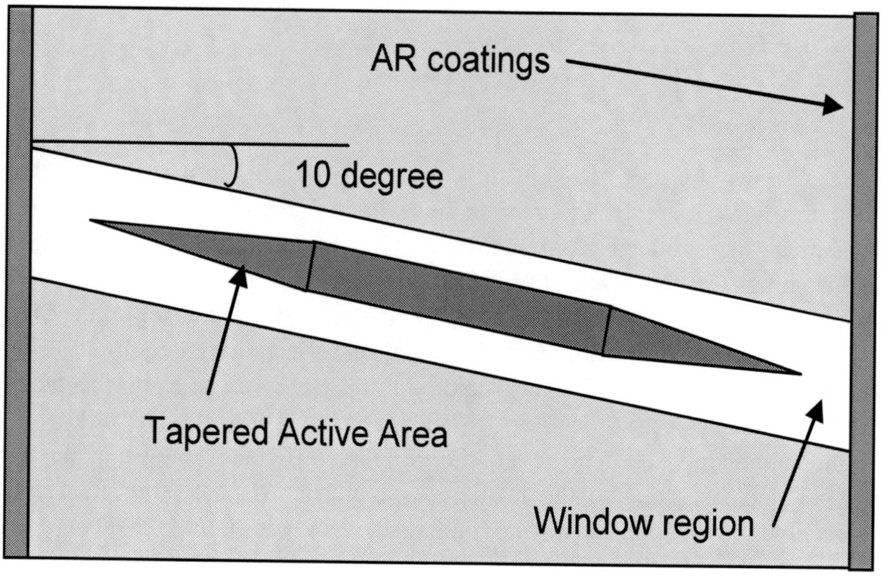

Figure 4. Schematic of the SOA top-view, with slant active area, mode expanders.

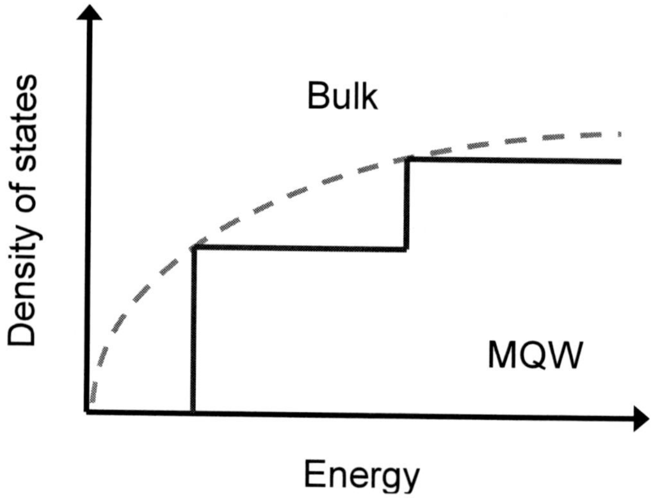

Figure 5. Density of states functions for bulk and multi-quantum wells SOA.

Since SOAs are usually used as in-line or pre-amplifier, or reflective modulator in the ONU, polarization insensitive is desirable. The polarization sensitivity of the device is due to the difference between the optical conferment factors in transverse electric (TE) Γ_{TE} and transverse magnetic (TM) Γ_{TM} modes. In order to reduce the polarization dependent gain (PDG), equalizing the Γ_{TE} and Γ_{TM} by using the square cross section waveguide [45] or using strained materials (such as tensile strained multi-quantum-wells (MQWs) [46]) in the active area to increase the gain coefficient of the TM mode with respect to TE mode.

The active area of the SOA can use bulk or quantum-well structure. The gain bandwidth of the MQWs SOA is usually greater than that of a bulk SOA. This is because in bulk SOA, the density of states is a continuous function of energy, while the density of states in a MQWs SOA shows independency at some energy levels (Figure 5). Hence, this decreases the dependence on energy providing a flatter gain bandwidth. Besides, as mentioned above, we can use strained MQWs to reduce the polarization dependency of the SOA. Hence, MQWs SOA are widely deployed in different areas in optical communication.

Usually, high saturation output power (P_{sat}) is desirable for SOA applications. The rate of change of carrier density inside a SOA can be written as the rate equation:

$$\frac{dn}{dt} = \frac{J}{ed} - \frac{n}{\tau} - a_1(n - n_0)\frac{I_S}{hv} \tag{4}$$

where J is the current density, e is the electronic charge, d is the active area thickness, τ is the carrier life time, a_1 is the differential gain coefficient, n_0 is the transparent carrier density, I_s is the signal intensity, h is the Plank constant and v is the optical frequency. At steady state conditions, the differential in (4) is zero, hence:

$$\frac{n}{\tau} + a_1 n \frac{I_S}{hv} = \frac{J}{ed} + a_1 n_0 \frac{I_S}{hv}$$

$$n = \frac{\tau J}{ed}\left(\frac{I_{sat}}{I_{sat} + I_S}\right) + \frac{n_0 I_S}{I_{sat} + I_S} \tag{5}$$

where the saturation intensity I_{sat} is given by:

$$I_{sat} = \frac{hv}{a_1 \tau} \tag{6}$$

We can define saturation output power by:

$$P_{sat} = \frac{AI_{sat}}{\Gamma} \tag{7}$$

where A is the cross section of the active area. From (6) and (7), we can see that by increasing I_{sat} or decreasing τ, the saturation output power of the SOA can be increased. τ is inversely proportional to the carrier density, so applying higher bias current can increase the SOA saturation power. Besides, choosing different materials can change a_1, so the SOA saturation power can be affected.

3. Characteristics of OSA

The gain spectrum peak and the wavelength of the InGaAsP based SOA is a function of carrier density, such that:

$$g_m = a_1(n - n_0) \tag{8}$$

where a_1 is the differential gain coefficient, n_0 is the transparent carrier density. Usually $a_1 = 6.7 \times 10^{-20}$ m^2 and $n_0 = 1.2 \times 10^{24}$ m^{-3} for InGaAsP material. Figure 6 shows the measured gain profile of a commercially available reflective SOA (RSOA). We can observe that the gain peak wavelength will shift to shorter wavelength when the bias current (or carrier density) increases. This is because when the carrier density increases, carriers will occupy higher energy states in the conduction band. Hence the emitted photon will have higher optical frequency or shorter wavelength.

Then, we study the on-off characteristics of the SOA. In the hybrid WDM-TDM PON, a group of ONUs shares the same wavelength in a TDM-PON. In order to prevent upstream signal collision, each ONU in the same group will transmit the upstream optical packet in different TDM time slots assigned by the network. When one ONU is sending the upstream packet, other inactive ONUs should be switched off. Hence, two issues arise: Can the SOA be switched on or off fast enough? Can the inactive ONU be switched off completely? If the SOA cannot be switched off completely, the inter-burst residual power will introduce in-band crosstalk and degrade the upstream signal performance particularly at high split-ratio TDM-PON.

To discuss the first issue, an experiment was performed by launching a continuous-wave (CW) optical signal with input power of -15 dBm into a commercially available SOA with modulation speed of about 1 GHz. The SOA has a frequency response of about 1 GHz, and it was gated by an applied electrical signal. Figure 7 shows the output optical time trace of the SOA, having the average optical output power of 2 dBm. The extinction ratio of the on-off is > 14 dB.

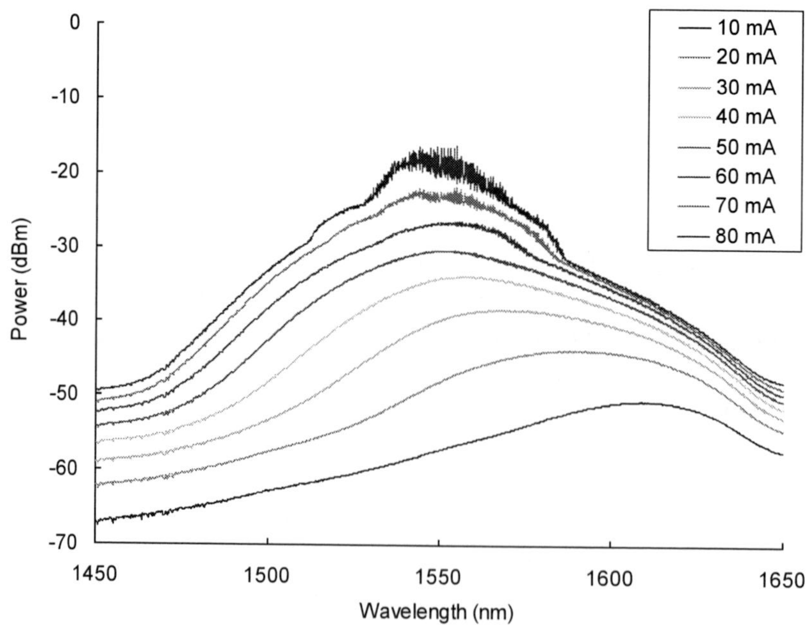

Figure 6. Measured ASE spectrum of a RSOA at different bias currents.

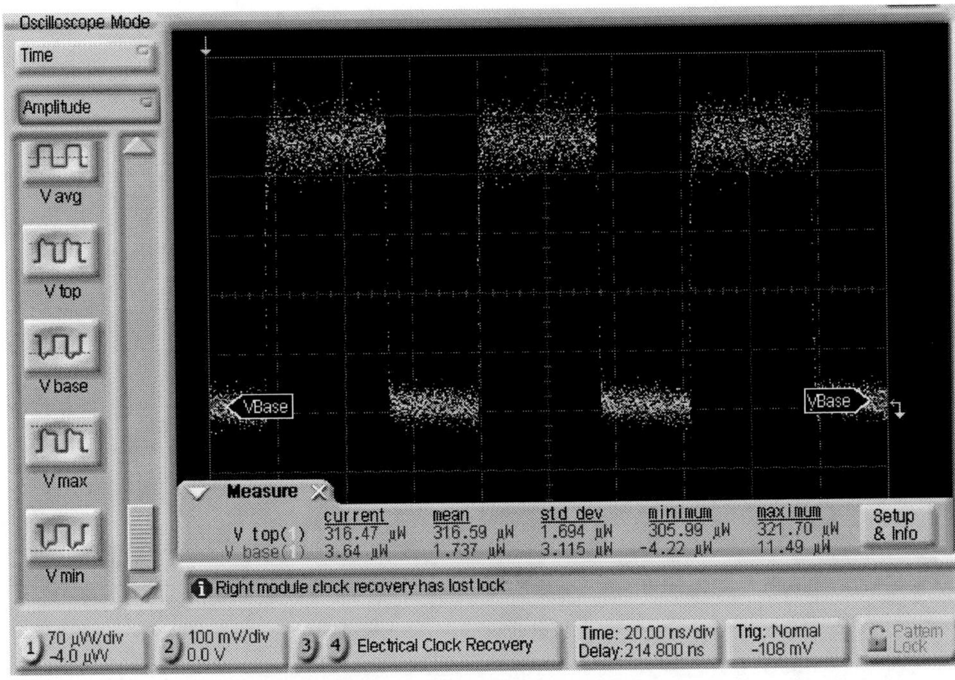

Figure 7. On-off characteristics of a SOA.

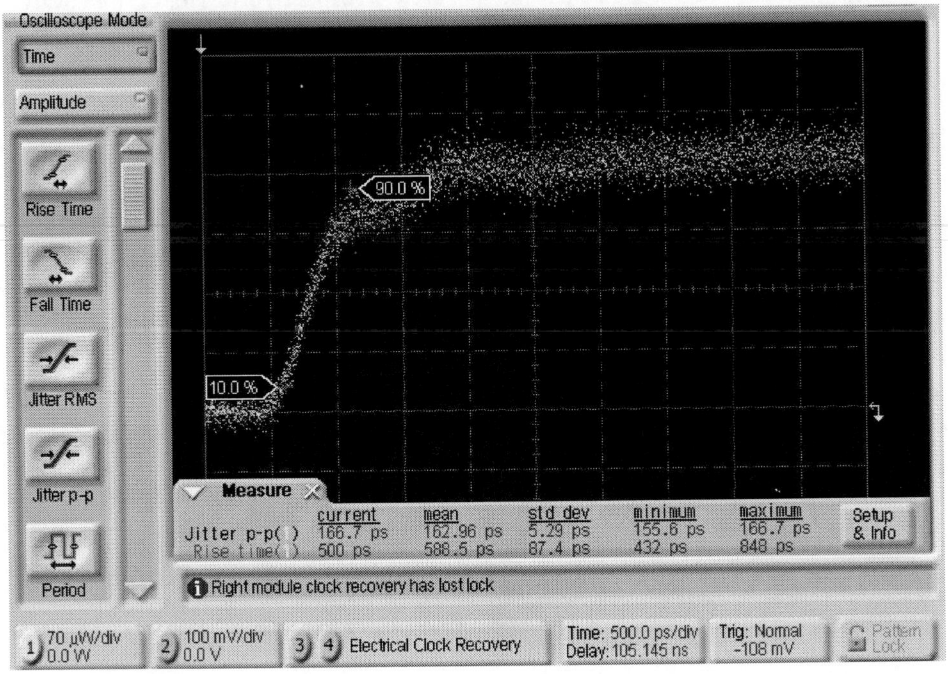

Figure 8. Rise time of a SOA.

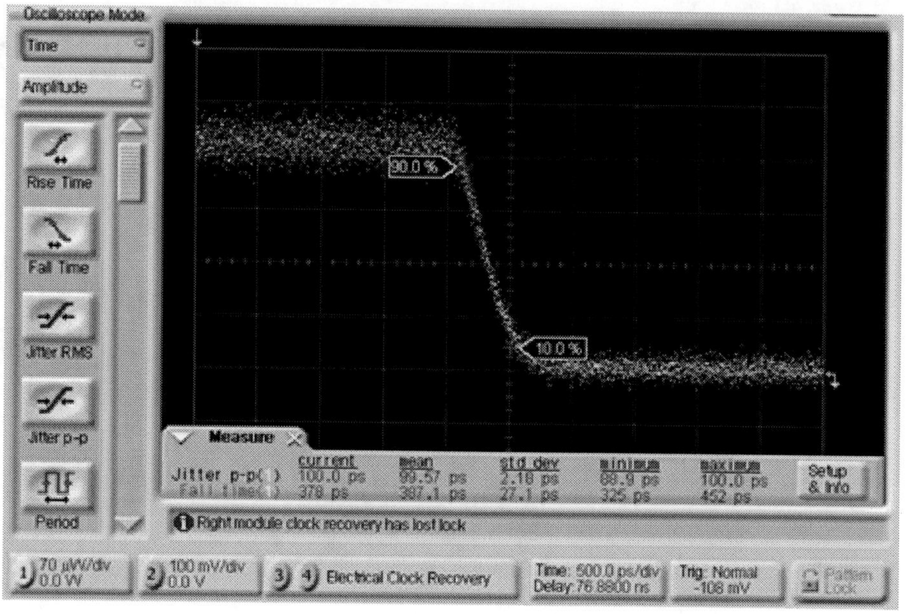

Figure 9. Fall time of a SOA.

We also studied the rise and fall times of the SOA. The rise and fall times are about 600 ps and 400 ps, as shown in Figure 8 and Figure 9, respectively. Hence the SOA can be operated at about 1 Gb/s. Operation at 10 Gb/s is also possible by using electronic equalizer processing at the receiver [47] or including a delay interferometer at the output [48].

Detail analysis of the OSNR penalty imposed by inter-bust residual power caused in-band crosstalk in the TDM-PON can be found in [49]. The imposed OSNR penalty can be defined as:

$$\delta_{OSNR} = -10\log_{10}\left(1 - \frac{(1+\sqrt{\alpha})^2(1+\alpha)(N-1)Q^2}{(1-\alpha)^2\beta\Delta L}\right) \tag{9}$$

where α is the ONU extinction ratio, β is the inter-burst suppression, N is the network split-ratio, and ΔL is the differential loss at a particular bit-error rate (BER). We can observe from (9) that the OSNR penalty can be reduced by decreasing the split-ratio of the network (hence the number of interferers is lower), or increasing the inter-burst suppression.

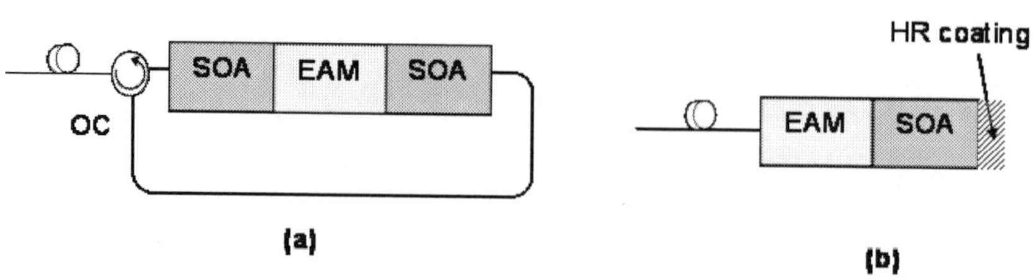

Figure 10. Schematics of RONU designs using (a) SOA-EAM-SOA and (b) R-SOA-EAM.

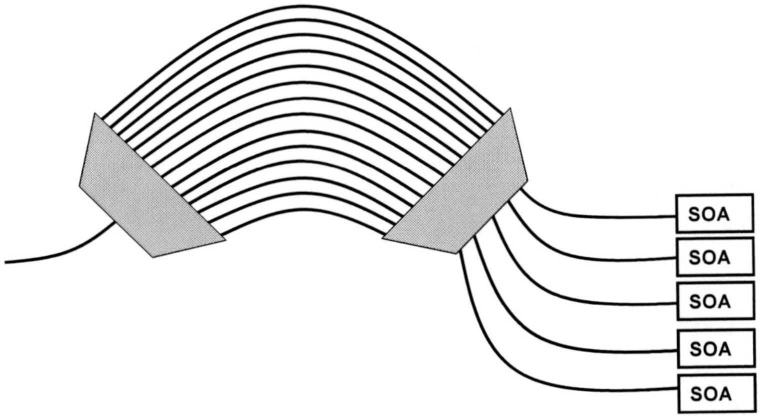

Figure 11. Schematic of integrated AWG and SOAs.

4. SOA INTEGRATED DEVICES FOR ACCESS NETWORKS

In this section, we will discuss the integrated devices using SOAs in the access network. And the SOAs in the devices are serving as the gain blocks. For high speed (> 10 Gb/s operation), using direction modulation of the RSOA is still technological challenging nowadays. Hence electro-absorption modulator (EAM) can be used as the colorless ONU for high speed operation [50-52]. However, the EAM has a relatively high insertion loss, and it is beneficial to monolithically integrate the EAM with SOA in the transmission or reflection mode devices, as shown in Figure 10 (a) and (b), respectively.

As described in [50], the R-SOA-EAM architecture is based on a buried heterostructure design with a butt-coupling interface for the SOA and the EAM sections. The SOA has a slant active area, mode expander and AR coating to minimize the facet reflectivity of $< 10^{-5}$. The length of the SOA section is 670 μm. In order to reduce the PDG, the active area is composed of tensile strained InGaAs superlattice. The EAM section has InGaAs/InAlAs MQW architecture. The length of the EAM section is 50 μm. A high reflection (HR) coating is applied to the rear facet for the reflective operation. A semi-insulating InP current blocking layer is used between the SOA and the EAM sections to provide high electrical resistance. The isolation resistance between the SOA and REAM was 36 kΩ when the applied bias was 2.5 V. The integrated R-SOA-EAM was capable of tolerating optical carrier power variations of up to 13dB when being operated in a 10Gb/s, 128-way split, 100km reach DWDM-TDMA PON [52].

A cascaded EAMs monolithically integrated with SOAs in a long waveguide was also demonstrated for 40 Gb/s data transmission [53]. The active areas of the EAMs and SOAs are MOCVD-grown InGaAsP MQWs (consisting of 6 wells) sandwiched by p- and n- InP layers. The lengths of EAM and SOA are 300 μm and 600 μm respectively. The width of optical waveguide is 3 μm. The isolation between EAM and SOA is formed by using H-ion implantation. The n- and p- contacts are made by evaporating Ni/AuGe/Au and Ti/Au metals.

An arrayed waveguide grating (AWG) is needed in a WDM-PON for distributing wavelengths to different ONUs. Integrating SOAs with an AWG acting as in-line amplifiers can be beneficial for compensating the transmission loss of the optical signals. Figure 11 shows the schematic diagram of an integrated SOA-AWG. Selective area growth metal organic vapor phase epitaxy (SAG-MOVPE) can be used for the fabrication of the device [54]. This method can tailor layout of different bandgaps with a single growth on patterned substrate. Since there are different distances between the ONUs and the OLT in the PON, the received powers at the OLT can vary due to different path loss. By adjusting the gain of the SOAs with different applied currents, the received power can be equalized and controlled within the dynamic range of the receiver.

5. RSOA as Colorless Modulator inside ONU

As mentioned before, WDM-PON is a promising high capacity access network. However, one great challenge is that wavelength tunable laser or wavelength specific laser is required at the cost-sensitive ONU. Wavelength tunable laser is still too expensive nowadays for the end-user. Using different wavelength specific lasers will create inventory issue. One of the cost effective WDM-PONs is to use colorless ONU, its wavelength can be assigned by the network. The simplest colorless ONU implement is to spectrum-slice the RSOA. As described in [55], the RSOA was directly modulated at 155 Mb/s and the output signal was spectrum-sliced by an AWG. The measured power margins were > 5.3 dB. This scheme is simple and inexpensive, however, the modulation speed is usually < 1 Gb/s and transmission distances are limited by the dispersion and slicing loss. Also spectrum-slicing with incoherent light (ASE) introduces excess intensity noise (EIN) generated by the spontaneous-spontaneous beat noise that falls within the receiver bandwidth.

For higher speed operation, ASE-injected RSOAs [56-58], CW-injected RSOAs [59] or self-injected RSOAs [60] can be deployed. In the ASE injection architectures, the broadband light source in the CO or remote node (RN) generates a broadband ASE for feeding into the RSOA at the ONU. The ASE will also be spectrum-sliced, hence allowing the injected RSOA operate in a single-mode light source with negligible crosstalk to other WDM channels in the WDM-PON. In the CW injection architecture, CW light source is used instead of the ASE light source. The CW injection scheme allows much higher data rate operation with extended transmission.

In the carrier distribution schemes (ASE or CW injections), usually the distributed optical carriers are un-modulated. The injected ASE or CW signals will then be modulated at the RSOA to generate the upstream signal, while the downstream signal of the WDM-PON will be carried in another wavelength. In the signal remodulation PON, the downstream signal will be modulated at the ONU to produce the upstream signal. This can further reduce the cost by wavelength reuse. In this case, a single wavelength can carry both downstream and upstream signals.

We proposed and demonstrated a signal remodulation 128 split-ratio WDM-TDM PON with RSOA-based ONU, using downstream differential phase shift keying (DPSK) and upstream on-off keying (OOK) [61] (Figure 12). A CW was encoded by DPSK data via a phase modulator (PM) at 10 Gb/s. The downstream signal was transmitted through 20 km SSMF. Dual-feeder fiber architecture was used to mitigate Rayleigh backscattering at the head-end receiver (Rx), while providing the advantages of using single distribution/drop fiber to reach users. At the ONU, 10% of optical power was taped and demodulated by a 1-bit delayed interferometer (DI). The rest of downstream optical signal was then launched into the RSOA, which was electrically driven by a 2.5 Gb/s non-return-to-zero (NRZ) data for signal remodulation. A high split-ratio of 128 can be achieved.

Figure 13 shows the BER measurements of the downstream and upstream signals at back-to-back and after 20 km transmission. Error free operations were observed in each case. 1.5 dB power penalties were measured in the downstream DPSK signal. Negligible power penalty was observed in the upstream OOK signal after the 20 km transmission. Less than 1 dB power penalty was observed if the downstream CW was used instead of the DPSK signal to launch into the RSOA for producing the upstream OOK signal. The corresponding eye diagrams are shown in the insets.

As mentioned in the introduction, LR-PON has been proposed to integrate the metro and access sections of the network. It can reduce the number of network elements and interconnection interfaces, and are predicted to significantly reduce the capital and operational cost. Although LR-PONs are as yet a research concept, the potential benefits they offer are illustrated that they are a very worthy research topic. We studied the possibility of using orthogonal frequency division multiplexed (OFDM) for signal remodulation in the LR-PON [62]. The inherent advantage of OFDM frequency diversity transmission allows effectively mitigation of chromatic dispersion. Also due to the highly spectral efficiency, low-bandwidth optical components can still be used.

Figure 14 shows the proposed signal remodulation LR-PON. A baseband digital signal processing (DSP) was used to produce the OFDM 16-quadrature amplitude modulation (QAM) signal.

An arbitrary waveform generator with 4 GHz sampling rate was then used to convert the digital to analog data, which was then applied to the EAM for downstream connection. The 4 Gb/s OFDM signal consisted of 16 subcarriers occupying about 1 GHz bandwidth. Each subcarrier was in 16-QAM format. The 1 G symbol/s OFDM signal occupied the radio frequency (RF) spectrum from 62.5 MHz to 1,125 MHz, with data pattern consisted of 8,192 OFDM symbols. The cyclic prefix was 1/32 symbol time. The signal transmitted through 90 km and 10 km of feeder and distribution/drop fibers respectively, via a pair of AWGs. At the ONU, 10% of downstream power was tapped out and detected.

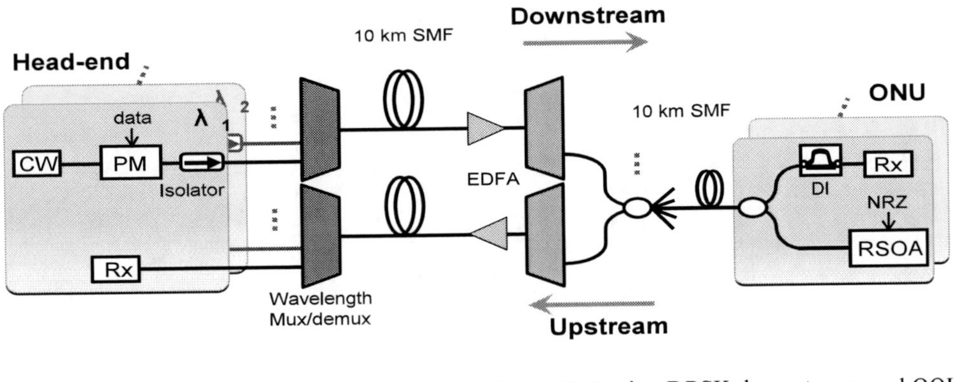

Figure 12. Architecture of the high split-ratio WDM-TDM PON using DPSK downstream and OOK upstream signals. PM: phase modulator, SMF: single mode fiber, DI: delayed interferometer, EDFA: erbium-doped fiber amplifier [61].

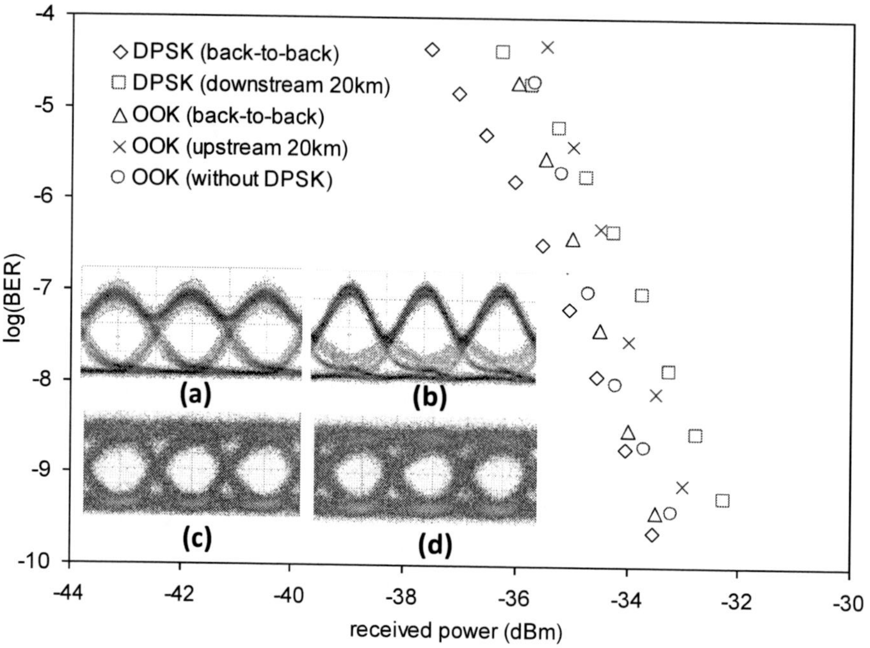

Figure 13. BER measurements of the WDM-TDM PON. Insets: eye diagrams of (a) demodulated DPSK at back-to-back, (b) DPSK after 20 km transmission, (c) OOK at back-to-back, (d) OOK after 20 km transmission [61].

An analog to digital converter, which is a real-time 10 GHz sampling oscilloscope, converted the downstream signal detected by the optically pre-amplified Rx to digital signal for demodulation. The demodulation was performed using computer software; and it consisted of a synchronizer to extract the carrier phase and aligned the OFDM symbol boundaries; a fast Fourier transform (FFT) to translate signal from time to frequency domain; a QAM decoder to analyze the symbol on each subcarrier to make the final decision. The measured error vector magnitude (EVM) was used to calculate the BER.

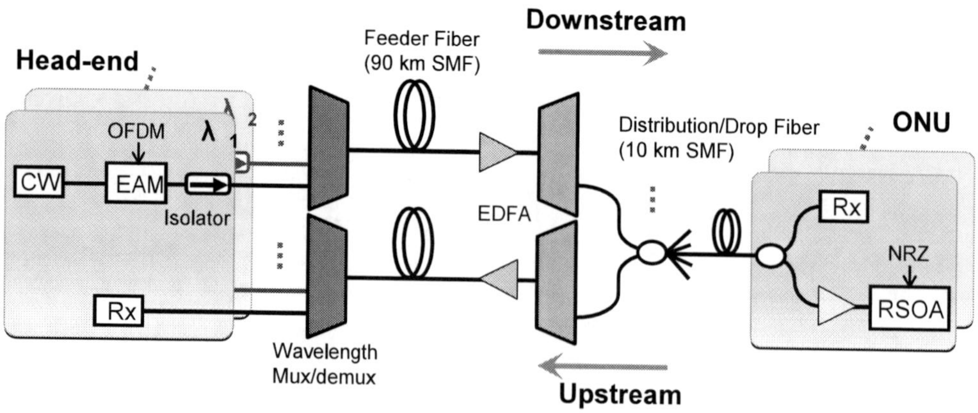

Figure 14. Architecture of the LR PON using OFDM downstream and OOK upstream signals. EAM: electro-absorption modulator, SMF: single mode fiber, EDFA: erbium-doped fiber amplifier [62].

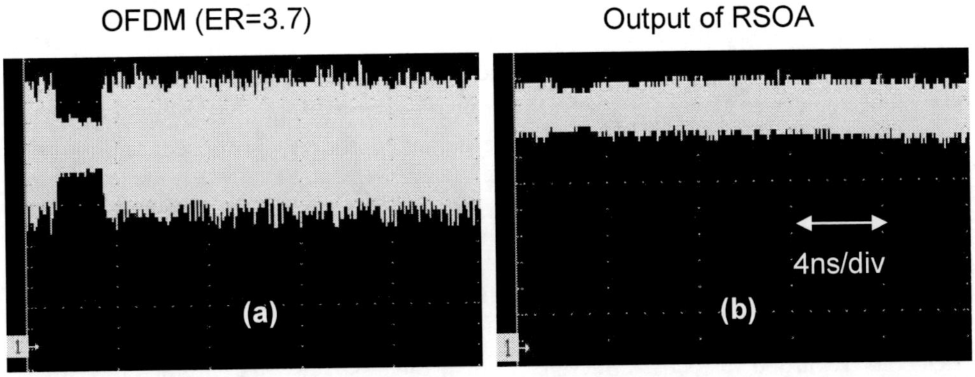

Figure 15. Time traces from the real-time oscilloscope of (a) the OFDM, output of (b) the RSOA when the ER of the input OFDM signal is 3.7 dB.

Figure 15 shows the time traces from the real-time oscilloscope of (a) the downstream OFDM (ER = 3.7 dB), and the (b) output of the RSOA. The electrical data applied to the RSOA was switched-off and only DC-bias was used. We can clearly observe that by using the gain saturation property of the RSOA, the downstream OFDM signal can be significantly suppressed, hence producing a better upstream OOK signal.

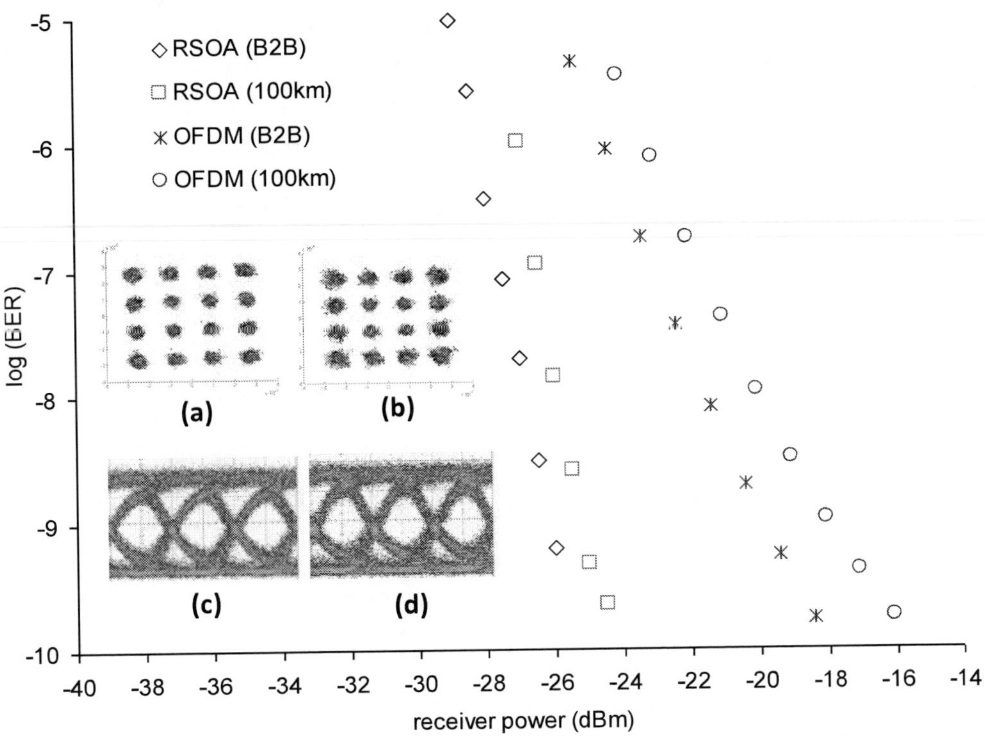

Figure 16. BER measurements of different signals. Insets: constellation diagrams of OFDM signal at ER = 3.7 dB at (a) B2B, (b) 100km, output remodulated NRZ eye diagrams of RSOA at (c) B2B, (d) 100km, when downstream OFDM at ER = 3.7 dB.

Figure 16 shows the BER measurements of each signal at B2B and after the 100 km LR transmission. Insets show the constellation diagrams of OFDM signal at ER = 3.7 dB, output remodulated NRZ eye diagrams of RSOA when downstream OFDM at ER = 3.7 dB. We can observe clear open eyes after 100 km SSMF transmission without dispersion compensation in each case. Error-free BER operation can be achieved for the proposed remodulation LR system.

Conclusion

PON is regarded as cost-effective optical access network architecture. The first generation of TDM-PON is now standardized and commercially available. They have already been deployed in Japan, North America, and Korea. As the bandwidth demand continues to grow, researchers are planning the next generation PONs, such as high data rate PONs (40 Gb/s – 100 Gb/s), LR-PONs and WDM PONs, which are currently under intensive studies. SOAs can be used as gain blocks to increase the power budget or as colorless modulators in the next generation access networks. SOA is compact, has low PDG and low power consumption. It is Integratable with other photonic devices and is potentially low cost. Due to the improvement in semiconductor fabrication technologies, SOA can be operated at high data rate (> 2.5 Gb/s) and in a wide temperature range, with high output saturation power. In this chapter, we have described the basic principle of SOA, followed by the SOA characteristics. Then, we presented some integrated devices using SOAs for next generation access networks. Finally, we proposed and demonstrated using RSOA as the colorless ONU for the next generation signal remodulation WDM-PON and signal remodulation LR-PON.

References

[1] D. B., Payne and R. P. Davey, "The future of fibre access systems?" *BT Technology Journal,* vol. 20, pp. 104-114, 2002.

[2] S. Chi, C. H. Yeh, and C. W. Chow, "Broadband access technology for passive optical network," *Proc. of SPIE*, vol. 7234, 723408-1 (Invited Paper), 2009.

[3] http://www.ieee802.org/3/av/

[4] D. Qian, N. Cvijetic, Y. –K. Huang, J. Yu, T. Wang, "100km long reach upstream 36Gb/s-OFDMA-PON over a single wavelength with source-free ONUs," *Proc. ECOC*, 8.5.1, 2009.

[5] J. Yu, Z. Jia, P. N. Ji, T. Wang, "40-Gb/s wavelength-division-multiplexing passive optical network with centralized lightwave source," *Proc. OFC*, OTuH8, 2008.

[6] D. Qian, N. Cvijetic, J. Hu, T. Wang, "40-Gb/s MIMO-OFDM-PON using polarization multiplexing and direct-detection," *Proc. OFC*, OMV3, 2009.

[7] S. –G. Mun, J. –H. Moon, H. –K. Lee, J. –Y. Kim, and C. –H. Lee, "A WDM-PON with a 40 Gb/s (32 x 1.25 Gb/s) capacity based on wavelength-locked Fabry-Perot laser diodes," *Opt. Express*, vol. 16, 11361-11368, 2008.

[8] C. W. Chow, L. Xu, C. H. Yeh, H. K. Tsang, W. Hofmann, and M. C. Amann, "40 Gb/s upstream transmitters using directly-modulated 1.55 μm VCSEL array for high-split-ratio PONs," *IEEE Photon. Technol. Lett.*, vol. 22, pp. 347-349, 2010.

[9] C. H. Yeh, C. W. Chow, C. H. Wang, Y. F. Wu, F. Y. Shih and S. Chi, "Using OOK modulation for symmetric 40 Gb/s long reach time-sharing passive optical networks," *IEEE Photon. Technol. Lett.*, vol. 22, pp. 619-621, 2010.

[10] C. H. Yeh, C. W. Chow, C. -H. Hsu, "40 Gb/s time division multiplexed passive optical networks using downstream OOK and upstream OFDM modulations," *IEEE Photon. Technol. Lett.*, vol. 22, pp. 118-120, 2010.

[11] Z. Xu, T. Cheng, X. Cheng, Y. Yeo, Y. Wen, Y. Wang, and W. Rong, "Broadcast capable 40-Gb/s WDM passive optical networks," *Proc. ECOC*, P6.05, 2009.

[12] P. P. Iannone, K. C. Reichmann, C. R. Doerr, L. L. Buhl, M. A. Cappuzzo, E. Y. Chen, L. T. Gomez, J. E. Johnson, A. M. Kanan, J. L. Lentz, and R. McDonough, "A 40Gb/s CWDM-TDM PON with a cyclic CWDM multiplexer/demultiplexer," *Proc. ECOC*, 8.5.6, 2009.

[13] D. Qian, N. Cvijetic, J. Hu, T. Wang, "108 Gb/s OFDMA-PON with polarization multiplexing and direct detection," *J. Lightw. Technol.*, vol. 28, pp. 484, 2010.

[14] R. P. Davey, P. Healey, I. Hope, P. Watkinson, D. B. Payne, O. Marmur, J. Ruhmann, and Y. Zuiderveld, "DWDM reach extension of a GPON to 135 km" *Proc. OFC*, PDP35, 2005.

[15] D. P. Shea, A. D. Ellis, D. B. Payne, R. P. Davey, and J. E. Mitchell, "10 Gbit/s PON with 100 km reach and x1024 split," *Proc. ECOC*, We.P.147, 2003.

[16] D. Nesset, R. P. Davey, D. Shea, P. Kirkpatrick, S.Q. Shang, M. Lobel, B. Christensen, "10 Gbit/s bidirectional transmission in 1024-way split, 110 km reach, PON system using commercial transceiver modules, Super FEC and EDC," *Proc. ECOC*, Tu1.3.1, 2005.

[17] D. P. Shea, and J. E. Mitchell, "A 10-Gb/s 1024-way-split 100-km long-reach optical-access network," *J. Lightw. Technol.,* vol. 25, 685-693, 2007.

[18] P. D. Townsend, G. Talli, C. W. Chow, E. M. MacHale, C. Antony, R. Davey, T. De Ridder, X. Z. Qiu, P. Ossieur, H. G. Krimmel, D. W. Smith, I. Lealman, A. Poustie, S. Randel, H. Rohde, "Long reach passive optical networks," *IEEE LEOS Annual Meeting*, 2007.

[19] P. Ossieur, C. Antony, A. Naughton, A. Clarke, P. D. Townsend, H. G. Krimmel, T. De Ridder, X. Z. Qiu, C. Melange, A. Borghesani, D. Moodie, A. Poustie, R. Wyatt, B. Harmon, I. Lealman, G. Maxwell, D. Rogers, D. W. Smith, "A symmetric 320Gb/s capable, 100km extended reach hybrid DWDM-TDMA PON," *Proc. OFC*, NWB1, 2010.

[20] C. Antony, P. Ossieur, A. M. Clarke, A. Naughton, H.G. Krimmel, Y. Chang, A. Borghesani, D. Moodie, A. Poustie, R. Wyatt, B. Harmon, I. Lealman, G. Maxwell, D. Rogers, D. W. Smith, D. Nesset, R.P. Davey, P. D. Townsend, "Demonstration of a carrier distributed, 8192-split hybrid DWDM-TDMA PON over 124 km field-installed fibers," *Proc. OFC*, PDPD8, 2010.

[21] J. Prat, J. Lazaro, P. Chanclou, R. Soila, A. M. Gallardo, A. Teixeira, G. M. TosiBeleffi, I. Tomkos, "Results from EU project SARDANA on 10G extended reach WDM PONs," *Proc. OFC*, OThG5, 2010.

[22] C. W. Chow, C. H. Yeh, Y. T. Li, C. H. Wang, F. Y. Shih, Y. M. Lin, C. L. Pan, and S. Chi, "Demonstration of high spectral efficient long reach passive optical networks using OFDM-QAM," *Proc. CLEO*, CPDB7, 2008.

[23] C. W. Chow, C. H. Yeh, C. H. Wang, F. Y. Shih, C. L. Pan, and S. Chi, "WDM extended reach passive optical networks using OFDM-QAM," *Opt. Express*, vol. 16, 12096-12101, 2008.

[24] S. B. Park, D. K. Jung, D. J. Shin, H. S. Shin, I. K. Yun, J. S. Lee, Y. K. Oh, and Y. J. Oh, "Colorless operation of WDM-PON employing uncooled spectrum-sliced reflective semiconductor optical amplifiers," *IEEE Photon. Technol. Lett.,* vol. 19, pp. 248-250, 2007.

[25] A. D. McCoy, P. Horak, B. C. Thomsen, M. Ibsen, M. R. Mokhtar, D. J. Richardson, "Optimising signal quality in a spectrum-sliced WDM system using SOA-based noise reduction," *Proc. ECOC*, Tu4.6.4, 2004.

[26] P. Healey, P. Townsend, C. Ford, L. Johnston, P. Townley, I. Lealman, L. Rivers, S. Perrin and R. Moore, "Spectral slicing WDM-PON using wavelength-seeded reflective SOAs," *Electron. Lett.,* vol.37, pp. 1181-1182, 2001.

[27] H. S. Shin, D. K. Jung, H. S. Kim, D. J. Shin, S. B. Park, S. T. Hwang, Y. J. Oh, and C. S. Shim, "Spectrally pre-composed ASE injection for a wavelength-seeded reflective SOA in a WDM-PON," *Proc. ECOC*, We3.3.7, 2005.

[28] F. Payoux, P. Chanclou, N. Genay, "WDM PON with a single SLED seeding colorless RSOA-based OLT and ONUs," *Proc. ECOC*, Tu4.5.1, 2006.

[29] C. Arellano, C. Bock, J. Prat, and K. D. Langer, "RSOA-based optical network units for WDM-PON," *Proc. OFC*, OTuC1, 2006.

[30] K. L. Lee, and E. Wong, "Directly-modulated self-seeding reflective SOAs in WDM-PONs: performance dependence on seeding power and modulation effects," *Proc. ECOC*, Tu4.5.2, 2006.

[31] A. Borghesani, I. F. Lealman, A. Poustie, D. W. Smith, and R. Wyatt, "High temperature, colourless operation of a reflective semiconductor optical amplifier for 2.5Gbit/s upstream transmission in a WDM-PON," *Proc. ECOC*, Paper 6.4.1, 2007.

[32] E. K. MacHale, G. Talli, P. D. Townsend, A. Borghesani, I. Lealman, D. G. Moodie, and D. W. Smith, "Extended-reach PON employing 10Gb/s integrated reflective EAM-SOA," *Proc. ECOC*, Th.2.F.1, 2008.

[33] E. K. MacHale, G.. Talli, C. W. Chow, and P. D. Townsend, "Reduction of signal-induced Rayleigh noise in a 10Gb/s WDM-PON using a gain-saturated SOA," *Proc. ECOC*, Paper 7.6.3, 2007.

[34] L. Y. Chan, C. K. Chan, D. T. K. Tong, F. Tong, and L. K. Chen, "Upstream traffic transmitter using injection-locked Fabry-Perot laser diode as modulaotr for WDM access networks," *Electron. Lett.*, vol. 38, pp. 43-45, 2002.

[35] W. Hung, C. K. Chan, L. K. Chen, and F. Tong, "An optical network unit for WDM access networks with downstream DPSK and upstream remodulated OOK data using injection-locked FP laser," *IEEE Photon. Technol. Lett.,* vol. 15, pp. 1476-1478, 2003.

[36] G. Zeidler and D. Schicetanz, "Use of laser amplifiers in glass fibre communication systems," *Siemens Forch. u. Entwickl. Ber.*, vol. 2, pp. 227-234, 1973.

[37] S. D. Personick, "Applications for quantum amplifiers in simple digital optical communication systems," *Bell Syst. Tech. J.*, vol. 52, pp. 117-133, 1973.

[38] J. C. Simon, "GaInAsP semiconductor laser amplifiers for single-mode fibre communications," *J. Lightw. Technol.*, vol. 5, pp. 1286-1295, 1987

[39] M. J. Connelly, *Semiconductor Optical Amplifiers*, Kluwer Academic Publishers, 2004.

[40] A. Borghesani, "Semiconductor optical amplifiers for advanced optical applications," *Proc. ICTON*, pp. 119-122, 2006.

[41] M. Reed, T. M. Benson, P. C. Kendall and P. Sewell, "Antireflection-coated angled facet design," *IEE Proc. Pt. J. Optoelectronics*, vol. 143, pp. 214 -220, 1996.

[42] A. Vukovic, P. Sewell, T. M. Benson and P. C. Kendall, "Advances in facet design for buried lasers and amplifiers," *IEEE J. on Sel. Topics in Quantum Electron.*, vol. 6, pp. 175-184, 2000.

[43] I. Cha, M. Kitamura, H. Honmou and I. Mito, "1.5 μm band travelling-wave semiconductor optical amplifier with window facet structure," *Electron. Lett.*, vol. 25, pp. 1241-1242, 1989.

[44] N. A. Olsson, R. F. Kazarinov, W. A. Nordland, C. H. Henry, M. G. Oberg, H. G. White, P. A. Garbinski and A. Savage, "Polarisation-independent optical amplifier with buried facets," *Electron. Lett.*, vol. 25, pp. 1048-1049, 1989.

[45] B. Mersali, G. Gelly, A. Accard, J. L. Lafragette, P. Doussiere, M. Lambert and B. Fernier, "1.55μm high-gain polarisation-insensitive semiconductor travelling wave amplifier with low driving current," *Electron. Lett.*, vol. 26, pp. 124-125, 1990.

[46] M. Joma, H. Horikawa, C. Q. Xu, K. Yamada, Y. Katoh, and T. Kamijoh, "Polarisation insensitive semiconductor laser amplifiers with tensile strained InGaAsP/lnGaAsP multiple quantum well structure," *Appl. Phys. Lett.*, vol. 62, pp. 121-122, 1993.

[47] K. Y. Cho, Y. Takushima, and Y. C. Chung, "10-Gb/s operation of RSOA for WDM PON," *IEEE Photon. Technol. Lett.*, vol. 20, pp. 1533, 2008.

[48] H. Kim, "10-Gb/s operation of RSOA using a delay interferometer," *IEEE Photon. Technol. Lett.*, vol. 22, pp. 1379, 2010.

[49] B. Baekelandt, C. Mélange, J. Bauwelinck, P. Ossieur, T. de Ridder, X. –Z. Qiu, and J. Vandewege, "OSNR penalty imposed by linear in-band crosstalk caused by interburst residual power in multipoint-to-point networks," *IEEE Photon. Technol. Lett.*, vol. 20, pp. 587, 2008.

[50] D. Smith, I. Lealman, X. Chen, D. Moodie, P. Cannard, J. Dosanjh, L. Rivers, C. Ford, R. Cronin, T. Kerr, L. Johnston, R. Waller, R. Firth, A. Borghesani, R. Wyatt and A. Poustie, "Colourless 10Gb/s reflective SOA-EAM with low polarization sensitivity for long-reach DWDM-PON networks," *Proc. ECOC*, 8.6.3, 2009.

[51] D. W. Smit, "Reducing the optical component cost for future fibre access," *Proc. ECOC*, 4.7.4, 2009.

[52] E. K. MacHale, G. Talli, P. D. Townsend, A. Borghesani, I. Lealman, D. G. Moodie, and D. W. Smith, "Extended-reach PON employing 10Gb/s integrated reflective EAM-SOA," *Proc. ECOC*, Th.2.F.1, 2008.

[53] T. H. Wu, J. P. Wu, H. J. Yan, and Y. J. Chiu, "40Gb/s SOA-integrated EAM using cascaded structure," *Proc. IPRM*, 2010.

[54] A. Al Amin, X. Song, K. Sakurai and M. Sugiyama, "Integration of semiconductor optical amplifiers with an arrayed waveguide grating demultiplexer by MOVPE selective area growth," *Proc. IPR*, IFB4, 2004.

[55] P. Healey, P. Townsend, C. Ford, L. Johnston, P. Townley, I. Lealman, L. Rivers, S. Perrin, and R. Moore, "Spectral slicing WDM-PON using wavelength-seeded reflective SOAs," *Electron. Lett.*, vol. 37, pp. 1181-1182, 2001.

[56] H. S. Shin, D. K. Jung, H. S. Kim, D. J. Shin, S. B. Park, S. T. Hwang, Y. J. Oh, and C. S. Shim, "Spectrally pre-composed ASE injection for a wavelength-seeded reflective SOA in a WDM-PON," *Proc. ECOC*, We3.3.7, 2005.

[57] H. S. Shin, D. K. Jung, D. J. Shin, S. B. Park, J. S. Lee, I. K. Yun, S. W. Kim, Y. J. Oh, and C. S. Shim, "16 x 1.25 Gbit/s WDM-PON based on ASE-injected R-SOAs in 60°C temperature range," *Proc. OFC*, OTuC5, 2006.

[58] F. Payoux, P. Chanclou, and R. Brenot, "WDM PON with a single SLED seeding colorless RSOA-based OLT and ONUs," *Proc. ECOC*, Tu4.5.1, 2006.

[59] C. Arellano, B. Bock, J. Prat, and K. D. Langer, "RSOA-based optical network units for WDM-PON," *Proc. OFC*, OTuC1, 2006.

[60] K. L. Lee, and E. Wong, "Directly-modulated self-seeding reflective SOAs in WDM-PONs: performance dependence on seeding power and modulation effects," *Proc. ECOC*, Tu4.5.2, 2006.

[61] C. W. Chow, C. H. Yeh, C. H. Wang, F. Y. Shih and S. Chi, "Signal remodulation high split-ratio hybrid WDM-TDM PONs using RSOA-based ONUs," *Electron. Lett.*, vol. 45. pp. 903-905, 2009.

[62] C. W. Chow, C. H. Yeh, C. H. Wang, F. Y. Shih and S. Chi, "Signal remodulation of OFDM-QAM for long reach carrier distributed passive optical networks," *IEEE Photon. Technol. Lett.*, vol. 21, pp. 715-717, 2009.

In: Optical Amplifiers
Editor: Galina Nemova

ISBN 978-1-61209-835-7

Chapter 3

NONLINEAR DYNAMICS OF SEMICONDUCTOR OPTICAL AMPLIFIERS

Ramón Gutiérrez–Castrejón
Institute of Engineering,
Universidad Nacional Autónoma de México, UNAM,
Cd. Universitaria, Coyoacán 04510 Mexico D.F., Mexico

Abstract

The semiconductor optical amplifier (SOA) is a versatile device that has found application at the core of all–optical signal processing sub–systems. These sophisticated structures are expected to replace current electronic data processing elements in next–generation reconfigurable photonic telecommunication networks. Besides, there has also been a renewed interest to use SOAs to cost–effectively extend the power budget margins of current metro and access optical networks. However, in contrast to Er–doped fiber amplifiers, whose excited level has a long lifetime of about 10 ms, SOAs exhibit carrier lifetimes in the order of picoseconds and a nonlinear response that is orders of magnitude higher as compared to optical fibers. Therefore, these minute active devices normally become affected by signals modulated at Gb/s rates. As a result of this characteristic, a clear understanding of the nonlinear dynamical behavior of the light–matter interactions that take place within active semiconductor waveguides is necessary. This helps in the design and engineering of optical systems where SOAs play a significant role. In this contribution we first review some of the most relevant nonlinear dynamical phenomena found in SOAs. Then, a discussion is presented on how the finite response time associated to these phenomena affects some applications employed by the optical telecommunications industry, such as in–line amplification and wavelength conversion. Finally, using a well–tested photonic circuits and systems simulator, novel solutions to enhance the performance of the aforementioned SOA–based applications are demonstrated. Narrow–band optical filtering and the combination of several devices in ingenious architectures are among them. The former approach is mainly analyzed through the use of simulated response curves, while the latter is evaluated via eye pattern analysis. In particular, a significant reduction of data pattern–dependent amplitude jitter is numerically demonstrated using an original method. The novel approach consists in combining the effect of interferometer (in this case a Mach-Zehnder interferometer) unbalance and the injection of a modulated holding beam that constitutes the binary complement of the processed signal. This chapter, therefore, not only represents an aid for the neophyte scientist or engineer to become involved in the nonlinear dynamical behavior of semiconductor optical amplifiers, but it provides with new insights in SOA–induced patterning mitigation to the well-informed reader.

1. Introduction

Fiber amplifiers, like Er–doped [1] and Raman amplifiers [2, 3], have pervaded today's photonic telecommunication networks. This phenomenon can be explained from the high gain, high output saturation power, low coupling losses and relatively low noise figure (NF) that these technologies exhibit. Their use has enabled wavelength division multiplexing (WDM), a multi–channel technology that represents the cornerstone for the implementation and management of modern photonic networks [4]. Next–generation systems, however, will also take advantage of other contending technologies to fulfill new amplification requirements. One amplification technology that is worth to be considered is the semiconductor optical amplifier (SOA), also known as semiconductor laser amplifier. This versatile device is attractive because it offers potentially low production costs, wider and more flexible amplification bandwidth as compared to rare–earth–doped amplifiers, and monolithic integration with other components in a compact structure. Recent models have shown high performance, exhibiting an output saturation power above 19 dBm, a NF of 6 dB or lower, and a gain bandwidth of at least 100 nm [5]. So far, however, the use of SOAs as in–line amplifiers has been hindered due to its nonlinear response, which is four orders of magnitude higher than that observed in silica fibers. Moreover, in contrast to Er–doped fiber amplifiers, whose excited level has a long lifetime of about 10 ms, SOAs exhibit carrier lifetimes in the order of picoseconds. Consequently, these active devices normally become affected by optical signals modulated at Gb/s rates, commonly employed in current fiber optic telecommunication systems. In spite of these disadvantages, SOAs are expected to soon penetrate the in–line amplification market at the metro/access network level, where average signal strengths are relatively low and cost–effective technologies are essential to extend power budget margins [6]. Longer span lengths with high channel counts will only be reached at the expense of more advanced associated technologies. Among them, it can be mentioned the use of more sophisticated internal nanometric SOA structures referred to as quantum dots [7]; the implementation of advanced modulation formats [8, 9]; and other approaches treated in this chapter.

SOAs are not only potentially useful as in–line amplifiers. Actually, the main application of SOAs relies on the use of their (apparently deleterious) nonlinear behavior to turn this minute amplifiers into photonic switches. Under high saturation conditions, the SOA can operate as wavelength converter, Boolean logic gate (OR, XOR, AND, etc.) and optical time division demultiplexer, among other possibilities [10]. The functionality of SOAs in next–generation reconfigurable photonic networks is thus expected to span from the optical fiber links (as optical amplifiers) to the core of the all–optical signal processing elements located at the network nodes. The latter elements will eventually substitute current electronic data–processing equipment, thus avoiding the optic–electronic–optic pathway that nowadays limits bit rate, transparency and power saving.

This chapter is set out in the following manner. The next section represents an introduction to the bulk semiconductor optical amplifier, mainly from a physical standpoint. Section

3. provides a brief description of some nonlinear dynamical phenomena observed in these active devices. It mainly refers to inter– and intra–band interactions and their associated characteristic times. Four-wave mixing in SOAs is addressed in section 3.2.. Section 4. is devoted to more practical considerations. In particular, it presents the concept of data patterning, relevant when optical data streams, instead of pulse trains or continuous wave signals, are of interest. Mitigation of data–pattern effects is discussed in section 5., where two different approaches are presented. The first one deals with optical filtering. Both, linear and nonlinear filters are analyzed. The use of simulated response curves uncovers the effectiveness of this technique. The second one, discussed in section 5.2., refers to compensating architectures that lessen the amount of amplitude jitter. Among these schemes, the use of modulated and unmodulated holding beam techniques are of particular interest. Finally, in section 6., the main results of this research work are summarized.

2. The Bulk Semiconductor Optical Amplifier

An SOA consists of a forward–biased p–n hetero–junction, usually fabricated of InP/InGaAsP material, through which current flows; and optical waveguides to confine the incoming light to the active region, which provides amplification [11]. The SOA is therefore similar to a laser diode where the reflectivity of both facets is reduced to prevent optical feedback [12]. When a sufficiently high current is injected into the semiconductor device, population inversion can be achieved. An incoming beam having a photon energy slightly larger than the material band–gap can then be amplified through stimulated emission occurring in the active region. If the beam is powerful enough, recombination depopulates the conduction band and fills the valence band, leading to saturation of the amplification process [13]. However, if the amplifier is operated in the small–signal regime, its gain coefficient can be assumed to be independent of z, where z represents position in the longitudinal axis of the active waveguide. The power of the optical waveform propagating along the SOA, $P(z)$, can hence be written as

$$\frac{dP}{dz} = g_0 P(z) - \alpha_{int} P(z)\,, \tag{1}$$

leading to $P = P_{in}\exp((g_0 - \alpha_{int})z) = P_{in}G(z)$. Here, G represents the amplifier gain, g_0 represents the small–signal gain coefficient, and α_{int} stands for the amplifier internal losses. Typical values of the latter parameter in SOAs are around $\alpha_{int} \approx 4000\ \mathrm{m}^{-1}$, implying a total loss of $10\log(\exp(-\alpha_{int}L)) \approx -8.7\mathrm{dB}$ in an SOA $L = 0.5$ mm long.

In SOAs, a variation of the material gain is correspondingly accompanied by a change in the refractive index. This occurs because both are coupled through the semiconductor carrier density, N. This coupling, consequence of the complex nature of the electric susceptibility, is usually modeled via either the Kramers–Kronig relations [14] or, more regularly, the linewidth enhancement factor [15], given by

$$\alpha_N(N,\lambda) = -\frac{4\pi}{\lambda}\frac{\partial n/\partial N}{\partial g/\partial N}\,, \tag{2}$$

where λ represents the optical wavelength, g is the material gain, and n is the refractive index. To a good approximation, the refractive index is proportional to the carrier density

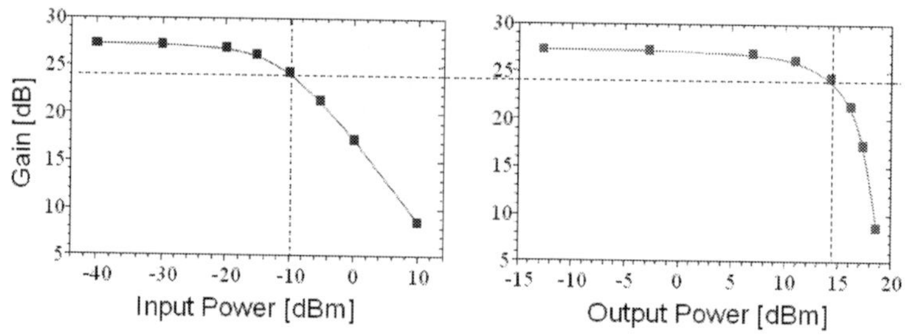

Figure 1. Nonlinear SOA gain saturation for continuous wave beams as a function of input (left) and output (right) power.

for wavelengths near the band edge, and hence to the optical intensity [16]. Under these considerations, the phase shift experienced by the propagating optical beam due to the SOA refractive index variation can be expressed as [15]:

$$\varphi(N) - \varphi(N_{ref}) = -\frac{L}{2}\alpha_N(N_{ref})(\Gamma g(N) - \Gamma g(N_{ref})), \tag{3}$$

where Γ is the, so–called, confinement factor, denoting the fraction of the waveguide–mode distribution that overlaps the amplifier active region, L is the amplifier length, and N_{ref} stands for a reference carrier density.

When a sufficiently intense beam is injected into an SOA, the amplifier gain becomes saturated, that is, the output power cannot longer be approximated by a linear function of the input power. Then, Eq. (1) takes the following form:

$$\frac{dP}{dz} = \frac{g_0 P(z)}{1 + P(z)/P_{sat}} - \alpha_{int} P(z), \tag{4}$$

where P_{sat} stands for the well–known saturation power [17]. This parameter can be understood as the power that leads to a reduction in the gain coefficient to half of its small–signal value, g_0. In practice, it is more convenient to use the input/output saturation power. It is defined as the input/output power at which the gain becomes reduced by 3 dB from its small–signal (linear) value, G_0. An example is presented in Fig. 1 for a small–signal gain $G_0 = 27$ dB. The vertical dotted lines indicate the input and output saturation power values, which are -10 and 14.5 dBm, respectively.

3. Dynamic Response of SOAs

3.1. Time domain

Most of practical applications of SOAs involve the injection of time–varying signals rather than continuous beams. When the extent of the temporal variation of the input signal power is significant as compared to the input saturation power, the SOA carrier density may become time–dependent as well, provided that the SOA nonlinear response is fast enough. Under these conditions, the gain and refractive index will dynamically vary according to the input driving signal. A common approach to characterize the dynamic nonlinear response of an SOA consists in injecting a high–energy optical pulse. If the pulse duration is short enough (in the order of tens of ps) carrier density depletion will take place, and the amplifier gain will be compressed. Due to the limited duration of the disturbance, restoration of the carrier distribution will occur in the active region, and the gain and refractive index of the SOA will relax back to its equilibrium value. The time it takes for the amplifier to recover its original carrier density value will mainly be determined by its effective carrier lifetime, which can be understood to be governed by two terms: the rate of spontaneous emission, R(N), and the rate of stimulated emission. The former is characterized by a differential carrier lifetime, $\tau_{diff} = (\partial R(N)/\partial N)^{-1}$, that usually is approximated by a constant value, τ, in the order of hundreds of picoseconds [18]. A more accurate description of the rate of spontaneous emission, especially when the SOA is operated in a wide interval of saturation levels, is based on the following polynomial expansion:

$$R(N) = AN + BN^2 + CN^3 . \tag{5}$$

Here, A is the so–called Schockley–Read–Hall (or leakage) coefficient. Its value is small and usually can be neglected [19]. B is the bimolecular radiative rate coefficient, and for high carrier densities becomes N–dependent [20]. And C is known as the Auger coefficient. It can be considered as the main non–radiative recombination process. According to this approach, the differential carrier lifetime is calculated as

$$\tau_{diff} = \frac{1}{A + 2BN + 3CN^2} . \tag{6}$$

When an SOA is operated far from saturation (small–signal regime) the rate of spontaneous emission exceeds the rate of stimulated emission. At high saturation, the latter is predominant, and a description of the amplifier response through the differential carrier lifetime is no longer enough. The effective carrier lifetime, which includes contributions from both spontaneous and stimulated recombinations, can be defined as [19]:

$$\tau_{eff}^{-1} = \frac{\partial}{\partial N}\left(\frac{N}{\tau} + \frac{g(N)P}{P_{sat}a_0\tau}\right) = \frac{1}{\tau} + \frac{\partial g(N)}{\partial N}\frac{P}{P_{sat}a_0\tau} , \tag{7}$$

where a constant differential lifetime has been assumed. a_0 is the differential gain. Since $a_0 = \partial g(N)/\partial N$, Eq. (7) can be recast as:

$$\tau_{eff}^{-1} = \frac{1 + P/P_{sat}}{\tau} . \tag{8}$$

According to this equation, when a weak beam is injected into the amplifier ($P/P_{sat} \approx 0$), the effective and differential carrier lifetimes are practically the same. On the other hand, when a strong beam is injected into the amplifier ($P/P_{sat} > 1$), the effective carrier lifetime can be significantly reduced, thus speeding up the response of the device. It must be noted that Eq. (8) was derived assuming a lumped SOA, and therefore it can only be regarded as an aid to understand the behavior of SOA dynamics.

The effective carrier lifetime is closely related to the SOA recovery time, a measurable quantity that provides a good indication on how fast an SOA can be optically driven. For many applications, both parameters are equivalent. As already mentioned, the recovery time can be calculated from the response of the amplifier to an injected optical pulse. Ideally, it should be measured from the gain, and it is defined as the time needed for the gain to

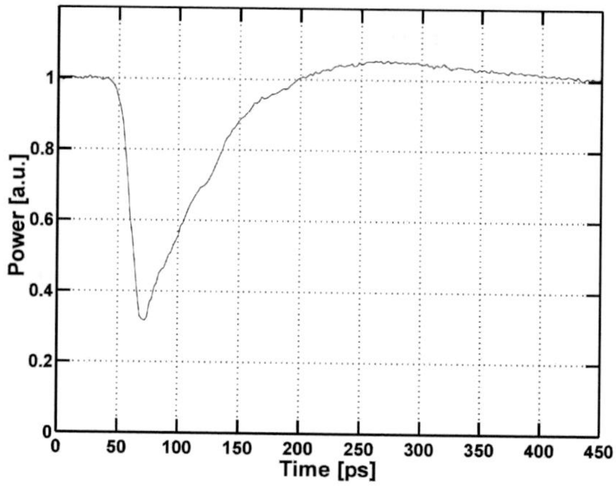

Figure 2. Measurement of SOA nonlinear dynamics in response to a strong optical disturbance. The probe beam power is assumed to be proportional to the carrier density level.

recover (10% to 90%) to steady–state after being compressed by an optical pulse. In practice, however, a probe CW beam is coupled with the pump pulse to read out the variation experienced by the amplifier gain. Pump and probe beams should have different frequencies or polarization states, in such a manner that the latter can easily be differentiated at the amplifier output. Fig. 2 shows the probe beam evolution when the SOA is affected by a short high–energy optical pulse. A fast depletion of the gain is observed. It is followed by a slow recovery of the carrier density to its original steady–state level. The recovery time in this case can be estimated to be about 80 ps. The recovery time can be shortened by increasing the driving current of the SOA [21], and by using a longer SOA [22]. In this example a 1 mm long SOA having an injection current of 400 mA was employed.

If the characteristic time–scale of the disturbance (i.e the width of an injected pulse) is comparable or shorter than the sub–picosecond scattering times on which a quasi–equilibrium carrier distribution is established in the SOA, the detailed evolution of the carrier distribution, and not only the overall carrier concentration, becomes important to determine the device recovery time [23–26]. There are mainly two intra–band nonlinear effects that produce carrier energy redistribution within each band: carrier heating (CH) and spectral hole–burning (SHB). Because of the complex nature of the electrical susceptibility, such processes affect both the gain and refractive index of the amplifier, causing ultra–fast gain compression and phase–shifting, respectively.

In thermal equilibrium at temperature T, the carrier density in the conduction band of a bulk semiconductor can be understood as the product of the density of states per unit volume, which has a parabolic shape, and a probability distribution function, which indicates what the likelihood of occupancy is for electron states with energy E_c. Since electrons are Fermions, their distribution function is given by:

$$f_c(E_c) = \frac{1}{1+\exp([E_c - E_{fc}]/k_B T)}, \tag{9}$$

where E_{fc} is the quasi–Fermi level of the electrons and k_B is Boltzmann constant. The

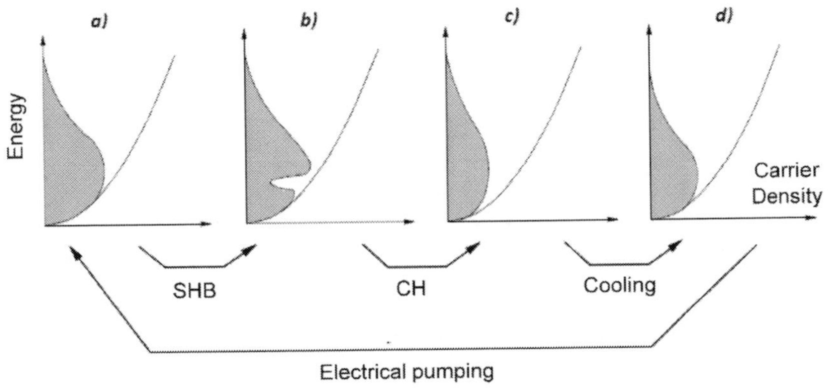

Figure 3. Temporal evolution of the free carrier density distribution in the conduction band of a semiconductor optical amplifier induced by a short optical pulse [27].

distribution of electrons in the conduction band can then be schematically represented as shown in Fig. 3.a.

When a short optical pulse is injected into the SOA, it stimulates carriers to recombine in only a narrow neighborhood of energies corresponding to the pulse photon energy. This process may create a spectral hole, that is, a local deviation of the carrier distribution function from the original Fermi–Dirac distribution presented in Eq. (9). This phenomenon is schematically shown in Fig. 3.b, and it is called SHB. Note that, due to stimulated emission, the total carrier density (and corresponding gain) within the band becomes reduced. Relaxation back to the quasi–equilibrium Fermi distribution occurs via carrier–carrier scattering in time scales of about 50 to 100 fs for InGaAsP amplifiers [28].

At this point of the light–matter interaction, the carrier temperature is higher than the lattice temperature, and that is why this phenomenon is called CH. Additionally, free–carrier absorption can also contribute to CH. In this event, a free carrier absorbs a photon and is promoted to a higher energy level within the same band. Similarly, due to the high photon density in the active region, two–photon absorption can also take place. There, two photons are absorbed and an electron is transferred from the valence band to a high–energy level in the conduction band. Furthermore, stimulated emission may remove carriers from low–energy states within the conduction band, leading to an increase in the overall average thermal energy [19]. The result of CH is schematically depicted in Fig. 3.c. The carrier distribution is Fermi–Dirac again, but it has a higher average temperature than that observed in Fig. 3.a. After about 700 fs [29, 30], cooling of the electron–hole plasma to its original value takes place via optical–phonon emission. This is illustrated in Fig. 3.d. Finally, when electrons are re–supplied to the active region of the SOA via the pumping mechanism (driving current), the original steady–state density, shown in Fig. 3.a, is achieved again.

From the explanation above, it is not difficult to understand why SHB induces a reduction of the gain. However, it must be noted that even after the Fermi–Dirac distribution is recovered, the gain compression persists. Indeed, if one compares the integrated area under the curves in Figs. 3.c and 3.a, which are proportional to the electron density, it can be observed that the heated hole–electron plasma will have a reduced gain as compared to the steady–state plasma. Such compression will recover fast, mainly mediated by the heated

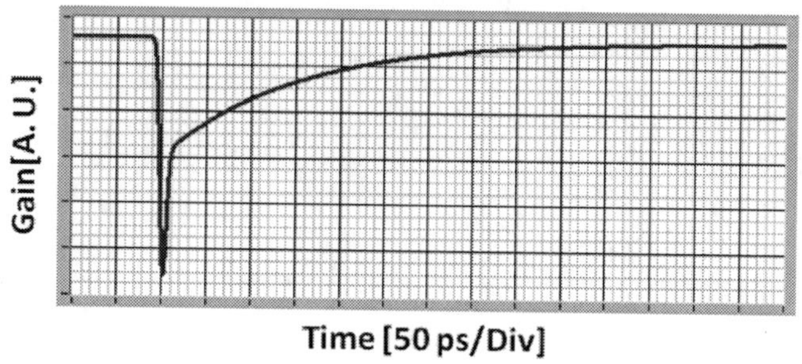

Figure 4. Gain recovery simulation after injection of a 4 ps Gaussian pulse. Fast and slow recovery components are visible. The 10–90% recovery time of the latter is about 400 ps.

carriers that cool down to the lattice temperature. A time constant, τ_{CH}, can be associated to this recovery. This will be followed by a second, much slower gain recovery that results from the effect of electrical pumping. With sufficient time–resolution, this recovery phenomenon can in principle be experimentally observed. The corresponding simulation is presented in Fig. 4. It is the result of injecting a 4–ps wide Gaussian pulse into a 0.5 mm–long SOA. The fast and slow recovery components are clearly visible[1].

To the gain recovery illustrated in Figs. 2 and 4 corresponds a phase recovery. This can be deducted from Eq. (3). Although more technically challenging, experimental analyses of phase recovery in SOAs have been successfully carried out [32]. In that research work, phase–shifts exceeding 2π have been measured for 1.5 mm–long amplifiers that are excited by a strong optical pulse. These experiments have also demonstrated that, in general, phase–shifting increases as a function of current density. Moreover, high resolution measurements have shown that large intra–band gain compressions (such as that shown in Fig. 4) are accompanied by hardly discernible corresponding phase–shifts. The linewidth enhancement factors associated with CH and SHB can thus be considered almost negligible. This has been explained using a simple parabolic band model and Kramers–Kronig analysis [33]. The situation is different for the linewidth enhancement factor associated with band–filling, where α_N values between 6 and 10 (N and λ–dependent) are common in bulk devices [15, 17]. These values lead to a dynamical phase response similar to that presented in Fig. 4 for the gain, but disregarding the ultra–fast component. A recovery time associated with the phase can thus be defined. From a practical standpoint, however, the use of the (10–90%) recovery time is not entirely adequate. This is because in most applications, only a fraction of the complete recovery of the phase is of interest. The X–full recovery time is alternatively defined as the period during which the phase recovers by X from its maximum compression value after being excited by a short optical pulse [34]. The value of X will depend on the application. For instance, in optical switching, only an X = π phase–shift matters [35], and then the use of the π–full recovery time makes sense. The dynamic behavior of the phase is particularly important for its impact on interferometric arrangements [32].

[1]The simulations in this work were carried out using an advanced and efficient SOA model [31] built around a system of rate equations for the carrier densities, derived from a density matrix formalism, and complemented with the corresponding waveform propagation equation within the slowly varying envelope approximation, thus following a semi–classical approach.

3.2. Frequency domain

The already explained dynamical phenomena also lead to four–wave mixing (FWM) in SOAs. This nonlinear process can more easily be understood if viewed in the frequency representation domain. It occurs when two sufficiently powerful optical beams of different central frequencies, ω_0 and $\omega_0 + 2\pi\Omega$, are injected into the active device. Interaction of the signals in the nonlinear medium will produce modulation at the beating frequency, leading to the appearance of gain and refractive index gratings. The relative weight of gain and index gratings is related to the linewidth enhancement factor [36]. Both input signals will scatter off the induced gratings, producing side bands at higher and lower frequencies from the inputs [19]. These are the so–called wave–mixing products. Theoretical and experimental research on the nonlinear behavior of bulk SOAs [28, 37] have shown that the nonlinear susceptibility that models the wave mixing contribution to the polarization of the medium can be expressed as a sum of contributions from carrier density pulsation (CDP) and CH, as $\chi = \chi^{CDP} + \chi^{CH}$. In particular, the wave mixing product at frequency $\omega_0 - 2\pi\Omega$ is characterized by a response function $R(\Omega)$ [38] composed of CDP and CH contributions, as follows [39]:

$$R(\Omega) = R_{CDP}(\Omega) + R_{CH}(\Omega)\,, \tag{10}$$

where

$$R_{CDP}(\Omega) = \frac{q_1}{1 - 2\pi i \Omega \tau}\,, \tag{11}$$

$$R_{CH}(\Omega) = \frac{q_2}{1 - 2\pi i \Omega \tau_{CH}}\,, \tag{12}$$

with $q_{1,2}$ being a proportionality factor. Depending on the frequency separation of the injected beams, Ω, either carrier–density modulation (inter–band dynamics and Eq. (11)) or carrier–occupation probability modulation (intra–band dynamics and Eq. (12)) will dominate the interaction. For a detuning $\Omega < 1$ GHz, χ^{CDP} dominates. However, when Ω is large enough as compared to the inverse of τ, the carrier density cannot longer respond to the beat frequencies Ω, 2Ω, etc. Thus, little or no χ^{CDP} is generated [40], and the FWM efficiency falls off rapidly with increasing detuning, at a rate of 20 dB per decade [19]. On the other hand, at large detuning, in the order of hundreds of GHz, $R_{CH}(\Omega)$ is the dominant term. For instance, for τ_{CH} = 0.8 ps and Ω = 200 GHz, $2\pi\Omega\tau_{CH} = 1$; whereas for τ = 400 ps, $2\pi\Omega\tau = 502$.

Current dense wavelength division multiplexed (DWDM) systems [41] operate with a channel spacing between 50 and 400 GHz. Amplification of DWDM systems using SOAs might therefore lead to the appearance of CH–based FWM tones at the amplifier output. This is an undesirable effect because of the resulting nonlinear in–band cross–talk, which degrades the quality of the transmitted data. Fig. 5.a shows a simulation of the effect of FWM when two signals with Ω = 200 GHz are injected into an SOA having τ = 400 ps and τ_{CH} = 0.8 ps. Both signals are modulated at 25 Gb/s in NRZ format, and each has an average power of -5 dBm at the SOA input. The spectrum on the left hand side (LHS) displays FWM–generated sidebands and even higher order components that are hardly visible. According to the figure, the first order components are 27 dB weaker than the injected signals. Their influence on the system performance is, in this case, very limited, and cross–gain modulation becomes a more relevant deleterious effect [42]. This is not necessarily the situation in multi–channel systems, where the number of FWM tones increases

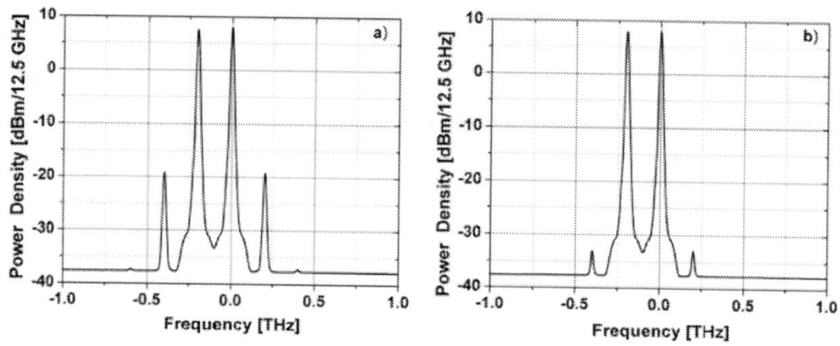

Figure 5. Calculated spectrum of two 25 Gb/s signals after being injected into an SOA. FWM–generated sidebands are visible. a) Carrier heating enabled. b) Carrier heating artificially disabled.

as $(N-1)N^2/2$, where N is the number of WDM channels [43]. Fig. 5.b shows the same spectrum presented in Fig. 5.a but this time measured from simulations where the CH effect has been artificially disabled. Under these circumstances, the FWM–generated products are greatly reduced. This confirms that CH is the main responsible for the appearance of dynamical gratings in the amplifier at the analyzed detuning value, which in turn induces the appearance of the FWM tones.

4. Practical Considerations

When an SOA is operated within the saturation regime, the recovery time becomes an important parameter, because it limits the bit rate[2] of the optical signal to be processed. This is true whether the SOA is employed as in–line amplifier or as an optical signal processor. The most common modulation format [44] in the telecommunications industry is still on–off keying, where a mark (or an optical pulse) is commonly associated with the bit "one", whereas a space (or absence of a pulse) is commonly associated with the bit "zero". In this manner, binary data can be processed and transmitted along hundreds of kilometers using optical fibers. In "slow" SOAs, having a long recovery time as compared to the data signal bit period, data–patterning arises. In this context, patterning is understood as the pattern–dependent wandering experienced by the power of the signal processed by the amplifier when it consists of random binary data. Indeed, for an SOA with a long recovery time, the gain only fully recovers when a sufficiently large number of spaces follows a mark; otherwise the successive pulse (mark) carves in an already saturated gain, leading to an amplified or processed signal whose power depends on the preceding data–pattern [10]. This behavior may lead to serious distortions in the processed optical signal. Data–patterning effects not only affect the power of the signal, but according to Eq. (3), they affect the signal phase as well.

Fig. 6 illustrates a simulation of the effect of patterning in an amplified optical signal. The input signal consists of binary data in NRZ format at a bit rate of 25 Gb/s (bit period of 40 ps) and having an average input power of 0 dBm. The signal is injected into an SOA

[2]The data bit rate is defined as the number of bits per time unit. The bit period is the inverse of the bit rate.

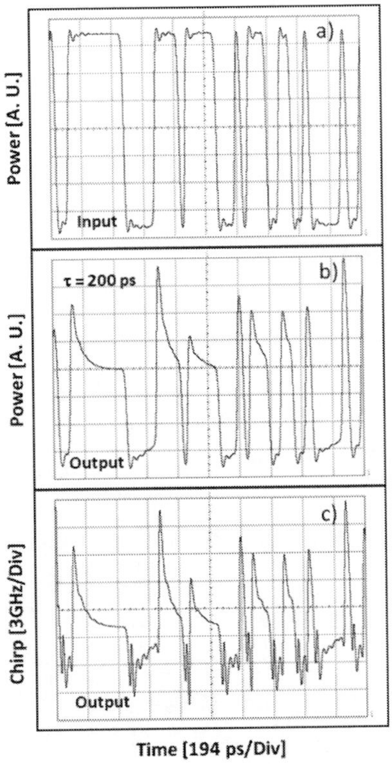

Figure 6. Patterning produced by an SOA in a 25 Gb/s pseudo–random binary optical signal. (a) Input data signal. (b) Amplified output signal. (c) Chirp of the output signal. Only a segment 50 bits long is shown. The SOA carrier lifetime was set to 200 ps.

that is modeled with a constant differential carrier lifetime of $\tau = 200$ ps, and an output saturation power of 8 dBm. The input optical signal, shown in Fig. 6.a, is therefore strong enough to saturate the device, whose small signal gain is 20 dB. Since the signal bit period is five times smaller than the SOA carrier lifetime, pattering arises. This is evident in the amplified signal shown in Fig. 6.b. In particular, it can be observed that a strong output pulse follows a relatively long sequence of spaces (zeroes). However, when only one or two spaces are present, the subsequent mark (one) is weaker. This occurs because the amplifier carrier density has not fully recovered when the new pulse arrives. Therefore, the pulse (mark) experiences a lower gain (whose coefficient g is proportional to the carrier density). Patterning then refers to the difference in power values observed in the output pulses produced by the random nature of the data. For completeness, the chirp of the output signal[3] is also presented in Fig. 6.c. The simulation shows how the phase also suffers from patterning. For the simulation, a linewidth enhancement factor, $\alpha_N = 9$, was utilized.

Fig. 7 shows the signal resulting from simulations carried out with shorter differential carrier lifetimes: $\tau = 100$ ps (top) and $\tau = 40$ ps (bottom). As already explained, when “faster” SOAs are employed, the SOA recovery time decreases, and then the patterning–induced power difference between the strongest and weakest output pulses becomes nar-

[3]The chirp is here defined as $\Delta\nu = \frac{-1}{2\pi}\frac{\partial\varphi}{\partial t}$.

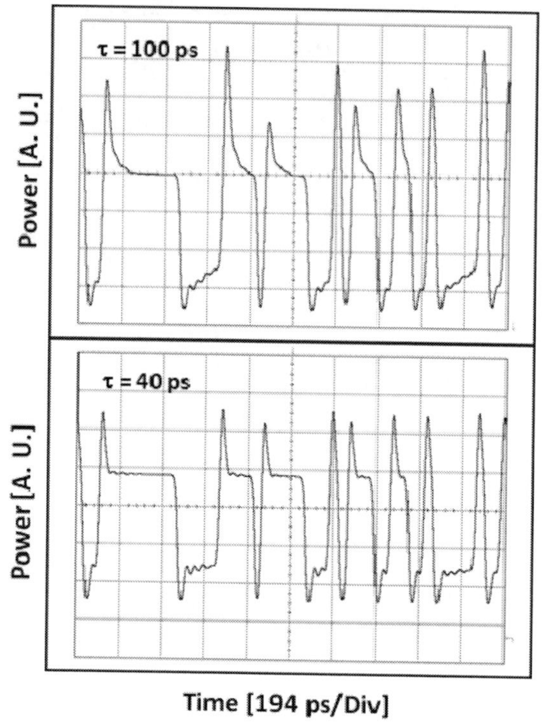

Figure 7. Output of a 25 Gb/s signal after being amplified by an SOA having a carrier lifetime of τ = 100 ps (top) and τ = 40 ps (bottom). In the latter, data pattern effects are practically absent.

rower. This is because the excursion of the carrier density fluctuations also diminishes, as can be observed by comparing Fig. 6.b and Fig. 7 (top). When the carrier lifetime becomes equal to the bit period of the optical signal (in this case 40 ps), pattern effects are greatly reduced. This phenomenon is shown in Fig. 7 (bottom). In this case, even after the occurrence of a single space (zero) that is immersed within a sequence of marks, the gain is able to recover. Consequently, all pulses become amplified with practically the same gain level.

From the discussion above, it is clear that adequate testing of pattern effects in an SOA, having a recovery time τ_{rec}, requires the use of a pseudo–random bit sequence (PRBS) with a bit period $T < \tau_{rec}$. Moreover, the PRBS must include a sequence consisting of k spaces, where k is an integer greater than τ_{rec}/T. Usually, the utilized PRBS has a length of 2^k bits that cycles through all possible sequences of k bits [45]. Evaluation of the patterning level in an optical signal can be accomplished through a parameter called *fluctuations in the mark power level* (FMPL), which is defined as FMPL = 100 × (maximum mark level - minimum mark level)/ minimum mark level [46]. Clearly, a low value of FMPL corresponds to a low patterning signal. The FMPL is usually measured from the signal's eye diagram, where all the bits that form the testing sequence are superimposed. The minimum and maximum mark levels can then be easily indentified. Fig. 8 shows the eye diagram resulting from an all–optical wavelength conversion process [47] from λ = 1540 nm to λ = 1580 nm, where a 2^9=512–bits long PRBS, running at 160 Gb/s, has been employed. A

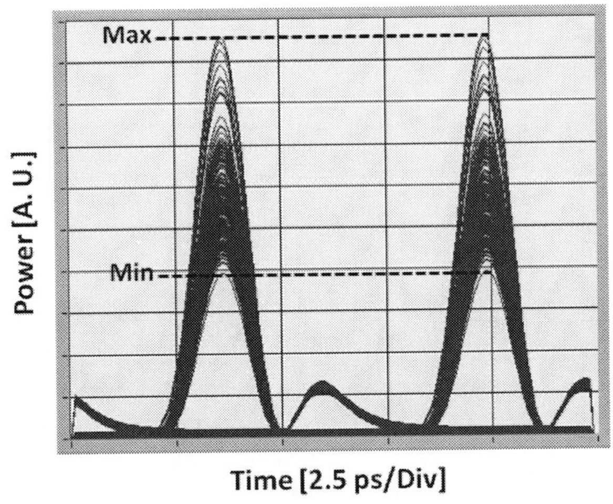

Figure 8. Eye diagram of a simulated wavelength conversion process at 160 Gb/s. The maximum and minimum mark levels are shown. The calculated FMPL is 144.8%.

high FMPL of nearly 145% can be measured. This marked patterning level is consequence of the very high bit rate used in the simulation. Wavelength conversion is an example of an all–optical nonlinear signal process that is affected by pattern effects. It is achieved in this case through the use of a Mach–Zehnder interferometer [10], where phase fluctuations produced by cross–phase modulation in two different SOAs form a phase transmission window via coherent waveform recombination. Data–induced phase wandering becomes amplitude fluctuation, leading to amplitude jitter, or, in other words, patterning. The same occurs in other all–optical signal processes based on sub–systems built with SOAs. Among them, we can mention all–optical Boolean gating via active interferometers [10], ultra–high capacity optical packet routing [48], all–optical error counting, and photonic header processing [49].

5. Mitigation of Data Pattern Effects

Mitigation of pattern effects in SOAs is currently being investigated following two paths. On the one hand, we have the development of more sophisticated internal nanometric structures to achieve faster relaxation effects. Quantum dots [50] exhibit ultrafast gain recovery [51, 52], thus enabling patterning–free high–speed signal amplification [7, 53]. They also present reduced chirp due to a low linewidth enhancement factor [54]. On the other hand, we have the use of linear [55] and nonlinear filters [56], and the design of ingenious integrated structures combining two or more bulk amplifiers. The analysis of quantum dots is beyond the scope of this chapter. We therefore refer the reader to the vast number of references on this topic.

5.1. Filtering

The use of a narrow–band blue–shifted optical bandpass filter, placed after a single SOA, has demonstrated a bandwidth enhancement effect that decreases patterning. This technique has been used, for instance, to demonstrate error–free wavelength conversion at 160 Gb/s

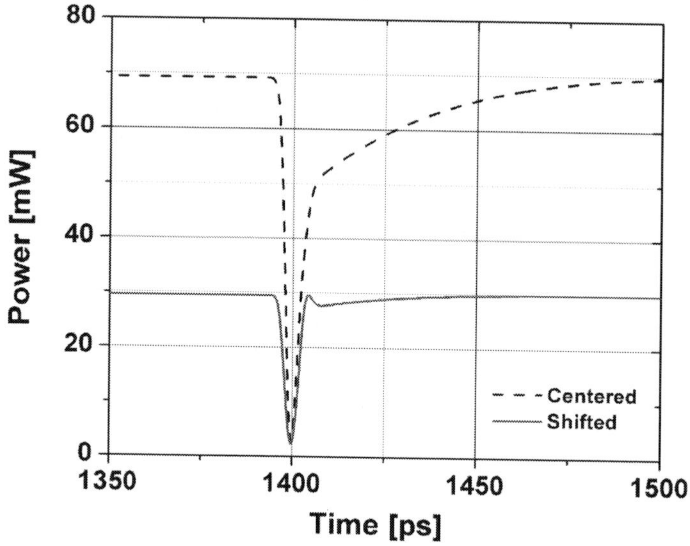

Figure 9. Probe beam response to an optical excitation in an SOA. The probe is decoupled via a narrow–band optical filter that is centered (dashed) or shifted by 100 GHz (solid) from the probe frequency.

[57]. More sophisticated schemes, which pass and recombine both red and blue–shifted parts of the signal spectrum, have also been reported [58]. Their drawback is a reduction in the optical signal–to–noise ratio of the converted signal.

Fig. 9 graphically presents, through the use of simulations, the improvement in gain recovery time obtained when a shifted filter is used. The dashed line depicts the output power of a probe (originally CW) beam that is launched together with a 2.2 ps Gaussian pulse into a 1 mm–long SOA. The probe beam at λ_1 = 1577.85 nm is decoupled from the pump pulse, centered at λ_2 = 1545.32 nm, through the use of a Gaussian optical filter 180 GHz wide, and centered at λ_1. In agreement with Fig. 4, the dashed curve shows a strong depletion of the gain, followed by a recovery with fast and slow components. As already explained, they correspond to the time it takes for the electron–hole plasma to cool down to the lattice temperature, and the time the pumping mechanism needs to re–supply electrons to the active region, respectively. The solid curve results from shifting the same optical filter by 100 GHz. The SOA gain response now shows a single time component, leading to full recovery in less than 6 ps. The improvement in recovery time gained with the use of this novel technique is, however, eclipsed by a reduced modulation depth. This means that, for instance, a wavelength–converted signal obtained through the use of the shifted–filter technique will exhibit lower patterning, but a possibly degraded extinction ratio as well. This can be confirmed from a wavelength conversion simulation via cross–gain modulation (XGM), from λ_2 to λ_1, of a 100 Gb/s PRBS sequence, 256 bits long. Bits are represented by Gaussian pulses with a full width of 2.8 ps. The resulting eye diagrams are shown in Fig. 10. They exhibit inverse polarity due to the XGM process. The diagram in Fig. 10.a was obtained with the 180 GHz optical filter centered at the probe (or target) wavelength,

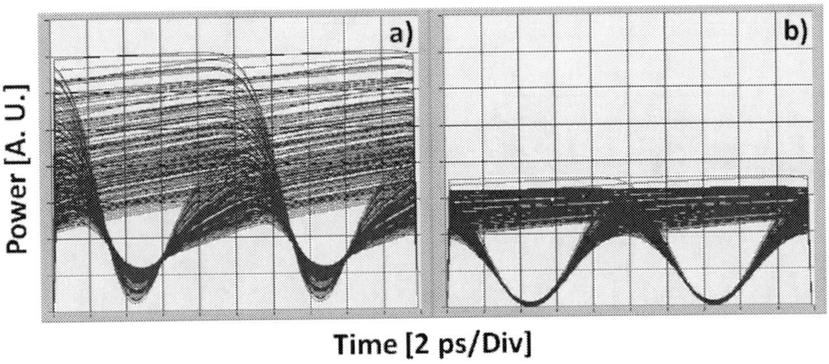

Figure 10. Eyes resulting from XGM–based wavelength conversion at 100 Gb/s for an optical filter that is centered (a) or shifted by 100 GHz (b) from the target wavelength.

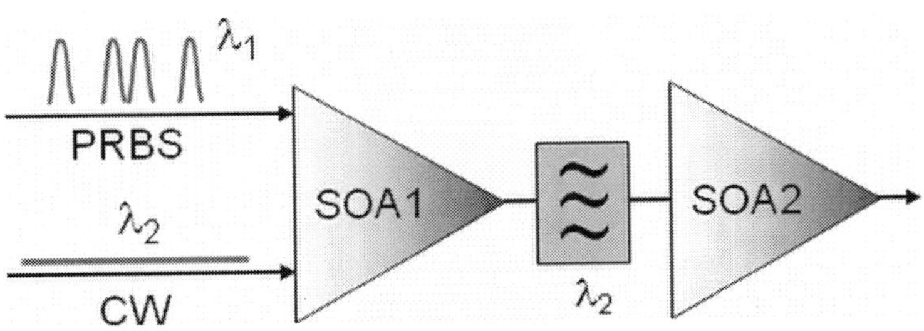

Figure 11. Schematic diagram of the turbo–switch [59].

λ_1. The diagram in Fig. 10.b corresponds to that obtained with the 100 GHz shifted filter. For the sake of comparison, both eyes are shown at the same scale. A reduced amplitude jitter is clearly observed in Fig. 10.b, leading to cleaner eyes. In spite of the extinction ratio degradation shown by some bits, the use of the shifted filter improves in this case the wavelength conversion process.

Nonlinear filtering of a wavelength–converted signal can be achieved through the use of the so–called *turbo–switch*, which is a recently introduced arrangement that increases the high–speed response of SOAs [59]. The turbo–switch can then be used to decrease the amount of patterning that results from the XGM nonlinear process. It is realized by cascading an extra SOA to the common wavelength conversion structure. A schematic of the turbo–switch is shown in Fig. 11. In contrast to the previously described linear filtering scheme, here an optical broad–band filter is placed between the SOAs. Its only purpose is to block the pump signal while allowing to pass the entire modulated spectrum of the probe (or target) beam. The bandpass filter together with the second SOA actually operate as a nonlinear filter. The somewhat complex action of the second SOA can be explained in terms of the self–gain dynamics of the second SOA, which act in opposition to the slow recovery of the amplitude and phase modulation transferred by the pump to the probe in the first SOA [56]. The result is an enhanced response of the SOA combination. This in turn leads to a converted signal with lower patterning and a higher optical signal–to–noise ratio as compared to the linear narrow filter scheme. A deeper insight into the turbo–switch dynamics has been gained using frequency–resolved optical gating techniques [60].

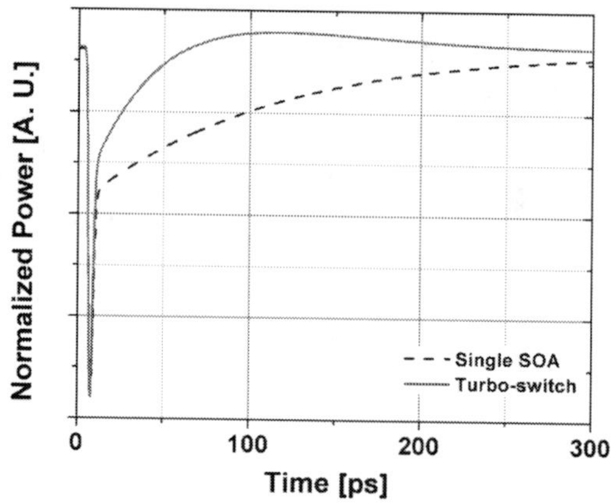

Figure 12. Simulated probe beam response at the output of a single SOA (dashed) and a turbo–switch (solid) when a strong Gaussian pulse, 2.2 ps wide, is injected.

Fig. 12 presents a comparison between the response of a single SOA and a turbo–switch to a 2.2 ps–wide Gaussian pulse. In all cases a 0.5 mm long SOA, having an injection current of 250 mA, was utilized. This is, in general terms, a relatively slow device. The turbo–switch optical filter was chosen to have a Gaussian shape and a broad band of 800 GHz. In the presence of the pulse and a CW probe beam, XGM takes place in the SOA. After an abrupt gain reduction, fast and slow recovery components are visible. This is evident in the dashed line, corresponding to the single SOA (non–turbo) response. The recovery obtained with the turbo–switch is faster, it amounts to about 60 ps for a full recovery, whereas the single SOA requires more than 300 ps to reach the same original gain level. Although the rate of gain recovery in the second SOA of the turbo–switch is similar to the one in the first SOA, the recovery occurs in the opposite sense because it is produced by the already modulated (and filtered) probe beam. The overall effect is a cancellation of the slow tail of the response of the first SOA and, as the solid curve shows, a more rapid gain recovery of the novel device [61].

The turbo–switch performance can be optimized in terms of pattern effect reduction by adequately tuning the optical power injected into the second SOA. This can be controlled by placing a variable optical attenuator immediately after the optical broad band filter. At high attenuations the response curve reverts to that of a single SOA, thus resembling that shown in Fig. 4. At low attenuation values, the gain response presents a marked overshoot. An optimum response results from moderate attenuation values [61]. When compared with the linear filter scheme, the main drawback of the turbo–switch is the use of two, instead of a single SOA. This clearly increases the structure cost and complexity. Another negative aspect of this scheme is the accumulation of amplified spontaneous emission. Error–free wavelength conversion demonstrations at 170 Gb/s have been reported using the turbo–switch scheme [59].

The dynamic behavior of the turbo–switch is not limited to the recovery of the gain, but also includes the phase response of the cross–modulated beam. This effect can be exploited in active interferometric structures to achieve non–inverted wavelength conversion and other switching–related functionalities that will be necessary on future all–optical packet–switched networks, such as Boolean gating [10]. Numerical simulations, however, have shown that, from a practical standpoint, the turbo–switch does not always represent the best solution to a specific need, since there are applications where the inclusion of the turbo–effect does not improve the sub–system performance and, in some cases, even introduces deleterious byproducts [62].

5.2. Novel architectures

For nearly fifteen years, ingenious SOA–based architectures, aimed for all–optical signal processing operations, have been proposed to reduce the gain and refractive index dependence of the amplifier on the specific signal bit pattern. Among them, cross–phase modulation–based active interferometric switches have been helpful to reduce the associated distortions. In reference [63], for instance, the input dynamic range of a semiconductor optical in–line amplifier has been demonstrated to be increased through the use of an active Mach-Zehnder interferometer (MZI), similar to that shown in Fig. 15. The novel approach consists in injecting unequal optical powers, derived from the same input data signal, to each of the two SOAs. Compensation occurs because the leading part of the signal pulse is attenuated by incomplete destructive interference, while the trailing part of the pulse experiences a higher transmittance owing to a nonlinear phase shift caused by carrier depletion. When operated at 10 Gb/s, the component shows an extension of 7 dB in dynamic range (sensitivity measured at BER=10^{-9}) as compared to a single SOA [63]. Similar schemes based on interferometer unbalance produced by asymmetric current injection in active MZIs have also been investigated. This research work has led to a reduction in power penalty from 6.8 to 0.9 dB in a 16 x 10 Gb/s DWDM in–line amplifier system [64], and a marked improvement in eye opening and reduction of amplitude jitter in wavelength conversion at 40 Gb/s [65]. Pattern effect compensation can also be achieved by employing SOAs of different length to produce the required interferometer unbalance.

A different approach to reduce patterning in SOA–based sub–systems is amplifier cascadation. One instance of this particular implementation is the turbo–switch, mentioned in section 5.1.. Another similar approach is discussed in reference [66]. The proposed architecture is shown in Fig. 13. The first stage, composed of an SOA and a filter that blocks λ_1, is simply a XGM–based wavelength converter, like the one shown in Fig. 14.a. The following multi–mode interference coupler (MMI) combines the wavelength–converted signal, having inverse polarity (and centered at λ_2), with an extra CW beam centered at λ_3. They are injected into SOA2, where a new wavelength conversion process, this time between the inverted signal and the extra CW beam, takes place. After the second filter, an uninverted signal centered at λ_3 emerges, exhibiting low power fluctuations. Moreover, since the emerging signal at λ_3 is the result of a conversion process from the data signal to a low–noise probe beam, the sub–system also generates an improvement in the optical signal–to–noise ratio. Experimental demonstrations, using a data sequence running at 10 Gb/s, show a negative power penalty of 4.1 dB (BER=10^{-9}) after wavelength conversion of a noisy signal [66]. Note that the value of λ_3 can be set to λ_1, turning this system into an in–line amplifier with an extended input power dynamic range.

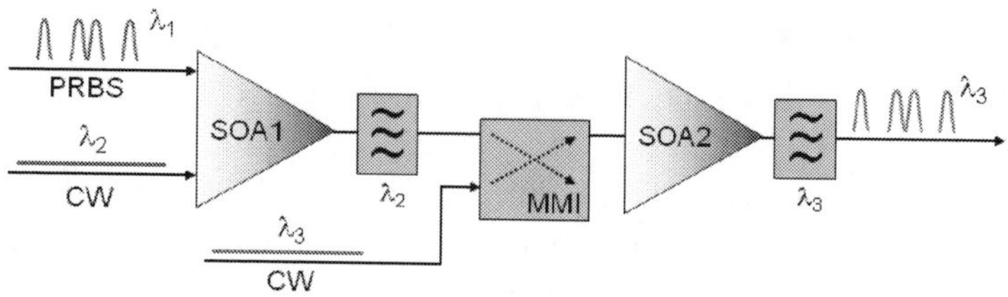

Figure 13. Schematic diagram of a cascaded wavelength converter to reduce patterning. MMI: multi–mode interference coupler.

Patterning mitigation in SOAs can also be achieved via cascadation of, not only amplifiers, but SOAs combined with electro–absorption modulators (EAMs) [67]. These two devices have almost opposite optical modulation characteristics. It is then envisaged that in a combined sub–system, formed by an SOA followed by an EAM, the latter may compensate the effect that the former induces in a probe beam when propagating together with a relatively strong optical pulse along the SOA–EAM system. Simulations have shown [68] that when an ultra–short pulse is injected together with a CW (probe) beam into an SOA, intra–band dynamics in the amplifier will produce a dark pulse at the probe wavelength, followed by a slow relaxation (see section 3. in this chapter). When injected into the subsequent EAM, cross–absorption modulation induced by the (now amplified) original pulse will produce the opposite slow recovery effect on the probe signal power, thus compensating the slow component of the previously formed dark pulse. The result at the EAM output is an undistorted pulse with inverse polarity. An optical–band pass filter placed at the sub–system output must be tuned to block the unwanted pump pulse that also emerges from the SOA–EAM system. According to this explanation, it can be anticipated that the cascaded system will chiefly be useful when ultra–short optical pulses running at very high bit–rates are of interest. Further simulations, including BER calculations, have confirmed this statement [68].

The use of an unmodulated (CW) holding beam to speed up the dynamic response of SOAs was proposed in 1994 [21]. It resorts to a short–wavelength CW beam that acts as an optical pump, bringing the carriers from the valence to the conduction band in a very short time once the gain is compressed. This technique received considerable attention, both experimentally [69] and theoretically [70]. A similar approach, useful for some applications involving the use of a CW beam, is to inject the holding beam at a high power level to drive the SOA into saturation. A schematic of such approach for the case of wavelength conversion via cross–gain modulation is shown in Fig. 14.a. A similar scheme, but having the filter centered at λ_1 allows for optical amplification of the PRBS. The use of the holding beam compresses the amplified spontaneous emission (ASE) noise and aids the amplifier gain to quickly approach the stationary level, thus reducing its recovery time. The assist light also increases the saturation power value. In muti–wavelength amplification, the use of an auxiliary CW holding beam considerably reduces inter–channel nonlinear crosstalk, produced by XGM and FWM. This has been experimentally demonstrated for an SOA–amplified, 4 x 10 Gb/s, WDM system having a channel–spacing of 100 GHz [71]. Measurement of the power penalty (BER=10^{-9}) produced by the SOA show that this parameter can be reduced up to an almost penalty–free level using the holding beam technique. Moreover, an enhancement

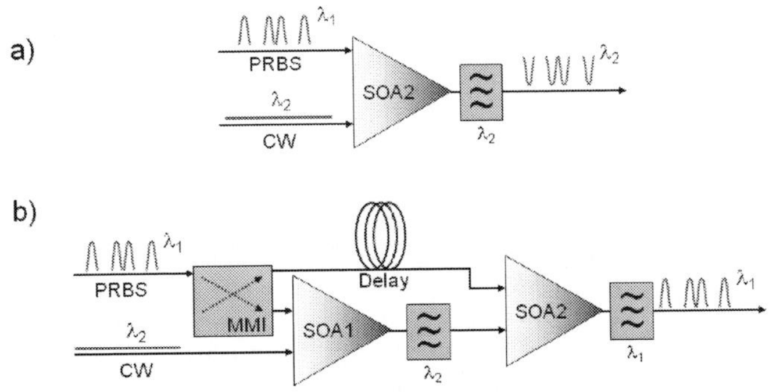

Figure 14. Schematic diagram of a) a wavelength coverter via cross–gain modulation, and b) a novel structure to amplify a data sequence with reduced patterning. MMI: multi–mode interference coupler.

of the input power dynamic range from 6 to 18 dB has also been demonstrated [71]. The use of this technique has also been suggested, and numerically compared with alternative schemes, to be used in the recently released IEEE 100 Gb/s Ethernet standard over single–mode fiber [72]. Unfortunately, amplifier gain reduction, together with the extinction ratio degradation it provokes [73], prevented this technique to become a serious contender at the applications level. An interesting alternative to reduce the data patterning effect is the employment of modulated holding beams.

The principle of operation of the structure shown in Fig. 14.b is based on the use of a cross–gain modulated holding beam, which is produced at SOA1 through the interaction of the PRBS centered at λ_1 and the CW beam at λ_2. After appropriate filtering (centered at λ_2), the modulated signal becomes an inverted replica of the original PRBS. This signal is then injected into SOA2 together with the original PRBS, derived from the MMI, which can be understood as a 50:50 splitter. The delay line is necessary to synchronize both input signals, which interact with each other in SOA2 via cross–gain modulation. The second optical broadband filter, centered at λ_1, isolates the signal carrying the amplified noninverted information. Patterning–reduced amplification takes place at SOA2, where the modulated holding beam helps to equalize the amplifier average gain level. Simulation results at 100 Gb/s have shown that, through adequate optimization of the CW input power, amplification of the data sequence above 20 dB can be achieved having an FMPL of about 13%, thus outperforming an unmodulated holding beam system operated under similar conditions. Moreover, relatively low extinction ratio degradation is also exhibited by the novel structure [46]. These results indicate the superior performance of the cross–gain modulated holding beam–based configuration, at the expense, of course, of a more complex structure.

Cross–gain modulated holding beams can also be conveniently used to reduce patterning in combination with interferometric structures, whose all–optical signal processing functionality depends on the cross–phase modulation effect. A practical interferometer for this purpose is the, already mentioned, MZI. Its structure is presented in Fig. 15. In the so–called differential scheme, the control pulse reaches SOA1 slightly later than it reaches SOA2. In both SOAs, the control pulse induces the same phase change into the signal beam via cross–phase modulation. However, since this occurs earlier in SOA2, a transmis-

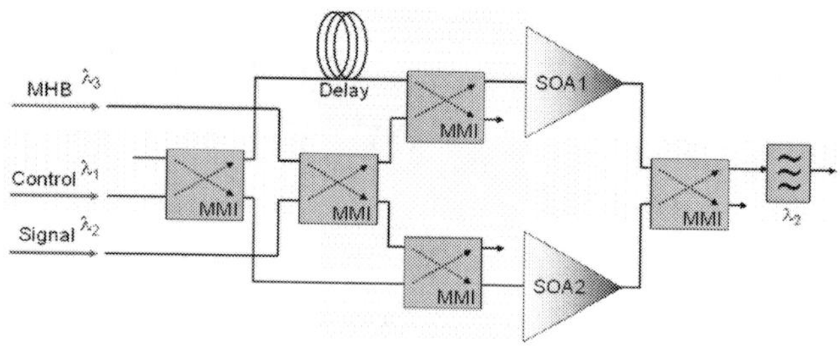

Figure 15. Schematic diagram of a differential Mach–Zehnder interferometer. MMI: multi–mode interference coupler. MHB: modulated holding beam.

sion window, produced by the corresponding interference that takes place in the rightmost MMI, is opened. Assuming that the interferometer delay is shorter than the recovery time of both SOAs and that a balanced interferometer is in use (same injection current in both SOAs), the delayed signal effectively cancels out the phase change caused by the lower arm, closing the already opened transmission window. In this way, the interferometer acts like a switch, with a (possibly ultra–fast) switching time determined by the interferometer delay and the control pulses duration [10]. For wavelength conversion operation, a data sequence is used as control pulse, whereas a CW beam centered at a different (target) wavelength, plays the role of signal beam. An optical filter is used to decouple both signals.

The performance of the MZI as wavelength converter can indeed be improved by injecting a cross–gain modulated holding beam. Its effect is similar to that explained with reference to the structure presented in Fig. 14.b. The time–varying holding beam at λ_3 can be produced with the structure shown in Fig. 14.a. The inverted and low–pass filtered version of the data signal is coupled into the MHB (modulated holding beam) port of the structure shown in Fig. 15, leading to a noninverted wavelength converted signal at the MZI output. The optical filter blocks λ_1 and λ_3, while the output signal is centered at λ_2. With the aid of the time–varying holding beam, the total power injected into the SOAs is temporally equalized, thereby reducing the degree of patterning [29]. By simultaneously injecting the modulated holding signal into both arms, the amplifier saturation level becomes somehow counterbalanced without significantly changing the phase difference that leads to the switching process.

Experimental work carried out with unmodulated CW holding beams has shown that amplitude jitter decreases when the holding beam power increases. This leads, however, to eye opening reduction. The effect can be explained from the carrier density compression that takes place in the SOA, resulting in a reduction of the gain and refractive index fluctuation and excursion. The situation is different when a modulated holding beam is utilized. In this case, as numerically demonstrated using PRBS sequences running at 160 Gb/s, the amplitude fluctuations are significantly reduced, but the extinction ratio is also improved [29]. These theoretical results position the modulated holding beam MZI architecture as a robust design to lessen temporal fluctuations in high–speed data processing operations.

A similar approach to reduce patterning using the structure shown in Fig. 15 consists in substituting the modulated holding beam with a logically complementary binary signal [74]. That means that if the control sequence consists of the following combination of marks and

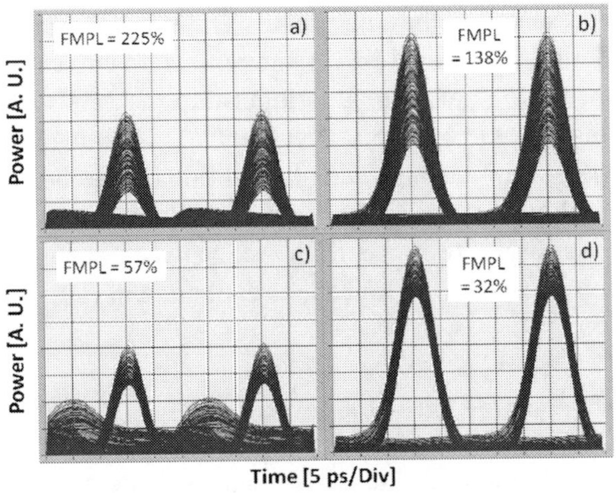

Figure 16. Simulated eye diagrams resulting from a 40 Gb/s wavelength conversion process using an MZI. The complementary signal technique is used in c) and d), leading to reduced patterning. The current injected into SOA1 is 399.5 mA, for b) and d), and 400.0 mA for a) and c). The current injected into SOA1 is 400.0 mA in all cases. The signals in a) and b) are re–scaled by 0.5 as compared to those in c) and d).

spaces: 1001110101010001, the complementary signal will be: 0110001010101110. In other words, whenever a pulse (representing a logical one) is not injected into the MZI through the control port, a compensating pulse will be input into the MHB port. In this manner, both SOAs in the MZI will receive nearly constant energy in every bit period, thus reducing patterning. This technique follows the same philosophy as the cross–gain modulated holding beam scheme mentioned in the previous paragraph. However, it takes it one step further, because instead of injecting a signal with a continuously varying power level (which results from the XGM process), a well–defined binary sequence is used to more accurately compensate the absence of energy during a bit period in the SOA. The drawback of the binary scheme stems from the complexity that producing the complementary sequence represents, at least as compared with the straightforward XGM process. A proposal is presented elsewhere [74].

Results from a *proof–of–concept* simulation are presented in Fig. 16. The simulated eye diagrams are derived from a wavelength down–conversion process at 40 Gb/s from λ_1= 1565.5 nm to λ_2= 1554.1 nm. The PRBS is composed of 256 bits in RZ format with Gaussian pulses having a duty cycle of 33%. After an optimization process, the complementary sequence, injected into port MHB of the MZI (see Fig. 15), is chosen to be three times more powerful than the data sequence. Identical, one millimeter–long, bulk amplifiers are used in both interferometer arms with standard parameters. The eye shown in Fig. 16.a was produced following the traditional MZI–based wavelength conversion approach with no holding beam [10]. It shows a high FMPL of 225%. Fig. 16.b, instead, was produced using an unbalanced interferometer, where the current injected into SOA1 was slightly lower (399.5 mA) than that injected into SOA2 (400.0 mA). As previously discussed in this section, this approach improves the output signal quality. This can be confirmed by the corresponding reduction of the FMPL value to 138%. An increase in eye opening can also be perceived, since both figures are presented at the same scale. The eye shown in Fig. 16.c is the result of a simulation considering a balanced MZI (same injection current in both SOAs), where the logically complementary binary signal is input into the MHB port. The reduction in data patterning as compared to Fig. 16.a is evident. This is quantitatively demonstrated with a reduction in FMPL from 225%, when the traditional approach is used, to 57%, when the novel scheme is employed. A marked after–pulse, however, is also visible. This is produced by an inadequate shifting, θ, in the difference between the phase change induced via cross–phase modulation in each SOA. Indeed, as explained in [10], the power at the output port of the MZI is given by

$$\frac{1}{4}\left(|A(t)|^2+|A(t-\tau)|^2\right)\left(1+M\cos\left(\phi(t)-\phi(t-\tau)+\pi+\theta\right)\right), \tag{13}$$

where $\phi(t-\tau)$ and $\phi(t)$ are the induced phase changes in SOA1 and SOA2, respectively. M stands for the modulation depth, which is close to unity in the analyzed situation; and $A(t)$ is the time–dependent slowly–varying envelope of the lightwave at the SOA output whose phase is given by $\phi(t)$. For an inadequate shifting θ in the argument of the cosine function, the phase window (i.e. $\cos(\phi(t)-\phi(t-\tau)+\pi+\theta)$) turns out to be "folded into itself", leading to the appearance of the aforementioned after–pulse. This can be corrected by adding the necessary phase shift to the argument of the cosine to "unfold" the phase window function. This can be accomplished by injecting different driving currents into the SOAs of the active interferometer, thus unbalancing the structure. The eye resulting from using a complementary binary signal together with the injection of a slightly lower current into SOA1 (399.5 mA) is presented in Fig. 16.d. When compared to Fig. 16.c, an increase in eye opening and a reduction in FMPL from 57% to 32% are clearly distinguished (figures c and d are presented at the same scale). From our numerical analysis, it is concluded that compensation via the injection of a complementary signal together with interferometer unbalance lead to improvement of the wavelength converter performance. A remarkable reduction of undesirable data pattern effects from FMPL = 225% to 32% is demonstrated. Further progress can be attained by more accurately optimizing the SOA parameters and input waveform characteristics. It can also be anticipated that these results are not exclusive of the wavelength converter, but expected to be observed in other MZI–based functional operations such as all–optical Boolean gating or in–line amplification.

6. Conclusion

In this chapter, a brief introduction on the nonlinear dynamical behavior of semiconductor optical amplifiers was presented. Time and frequency domain representations were addressed. The concept of data patterning was then introduced as the pattern–dependent wandering experienced by the power and phase of random data signals when nonlinearly processed by an SOA. Mitigation of data pattern effects then became the main subject of the chapter. With the aid of simulations, novel solutions to enhance the performance of SOA–based applications, such as in–line amplification and wavelength conversion, were demonstrated. In particular, the use of shifted bandpass filters and the turbo–switch were shown to decrease the gain recovery time of amplifiers, leading to a reduction in signal amplitude jitter at the output of the SOA. These techniques, however, also reduce the signal–to–noise ratio or accumulate amplified spontaneous emission noise. Such inconsistencies led us to investigate alternative approaches. A myriad of SOA–based novel architectures were discussed. It included unbalanced Mach–Zehnder interferometers via asymmetric current or power injection, cascadation of amplifiers or amplifiers plus absorbers, holding beam injection to speed up carrier repopulation in the conduction band or to drive the amplifier gain into saturation, and the use of modulated holding beams. Three different schemes were considered for the latter approach. The first one, being the simplest one, makes use of a cross–modulated holding beam that is derived from the original data signal. The beam, injected together with the amplifying signal, helps to equalize the average gain level of the amplifier, leading to practically patterning–free amplified signals. The second scheme also resorts to a cross–modulated holding beam. However, in contrast to the previous scheme, now a typical differential Mach–Zehnder structure is utilized. The main effect of the additional beam is to counterbalance the saturation level in both SOAs without significantly changing the phase difference that leads to the switching action. This reduces amplitude fluctuations and increases extinction ratio. The third scheme is also based on the use of a differential Mach–Zehnder interferometer, but instead of a cross–modulated holding beam, it employs the logical complement of the input data signal. In this arrangement, both SOAs receive nearly constant energy in every bit period, since there is a pulse constantly injected into the device. Although this scheme indeed reduces patterning, further improvement in the output signal quality can be gained when combined with interferometer unbalance. Simulations at 40 Gb/s, here presented for the first time, showed a reduction of data pattern effects from 225% in FMPL for a typical wavelength conversion process to 32% when using this original approach. Further investigations to optimize the architecture layout and amplifier material parameters are underway.

References

[1] E. Desurvire. *Erbium-doped fiber amplifiers: principles and applications* . John Wiley and Sons., 1994.

[2] M. N. Islam. Raman amplifiers for telecommunications. *Journal of Selected Topics in Quantum Electronics*, 8:548–559, 2002.

[3] J. Bromage. Raman amplification for fiber communications systems. *Journal of Lightwave Technology*, 22:79–93, 2004.

[4] M. J. O'Mahony, C. Politi, D. Klonidis, R. Nejabati, and D. Simeonidou. Future optical networks. *Journal of Lightwave Technology*, 24:4684–4696, 2006.

[5] S. Tanaka, S. Tomabechi, A. Uetake, M. Ekawa, and K. Morito. Record high saturation output power (+20 dBm) and low NF (6.0 dB) polarisation-insensitive MQW-SOA module. *Electronics Letters*, 42:1059–1060, 2006.

[6] T. Rogowski, S. Faralli, G. Bolognini, F. Di Pasquale, R. Di Muro, and B. Nayar. SOA-based WDM metro ring networks with link control technologies. *IEEE Photonics Technology Letters*, 19:1670–1672, 2007.

[7] A. Uskov, T. W. Berg, and J. Mørk. Theory of pulse-train amplification without patterning effects in quantum-dot semiconductor optical amplifiers. *IEEE Journal of Quantum Electronics*, 40:306–320, 2004.

[8] Z. Li, Y. Dong, J. Mo, Y. Wang, and C. Lu. 1050-km WDM transmission of 8x10.709 Gb/s DPSK signal using cascaded in-line semiconductor optical amplifier. *IEEE Photonics Technology Letters*, 16:1760–1762, 2004.

[9] J. D. Downie, J. Hurley, and Y. Mauro. 10.7 Gb/s uncompensated transmission over a 470 km hybrid fiber link with in-line SOAs using MLSE and duobinary signals. *Optics Express*, 16:15759–15764, 2008.

[10] R. Gutiérrez-Castrejón. Active interferometers for all–optical information processing. In D. Halsey and W. Raynor, editors, *Handbook of Interferometers. Research, Technology and Applications*, pages 793–831. Nova Science Publishers, 2009.

[11] D. Nesset, T. Kelly, and D. Marcenac. All-optical wavelength conversion using SOA nonlinearities. *IEEE Communications Magazine*, 36:56–61, 1998.

[12] Ch. Holtmann, P.-A. Besse, T. Brenner, and H. Melchior. Polarisation independent bulk active region semiconductor optical amplifiers for 1 .3 μm wavelengths. *IEEE Photonics Technology Letters*, 8:343–345, 1996.

[13] J. T. Verdeyen. *Laser Electronics*. Prentice Hall International, 1995.

[14] L. Kador. Kramers-Kronig relations in nonlinear optics. *Applied Physics Letters*, 66:2938, 1995.

[15] L. Occhi, L. Schares, and G. Guekos. Phase modeling based on the α factor in bulk semiconductor optical amplifiers. *IEEE Journal of Selected Topics in Quantum Electronics*, 9:788–797, 2003.

[16] R. J. Manning, A. D. Ellis, A. J. Poustie, and K. J. Blow. Semiconductor laser amplifiers for ultrafast all-optical signal processing. *Journal of the Optical Society of America B*, 14:3204–3216, 1997.

[17] R. Gutiérrez-Castrejón and M. Duelk. Modeling and simulation of semiconductor optical amplifier dynamics for telecommunication applications. In S. J. Bianco, editor, *Computer Physics Research Trends*, pages 89–124. Nova Science Publishers, 2007.

[18] R. Gutiérrez-Castrejón, L. Schares, L. Occhi, and G. Guekos. Modeling and measurement of longitudinal gain dynamics in saturated semiconductor optical amplifiers of different length. *IEEE Journal of Quantum Electronics*, 36:1476–1484, 2000.

[19] J. M. Wiesenfeld. Gain dynamics and associated nonlinearities in semiconductor optical amplifiers. *International Journal of High Speed Electronics and Systems*, 7:179–222, 1996.

[20] R. Olshansky, C. B. Su, J. Manning, and W. Powazinik. Measurement of radiative and nonradiative recombination rates in InGaAsP and AlGaAs light sources. *IEEE Journal of Quantum Electronics*, 20:838–854, 1984.

[21] R. J. Manning and D. A. O. Davies. Three-wavelength device for all-optical signal processing. *Optics Letters*, 19:889–891, 1994.

[22] F. Girardin, G. Guekos, and A. Houbavlis. Gain recovery of bulk semiconductor optical amplifiers. *IEEE Photonics Technology Letters*, 10:784–786, 1998.

[23] J. Mark and J. Mørk. Subpicosecond gain dynamics in InGaAsP optical amplifiers: experiment and theory. *Applied Physics Letters*, 61:2281–2283, 1992.

[24] K. L. Hall, G. Lenz, A. M. Darwish, and E. P. Ippen. Subpicosecond gain and index nonlinearities in InGaAsP diode lasers. *Optics Communications*, 111:589–612, 1994.

[25] P. Borri, S. Scaffetti, J. Mørk, W. Langbein, J. M. Hvam, A. Mecozzi, and F. Martelli. Measurement and calculation of the critical pulsewidth for gain saturation in semiconductor optical amplifiers. *Optics Communications*, 164:51–55, 1999.

[26] L. Occhi, Y. Ito, H. Kawaguchi, L. Schares, J. Eckner, and G. Guekos. Intraband gain dynamics in bulk semiconductor optical amplifiers: measurements and simulations. *IEEE Journal of Quantum Electronics*, 38:54–60, 2002.

[27] L. Occhi. *Semiconductor optical amplifiers made of ridge waveguide bulk InGaAsP/InP:Experimental characterisation and numerical modelling of gain, phase and noise*. PhD thesis, ETH Zürich, 2002.

[28] A. Uskov, J. Mørk, and J. Mark. Wave mixing in semiconductor laser amplifiers due to carrier heating and spectral hole burning. *IEEE Journal of Quantum Electronics*, 30:1769–1781, 1994.

[29] S. Bischoff, M. L. Nielsen, and J. Mørk. Improving the all-optical response of SOAs using a modulated holding signal. *Journal of Lightwave Technology*, 22:1303–1308, 2004.

[30] Y. Kim, H. Lee, S. Kim, J. Ko, and J. Jeong. Analysis of frequency chirping and extinction ratio of optical phase conjugate signals by four-wave mixing in SOAs. *IEEE Journal of Selected Topics in Quantum Electronics*, 5:873–879, 1999.

[31] R. Gutiérrez-Castrejón and M. Duelk. Uni–directional time–domain bulk SOA simulator considering carrier–depletion by amplified spontaneous emission. *IEEE Journal of Quantum Electronics*, 42:581–588, 2006.

[32] L. Schares, C. Schubert, C. Schmidt, H. G. Weber, L. Occhi, and G. Guekos. Phase dynamics of semiconductor optical amplifiers at 10-40 GHz. *IEEE Journal of Quantum Electronics*, 39:1394–1407, 2003.

[33] R. P. Giller, R. J. Manning, and D. Cotter. Gain and phase recovery of optically excited semiconductor optical amplifiers. *IEEE Photonics Technology Letters*, 10:1061–1063, 2006.

[34] R. Gutiérrez-Castrejón, L. Occhi, L. Schares, and G. Guekos. Recovery dynamics of cross-modulated beam phase in semiconductor amplifiers and applications to all-optical signal processing. *Optics Communications*, 195:167–177, 2001.

[35] R. J. Manning, D. A. O. Davies, and J. K. Lucek. Recovery rates in semiconductor laser amplifiers: optical and electrical bias dependencies. *Electronics Letters*, 30:1233–1235, 1994.

[36] G. P. Agrawal. Population pulsations and nondegenerate four-wave mixing in semiconductor lasers and amplifiers. *Journal of the Optical Society of America B*, 5:147–159, 1988.

[37] A. Uskov, J. Mørk, J. Mark, M. C. Tatham, and G. Sherlock. Terahertz four-wave mixing in semiconductor optical amplifiers: experiment and theory. *Applied Physics Letters*, 65:944–946, 1994.

[38] A. Mecozzi. Analytical theory of four-wave mixing in semiconductor amplifiers. *Optics Letters*, 19:892–894, 1994.

[39] S. Scotti, L. Graziani, A. D'Ottavi, F. Martelli, A. Mecozzi, P. Spano, R. Dall'Ara, F. Girardin, and G. Guekos. Effects of ultrafast processes on fequency converters based on four-wave mixing in semiconductor optical amplifiers. *IEEE Journal of Selected Topics in Quantum Electronics*, 3:1156–1161, 1997.

[40] K. Inoue. Observation of crosstalk due to four-wave mixing in a laser amplifier for FDM transmission. *Electronics Letters*, 23:1293–1295, 1987.

[41] G. P. Agrawal. *Fiber-optic communication systems*. John Wiley and Sons., 3rd edition, 2002.

[42] R. Gutiérrez-Castrejón, L. Schares, and M. Duelk. SOA nonlinearities in 4 x 25 Gb/s WDM pre-amplified system for 100-Gb/s Ethernet. *Optical and Quantum Electronics*, 40:1005–1019, 2009.

[43] R. W. Tkach, A. R. Chraplyvy, F. Forghieri, A. H. Gnauck, and R. M. Derosier. Four-photon mixing and high-speed WDM systems. *Journal of Lightwave Technology*, 13:841–849, 1995.

[44] Y. J. Wen, J. Mo, and Y. Wang. Advanced data modulation techniques for WDM transmission. *IEEE Communications Magazine*, 8:58–65, 2006.

[45] E. J. Watson. Primitive polynomials (mod 2). *Mathematics of Computation*, 16:368–369, 1962.

[46] R. Gutiérrez-Castrejón and A . Filios. Pattern–effect reduction using a cross–gain modulated holding beam in semiconductor optical in–line amplifier. *Journal of Lightwave Technology*, 24:4912–4917, 2006.

[47] T. Durhuus, B. Mikkelsen, C. Joergensen, S. L. Danielsen, and K. Stubkjaer. All-optical wavelength conversion by semiconductor optical amplifiers. *Journal of Lightwave Technology*, 14:942–954, 1996.

[48] J. Gripp, J. E. Simsarian, P. Bernasconi, J. D. Le Grange, L. Zhang, L. Buhl, D. Stiliadis, D. T. Neilson, and M. Zirngibl. Load-balanced optical packet router based on 40 Gb/s wavelength converters and time buffers. In *European Conf. on Optical Communications (ECOC 05)*, Post Deadline Paper Th 4.5.2, Glasgow, Scotland, U.K., September 2005.

[49] K. E. Zoiros. Semiconductor optical amplifier–based interferometric switches: overview and characteristic applications of two main types. In D. Halsey and W. Raynor, editors, *Handbook of Interferometers. Research, Technology and Applications*, pages 95–135. Nova Science Publishers, 2009.

[50] C. J. Murphy and J. L. Coffer. Quantum dots: a primer. *Applied Spectroscopy*, 56:16A–27A, 2002.

[51] V. Cesari, W. Langbein, P. Borri, M. Rossetti, A. Fiore, S. Mikhrin, I. Krestnikov, and A. Kovsh. Ultrafast gain dynamics in 1.3 μm InAs/GaAs quantum-dot optical amplifiers: The effect of p doping. *Applied Physics Letters*, 90:201103, 2007.

[52] J. L. Xiao, Y. D. Yang, and Y. Z. Huang. Investigation of gain recovery for InAs/GaAs quantum dot semiconductor optical amplifiers by rate equation simulation. *Optical and Quantum Electronics*, 41:613–626, 2009.

[53] T. Akiyama, N. Hatori, Y. Nakata, H. Ebe, and M. Sugawara. Pattern-effect-free semiconductor optical amplifier achieved using quantum dots. *Electronics Letters*, 38:1139–1140, 2002.

[54] J. M. Martinez, Y. Liu, R. Clavero, A. M. J. Koonen, J. Herrera, F. Ramos, H. J. S. Dorren, and J. Marti. All–optical processing based on a logic XOR gate and a flip–flop memory for packet–switched networks. *IEEE Photonics Technology Letters*, 19:1316–1318, 2007.

[55] M. L. Nielsen and J. Mørk. Increasing the modulation bandwidth of semiconductor optical amplifier-based switches by using optical filtering. *Journal of the Optical Society of America B*, 21:1606–1619, 2004.

[56] R. J. Manning, R. Giller, X. Yang, R. P. Webb, and D. Cotter. Faster switching with semiconductor optical amplifiers. In *International Conf. on Photonics in Switching (PS 07)*, pages 145–146, San Francisco, California, U.S.A, August 2007.

[57] Y. Liu, E. Tangdiongga, Z. Li, S. Zang, H. de Waardt, G. D. Khoe, and H. J. S. Dorren. Error-free all-optical wavelength conversion at 160 Gb/s using a semiconductor optical amplifier and an optical bandpass filter. *Journal of Lightwave Technology*, 24:230–235, 2006.

[58] J. Leuthold, D. M. Marom, S. Cabot, J. J. Jaques, R. Ryf, and C. R. Giles. All-optical wavelength conversion using a pulse reformatting optical filter. *Journal of Lightwave Technology*, 22:186–192, 2004.

[59] R. J. Manning, X. Yang, R. P. Webb, R. Giller, F. C. Garcia Gunning, and A. D. Ellis. The Turbo-Switch - a novel technique to increase the high-speed response of SOAs for wavelength conversion. In *Optical Fiber Conference (OFC 06)*, paper OWS8, Anaheim, California, U.S.A, March 2006.

[60] D. A. Reid, A. M. Clarke, X. Yang, R. Maher, R. P. Webb, R. J. Manning, and L. P. Barry. Characterization of a turbo-switch SOA wavelength converter using spectrographic pulse measurements. *IEEE Journal of Selected Topics in Quantum Electronics*, 14:841–848, 2008.

[61] R. P. Giller, X. Yang, R. J. Manning, R. P. Webb, and D. Cotter. Pattern effect mitigation in the turbo–switch. In *International Conf. on Photonics in Switching (PS 06)*, pages 1–3, Herakleion, Greece, October 2006.

[62] R. Gutiérrez-Castrejón. Turbo-Switched Mach-Zehnder Interferometer performance as all-optical signal processing element at 160 Gb/s. *Optics Communications*, 282:4345–4352, 2009.

[63] E. A. Patent, J. J. G. M. van der Tol, M. L. Nielsen, J. J. M. Binsma, Y. S. Oei, J. Mørk, and M. K. Smit. Integrated SOA-MZI for pattern-effect-free amplification. *Electronics Letters*, 41:549–551, 2005.

[64] Q. Xu, M. Yao, Y. Dong, W. Cai, and J. Zhang. Experimental demonstration of pattern effect compensation using an asymmetrical Mach-Zehnder interferometer with SOAs. *IEEE Photonics Technology Letters*, 13:1325–1327, 2001.

[65] R. Gutiérrez-Castrejón, M. Duelk, and P. Bernasconi. A versatile modular computational tool for complex optoelectronic integrated circuits simulation. *Optical and Quantum Electronics*, 38:1125–1134, 2006.

[66] H. Takeda and H. Uenohara. Novel optical signal regeneration system using cascaded wavelength converter based on cross-gain modulation in semiconductor optical amplifiers. *Japanese Journal of Applied Physics. Part 2*, 42(12A):L1446–L1448, 2003.

[67] K. Inoue. Technique to compensate waveform distortion in a gain–saturated semiconductor optical amplifier using a semiconductor saturable absorber. *Electronics Letters*, 34:376–378, 1998.

[68] E. Zhou, F. Ohman, C. Cheng, X. Zhang, W. Hong, J. Mørk, and D. Huang. Reduction of patterning effects in SOA–based wavelength convertes by combining cross–gain and cross–absorption modulation. *Optics Express*, 16:21522–21528, 2008.

[69] M. Usami, M. Tsurusawa, and Y. Matsushima. Mechanism for reducing recovery time of optical nonlinearity in semiconductor optical amplifier. *Applied Physics Letters*, 72:2657–2659, 1998.

[70] J. H. Kim, K. R. Oh, and K. M.Cho. Spectral characteristics of optical pulse amplifications with a holding light in semiconductor optical amplifiers. *Optics Communications*, 170:99–109, 1999.

[71] H. N. Tan, M. Matsuura, and N. Kishi. Enhancement of input power dynamic range for multiwavelength amplification and optical signal pocessing in a semiconductor optical amplifier using holding beam effect. *Journal of Lightwave Technology*, 28:2593–2602, 2010.

[72] R. Gutiérrez-Castrejón and M. Duelk. Electronic versus optical mitigation of nonlinearities in 4x25 Gb/s WDM system for 100 Gb Ethernet. In *Proceedings of the IEEE LEOS Summer Topical Meeting*, pages 153–154, Acapulco, Mexico, July 2008.

[73] J. J. Contreras-Torres and R. Gutiérrez-Castrejón. Performance analysis of an all-optical wavelength converter using a semiconductor optical amplifier simulator. In *2nd International Conf. on Electrical and Electronics Engineering (ICEEE)* , pages 97–100, Mexico City, Mexico, September 2005.

[74] R. P. Webb, J. M. Dailey, and R. J. Manning. Pattern compensation in SOA–based gates. *Optics Express*, 18:131502–131509, 2010.

In: Optical Amplifiers
Editor: Galina Nemova

ISBN 978-1-61209-835-7

Chapter 4

ZIRCONIA-BASED ERBIUM-DOPED FIBER AMPLIFIER

S. W. Harun,[1,*] N. A. D. Huri,[1] A. Hamzah,[1] H. Ahmad,[1] M. C. Paul,[2] S. Das,[2] M. Pal,[2] S. K. Bhadra,[2] S. Yoo,[3] M. P. Kalita,[3] A. J. Boyland,[3] and J. K. Sahu[3]

[1]Photonics Research Center, University of Malaya, Malaysia,
[2]Fiber Optics and Photonics Division, Central Glass and Ceramic Research Institute(CGCRI), India,
[3]Optoelectronics Research Center(ORC), Southampton University, UK

ABSTRACT

Extensive research has been done on erbium-doped waveguide and optical fiber amplifiers in materials including silica, alumina, telluride glass, phosphate glass, lithium niobate, silicon and others. Comparison shows that these materials demonstrate different qualities that have a significant impact on the overall performance and applicability of an optical amplifier. Some materials have wider emission bandwidths than others, which can be used to amplify more wavelength channels in wavelength division multiplexing (WDM) systems. Others allow higher erbium concentrations before detrimental effects such as concentration quenching and cluster formation occur, which can translate to equal gain in a more compact device. Extremely low waveguide loss is also possible with some materials and results in improved amplifier efficiency. In the choice of glass host, many researchers have focused on only high silica glass due to its proven reliability and compatibility with conventional fiber-optic components. In this chapter, a Zircornia-based erbium-doped fiber (Zr-EDF) is comprehensively reviewed as an alternative medium for wideband optical amplification with a compact design. With a combination of both Zr and Al, a high erbium doping concentration of more than 3000 ppm in the glass host has been achieved without any phase separations of rare earths. The Zr-EDF is fabricated using in a ternary glass host, zirconia-yttria–aluminum codoped silica fiber through solution doping technique along with modified chemical vapor deposition

* E-mail: swharun@gmail.com

(MCVD). A Zirconia-based erbium-doped fiber amplifier (Zr-EDFA) is capable to provide a wide-band amplification as well as flat-gain operation in both the C- and L-band regions using only a single gain medium. For instance, at a high input signal of 0dBm, a flat gain at average value of 13 dB is obtained with a gain variation of less than 2 dB within the wavelength region of 1530–1575 nm and using 2 m of Zr-EDF and 120 mW pump power. It was found that a Zr-EDFA can achieve even better flat-gain value and bandwidth as well as lower noise figure than the conventional Bismuth-based erbium-doped fiber amplifier. Using an advanced double-pass configuration, a flat gain of around 16 dB can be achieved with gain variation of approximately 2.5 dB throughout the wavelength range from 1530 nm to 1590 nm using 2 m of the Zr-EDF.

Keywords: Optical fiber amplifier, Zircornia-based erbium-doped fiber

1. Introduction

More and more research efforts have been directed towards developing highly efficient broad-band fiber amplifiers that will fully exploit the low-loss band of silica fibers in order to increase the transmission capacity of wavelength-division multiplexing (WDM) networks [1] due to the tremendous increase in communication traffic in recent years. A lot of research has been done on erbium-doped fiber amplifiers (EDFAs) using various host and co-dopant materials such as silica, alumina, telluride, phosphate, Bismuth and others, to improve the gain performance, compactness and cost of the devices [2-4]. Results show that these materials demonstrate different qualities that have a significant impact on the overall performance and applicability of an optical amplifier. Some materials have wider emission bandwidths than others, which can be used to amplify more wavelength channels in wavelength division multiplexing (WDM) systems. Others allow higher erbium concentrations before detrimental effects such as concentration quenching [5] and cluster formation [6] occur, which can translate to equal gain in a more compact device. Extremely low fiber loss is also possible with some materials and results in improved amplifier efficiency.

Efficient amplifiers in both C- and L-band regions have been proposed using various configurations such as double-pass, two-stage and etc [7]. In the choice of glass host, many researchers have focused on only high silica glass due to its proven reliability and compatibility with conventional fiber-optic components. The use of high concentration EDF such as Erbium/Ytterbium doped fiber is also proposed by many researchers to reduce the gain medium length. Recently, Lanthanum co-doped Bismuth-based erbium-doped fibers (Bi-EDFs) have been extensively studied for use in compact amplifiers with a short piece of gain medium [8]. However, this type of fiber is very difficult to splice with a standard single mode fiber (SMF) using the standard splicing machine due to the difference in melting temperature.

In this chapter, a new type of EDFA is proposed using a Zirconia-Yttria-Alumino silicate co-doped Erbium-doped fiber (Zr-EDF) as a gain medium. The Zr-EDF can be easily spliced with a standard SMF due to the similarity in the melting temperature. The performance of the Zr-EDF amplifier is investigated and compared with the conventional Bi-EDFA. An advanced double-pass configuration is also proposed to provide a wide-band amplification as well as flat-gain operation in both the C- and L-band regions using only a single gain medium. The next section describes the fabrication process of the EDF.

2. Fabrication of Crystalline Zirconia Yttria Alumino Silicate Erbium-Doped Fiber

This section describes the fabrication process of the crystalline zirconia yttria alumino silicate Erbium-doped fiber (Zr-EDF). Firstly, the new EDF performs were fabricated in a ternary glass host, zirconia-yttria–aluminum codoped silica fiber (Zr-Y-Al-EDF) using a MCVD process [9]. Doping of Er_2O_3 into Zirconia yttria-aluminosilicate based glass was done through solution doping process. Small amount of Y_2O_3 and P_2O_5 were added where both Y_2O_3 and P_2O_5 serve as a nucleating agent to increase the phase separation with generation of Er_2O_3 doped micro crystallites into the core matrix of optical fiber preform. The glass formers incorporated by the vapour phase deposition process involves SiO_2 and P_2O_5 along with glass modifiers Al_2O_3 ZrO_2, Er_2O_3 and Y_2O_3 incorporated by the solution doping technique using an alcoholic-water (1:5) mixture of suitable strength of $ErCl_3.6H_2O$, $AlCl_3.6H_2O$, $YCl_3.6H_2O$ and $ZrOCl_2$ $8H_2O$.

The preforms were fabricated by passing the gaseous chlorides vapour of $SiCl_4$ and $POCl_3$ through a rotating silica tube. The tube was heated by an external burner which moves along the tube. Due to the high temperature, the chlorides oxidise, forming particles which deposit on the inner wall of the tube. This deposited single porous layer after solution soaking process turns into a glassy layer when the burner passes over it (the temperature is around 1650°C). By this process Er_2O_3 doped preforms based on zirconia yttria alumino silicate glass were fabricated through deposition of single porous phospho-silica layer at optimum deposition temperature range within 1350-1400^{0}C. During this process, the pre-sintering temperature was set within a range of 1350 to 1400^{0}C. Then a suitable strength of the dopant precursors was used during solution doping process to get the optimized process parameters for making of fiber with NA around 0.17-0.20.

In the final stage, the tube was collapsed into a solid rod at a temperature higher than 2000°C. Generally in a bulk glass matrix pure zirconia exists in three crystal phases at different temperatures. At very high temperature above 2350°C ZrO_2 has a cubic structure. At intermediate temperatures between 1170 to 2350°C it has a tetragonal structure. At low temperatures around 1170°C the material transforms to the monoclinic structure.

The transformation from tetragonal to monoclinic is very rapid and is accompanied by a 3 to 5 percent volume increase that causes extensive cracking in the material. Such type of cracking within the doped core region was observed after fabrication of preform. This behaviour destroys the mechanical properties of fabricated preforms during cooling. However several oxides which dissolve in the zirconia crystal structure can slow down or eliminate these crystal structure changes. Commonly effective additives are used such as MgO, CaO, and Y_2O_3. In order to prevent such type of cracking, minor amount of Y_2O_3 is used. A fiber of 125 μm in diameter is drawn from the fabricated preforms at temperature of around 2000^{0}C using the conventional fibre drawing technique.

3. Optical Characteristics of the Preform and EDF

The morphology of the core region of some preform samples was studied using field-emission gun scanning electron microscopy (FEGSEM). The x-ray diffraction (XRD) curve of such type of fiber preform sample was also taken. The average dopant percentages of the fiber preform samples were measured by electron probe microanalyses (EPMA). The amplification characteristic of the drawn fiber was also investigated. In bulk zirconia-silicate glass phase-separation has been observed at temperatures below the onset of crystallization [10] which also results in structural homogeneity. Phase separation, or immiscibility, is a phenomenon that is known to exist in amorphous binary systems [11]. However in some ZrO_2–SiO_2 system, immiscibility exists even in the stable liquid phase above the melting point. The phase diagram of ZrO_2–SiO_2 system is shown in Figure 1 evaluated based on Fact-Sage software in which a stable immiscibility zone exists between 60–80 mole % SiO_2. Such stable immiscibility zone extent to temperatures lower than the melting point and gives a metastable immiscibility zone in a wide composition range where phase separation occurs normally in an amorphous state.

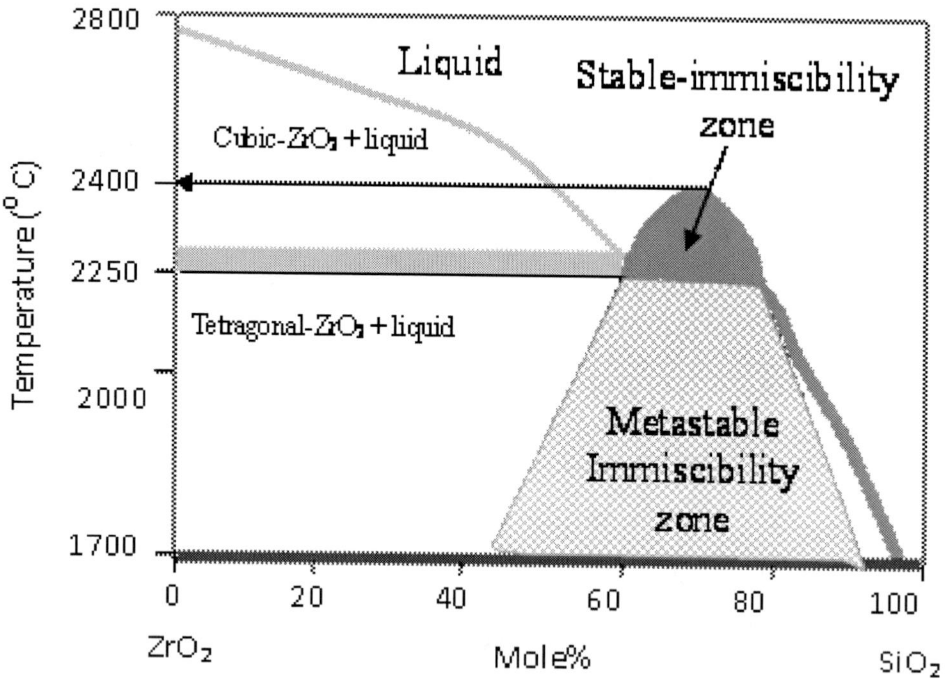

Figure 1. Phase-diagram of SiO_2-ZrO_2 system.

In zirconia-alumino silicate glass, it may be expected that the separated ZrO_2 and Al_2O_3 phases tends to mix together during heating at high temperature. Generally the homogeneous amorphous mixture of composition of $ZrAl_xO_y$ is thermodynamically more stable than the separated two phases. It may be expected that the two separated phases ZrO_2 and Al_2O_3 tends to mix into a homogeneous mixture before crystallization.

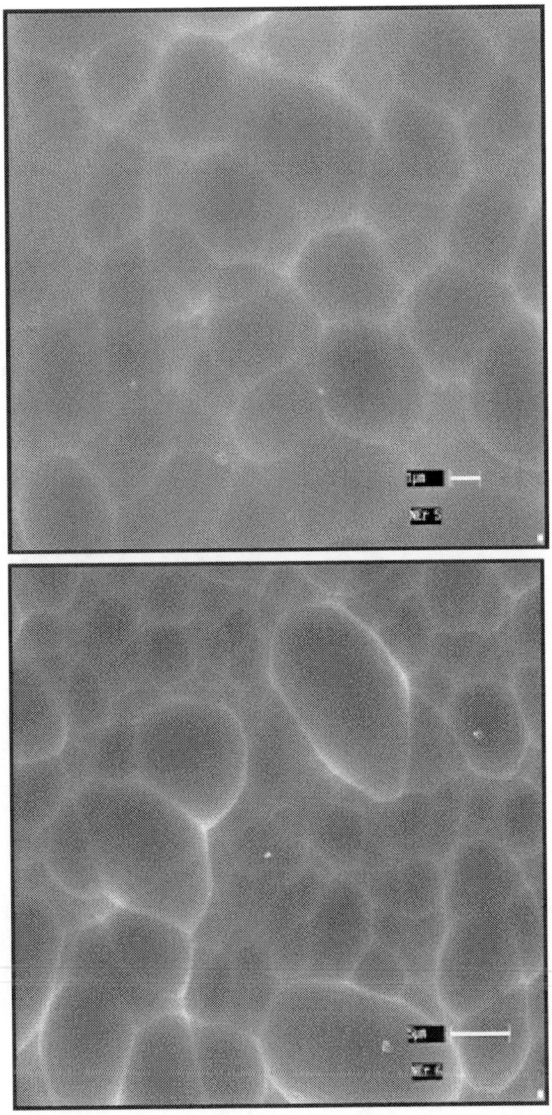

Figure 2. The microstructure of the core region of optical fiber performs (a) MEr-1 (b) MEr-2.

Three Er_2O_3 doped fibers (MEr-1, MEr-2 and MEr-3) were drawn from three different preforms. These preforms were fabricated by increasing the doping levels of Er_2O_3 and ZrO_2 while maintaining the same content of Al_2O_3. This was done by changing the composition of the soaking solution of a mixture of $ErCl_3.6H_2O$ and $ZrOCl_2.8H_2O$ during solution doping process. Since ZrO_2 sustain the crystalline nature at high temperature which becomes comparable to the collapsing temperature at the preform making stage and fiber drawing temperature, it may be expected retention of crystalline nature of some portion of ZrO_2 within the core glass matrix of preform as well as fiber. The microstructure of the doping region of two fiber preform samples MEr-1 and MEr-2, which are developed without any thermal treatment or annealing, are given in figure 2. The pictures clearly show some gain boundary within the region of silica based core glass matrix of optical fiber preforms.

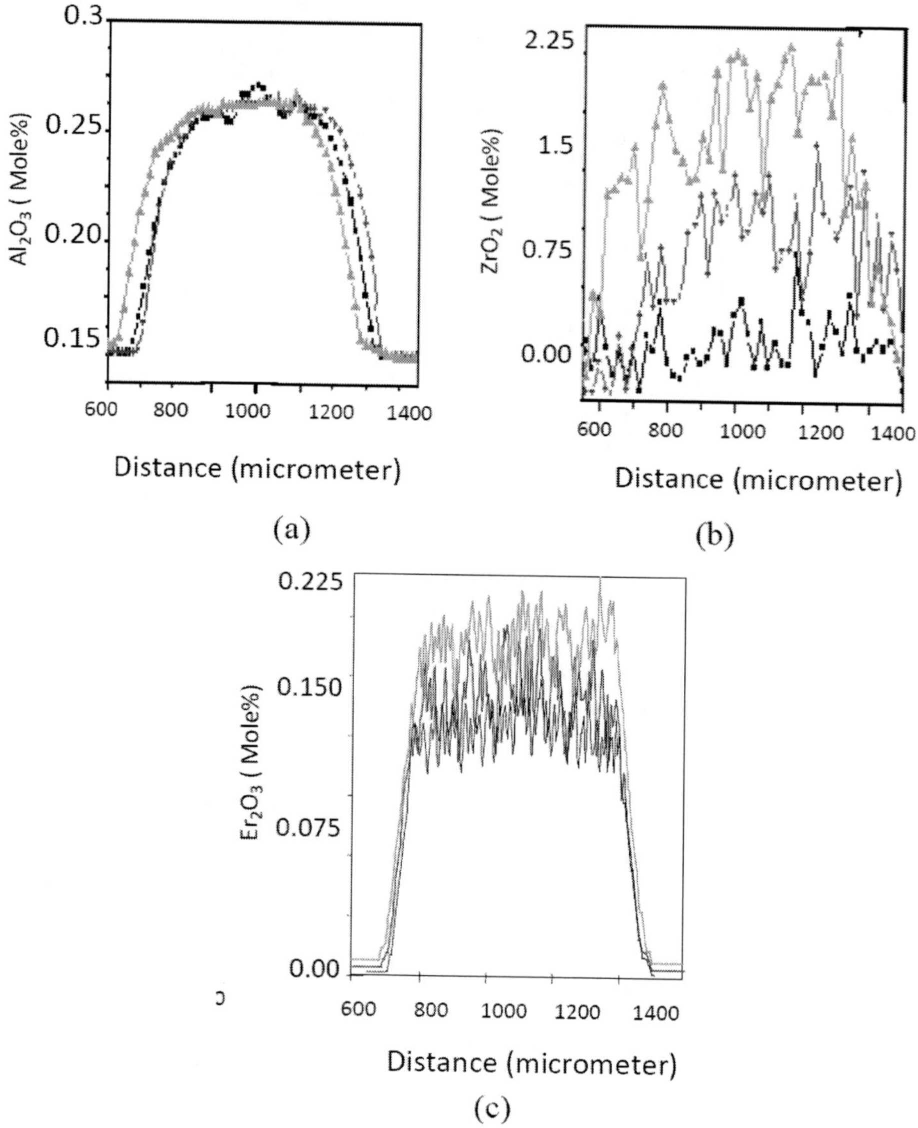

Figure 3. Distribution of (a) Al_2O_3, (b) ZrO_2 and (c) Er_2O_3 along the diameter of doped region of optical preforms.

The EPMA results of fiber preform measurements are shown in Figure 3. Figures 3 (a), (b) and (c) show the distribution of Al_2O_3, ZrO_2 and Er_2O_3, respectively along the diameter of doped region of optical preforms. All the three fibers contain almost the same doping level of Al_2O_3 around 0.24-0.26 mole% with increasing ZrO_2 from 0.65 to 2.21 mole% and Er_2O_3 from 0.155 to 0.225 mole%. The doping levels of Al_2O_3, ZrO_2 and Er_2O_3 evaluated from EPMA result is summarized in Table 1 and the fiber parameters in Table 2. The nature of the core glass matrix of optical fiber preform was evaluated from XRD study. The XRD curve is given in Figure 4. A small diffraction peak is observed at 2θ of 30°, which indicating the formation of tetragonal ZrO_2.

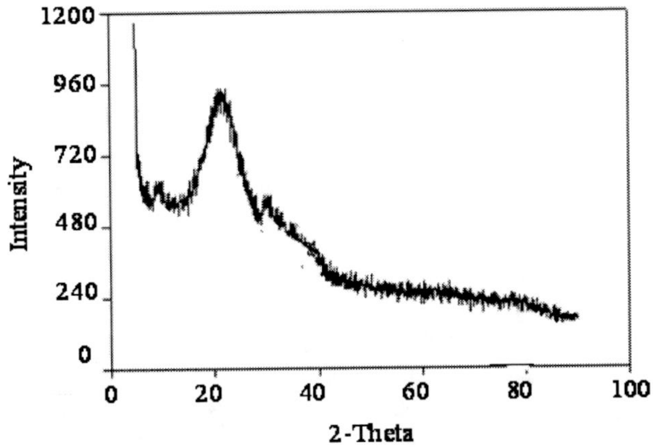

Figure 4. XRD curve of the doping region of the preform of MEr-3.

Table 1. Doping levels within the core region of preforms

Preform No	Al_2O_3 (mole%)	ZrO_2 (mole%)	Er_2O_3 (mole%)
MEr-1	0.25	0.65	0.155
MEr-2	0.26	1.47	0.195
MEr-3	0.24	2.10	0.225

The peak absorption of such three fibers at 978nm are found to be 14.5, 18.3 and 21.0 dB/m respectively which translates to the concentration of erbium ions of 2800, 3888 and 4320 ppm.wt, respectively. Both fibers (MEr-2 and MEr-3) show the same fluorescence life-time around 10.0 ms. However, the fiber MEr-1 shows slightly lower fluorescence life time around 8.0 ms. The amplification characteristics of these three fabricated fibers (MEr-1, MEr-2 and MEr-3) are also investigated.

Table 2. Fiber parameters

Fiber No	Core composition	NA	Core diameter (μm)	Fiber diameter (μm)	Erbium ion concentration (ppm wt)
MEr-1	SiO_2-Al_2O_3-Y_2O_3-ZrO_2-P_2O_5-Er_2O_3	0.17	10.5	125±3	2800
MEr-2	SiO_2-Al_2O_3-Y_2O_3-ZrO_2-P_2O_5-Er_2O_3	0.18	10.0	125±3	3888
MEr-3	SiO_2-Al_2O_3-Y_2O_3-ZrO_2-P_2O_5-Er_2O_3	0.20	10.0	125±3	4320

These three fibers have slight differences in core diameter, numerical aperture (NA) and erbium ion concentration. Core diameter for MEr-1, MEr-2 and MEr-3 are 10.5, 10.0 and 10.0 μm, respectively. The NA of the fibers varies from 0.17 to 0.20 as shown in Table 2.

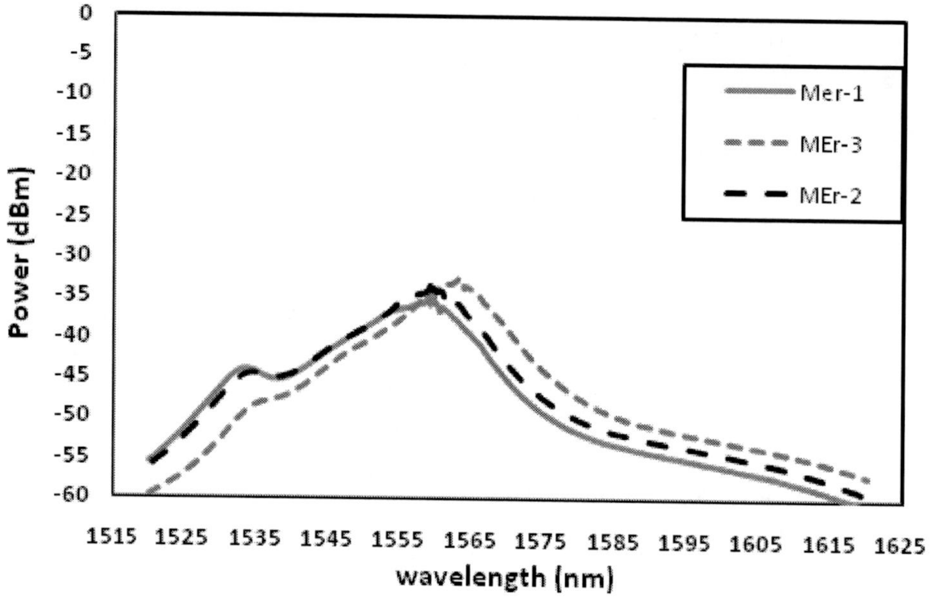

Figure 5. ASE spectrum of the forward pumped Zr-EDFA when it is pumped by 100 mW of 1480 nm pump powers.

4. Amplification Chacteristics of the Zr-Based EDFA

By pumping the fabricated Zr-EDF by 1480 nm laser diode, amplified Spontaneous Emission (ASE) is produced due to population inversion process. The ASE is an undesirable because its add noise during the amplification process. The ASE spectrum of the forward pumped EDFA is measured using an optical spectrum analyzer (OSA) and the output spectrum is shown in Figure 5 for various Zr-EDFs (MEr-1, MEr-2 and MEr-3). In the experiment, the 1480nm pump power and EDF length are fixed at 100 mW and 4 m, respectively. As shown in Figure 5, the ASE power is highest at L-band region but lowest at C-band region for MEr-3 EDF, which has the highest Erbium ion concentration. This shows that the ASE spectrum moves to a longer wavelength or L-band region as the Erbium ion concentration is increased from 2800 ppm to 4320 ppm. The higher Erbium ion increases the population inversion inside the gain medium, which emits light in C-band or 1550 nm region. The C-band ASE is then absorbed to emit light at the L-band region and therefore the gain spectrum shifts to a longer wavelength as the Erbium ion increases.

Figure 6 shows the ASE spectra at various EDF lengths for an EDFA with MEr-3. As shown in the figure, the overall ASE power increases and shifts toward L-band region with the increase of EDF length. The peak powers of -34 dBm, -33 dBm and -32 dBm are achieved by EDF lengths of 2, 3 and 4 m, respectively at wavelength region around 1560 nm. Some ripples are also observed at the peak power region due to the spurious reflections at splicing point, which cause lasing. The population inversion in the EDFA increases with the EDF length and this contributes to the wavelength shifting due to quasi-two level effect.

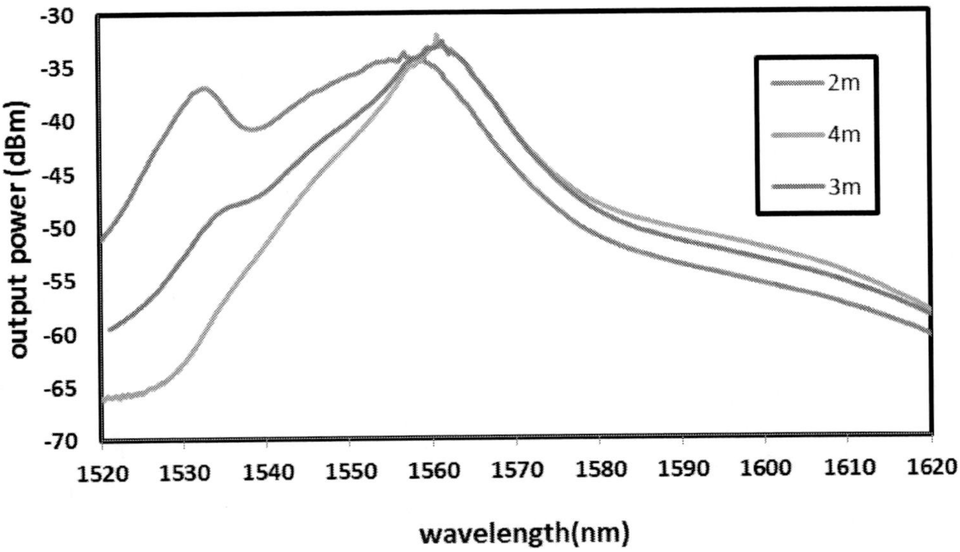

Figure 6. ASE spectrum with a various lengths of Zr-EDF.

The gain and noise Figure performances of the EDFAs are investigated at different length of MEr-3 EDF using a standard configuration with a forward pumping scheme. Tunable laser source (TLS) is used in conjunction with OSA for the measurements. In the experiment the input signal and the pump power were fixed to -30 dBm and 100 mW, respectively while the MEr-3 fiber length was varied from 2 to 4 m. The result is shown in Figure 7 where the gain spectrum moves toward a longer wavelength region as the MEr-3 fiber length is increased from 2 to 4 m. The maximum gain of 21.8 dB is obtained at 1560 nm with 2 m length of EDF. But the gain spectrum becomes flatter as the EDF length increases. The gain drops from 19.5 dB to 9.3 dB at 1550 nm but increases from 5.8 to 8.0 dB at 1600 nm as the length is changed from 2 to 4 m as shown in Figure 7. This is attributed to a quasi-two level system effect in gain medium which absorbs the shorter wavelength photons and emits at longer wavelengths. On the other hand, the noise figure varies from 5.5 to 10 dB within a wavelength region from 1530 to 1620 nm for all amplifiers. The shorter length EDF provides a relatively lower noise figure value especially at a shorter wavelength region. At EDF length of 3 m, over 10 dB signal gain and noise figure less than 9 dB was obtained from 1540 nm to 1580 nm.

Figures 8(a) and (b) show the measured gain and noise figure spectra at various 1480 nm pump power for input signal powers of -30 dBm and -4 dBm, respectively. In the experiment, the 3m long MEr-3 EDF is used as the gain medium and the 1480 nm pump power is varied from 60 mW to 160 mW. As shown in the figure, the gains increase as the pump power increases. For instance, at 1560 nm, the small signal gain is increased from 8.4 to 20.4 dB as the pump power is increased from 60 mW to 100 mW as shown in Figure 8(a). However, only a small gain increment of less than 1dB was observed when the pump power is increased from 100 mW to 160 mW. This shows that the optimum pump power for the 3 m EDF is around 100 mW. At high input signal of -4 dBm, a flat-gain at average value of 8.6 dB is obtained with a gain variation of less than 4.4 dB within a wavelength region from 1535 nm to 1605 nm. The noise figure is maintained below 9.6 dB at this wavelength region. At wavelength span from 1550 nm to 1580 nm, the signal gain is flattened around 10 dB and noise figure is measured to be less than 9 dB. As the pump power is increased to 160 mW, the gain enhancement can be achieved more than 10 dB at wavelength range from 1540 nm to 1580 nm.

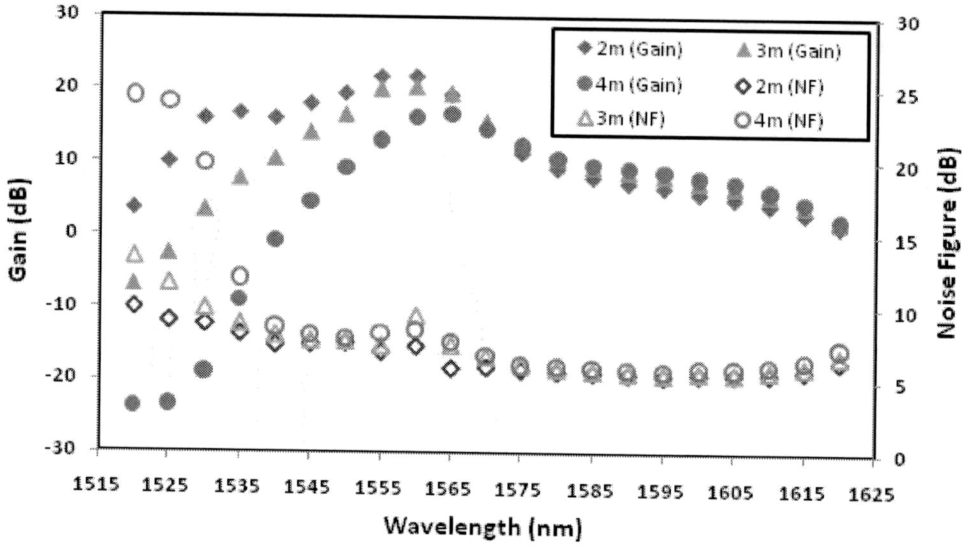

Figure 7. Optical gain and noise figure spectrum of the EDF at different lengths where the input signal is fixed to -30 dBm and 100 mW.

5. Performance Comparison with a Bismuth-Based EDFA

This section compares the performance of the proposed Zr-EDFA with the conventional Bi-EDFA. The configuration used is based on a forward pumped EDFA system. A piece of Zr-EDF or Bi-EDF is forward pumped by a 1480nm laser diode via a 1480/1550nm wavelength division multiplexing (WDM) coupler. Lanthanum/erbium co-doped Bi_2O_3-based fiber (Bi-EDF) is a commercial fiber from Asahi Glass Co. Ltd. This fiber was fabricated from a preform which was prepared by a melting method. The Erbium ion concentrations are 3250 wt-ppm and 6500 wt-ppm for C- and L-band operations, respectively. In this experiment, both Bi-EDF lengths are fixed at 0.5 m for both operations. The Bi-EDFs were fusion-spliced to high NA fibers (Corning HI980) using a commercial fusion-splicer and the average splice loss was estimated to be less than 0.5 dB/point. Angled-cleaving and splicing were applied to suppress the reflection due to the large refractive index difference between the Bi-EDF and silica fiber. The Zr-EDF is obtained from the fabrication process explained in the earlier section. It has an erbium ion concentration of around 2800 ppm and the optical characteristics are similar to MEr-1. The fiber lengths are fixed at 2 m and 3 m for C-band and L-band operations respectively.

Figure 9 shows the measured gain and noise figure of both Zr-EDFA and Bi-EDFA with the doped fiber length of 2 m and 0.5 m respectively, as the input signal and 1480 nm pump power is set at -30 dBm and 120 mW, respectively. As shown in the figure, Zr-EDFA illustrates more efficient gain and noise figure compared to Bi-EDFA. This is attributed to the optimized doped fiber length which is longer for the Zr-EDFA. Both EDFAs operates in C-band region with a relatively flat gain characteristic. The highest gain of 25.4 dB was obtained at 1560nm for the Zr-EDFA. Within wavelength region from 1530 to 1565 nm, overall gain for Zr-EDFA obtained was more than 22.5 dB with gain variation of less than 3 dB. The measured noise figure was less than 7.4 dB within this wavelength region. However, the maximum gain achieved for Bi-EDFA was 24.2 dB at 1535 nm with the corresponding noise figure of 7.6 dB.

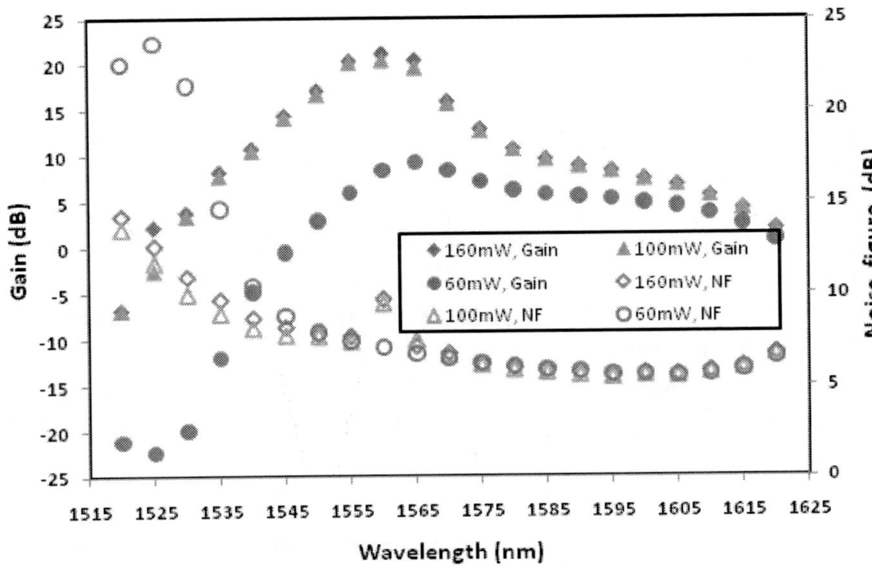

(a)-30 dBm

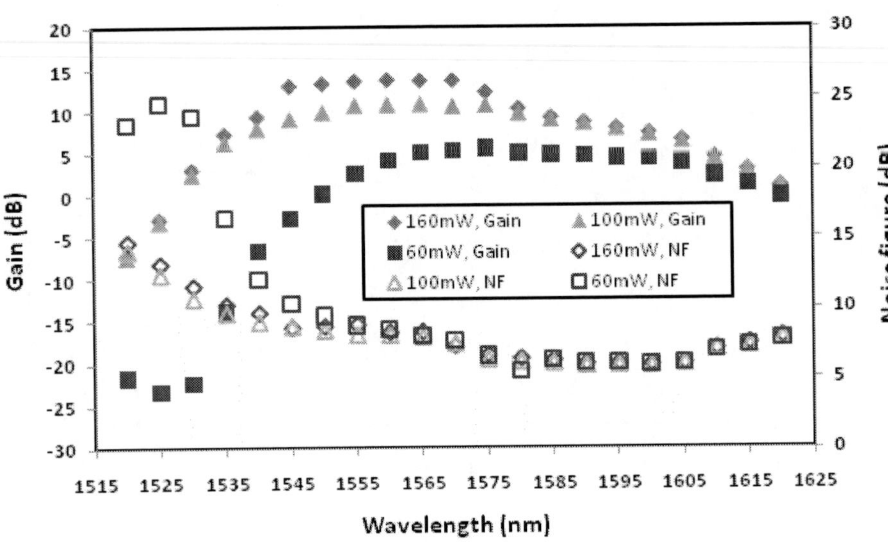

(b)-4 dBm

Figure 8. Optical gain and noise figure of the 3m MEr-3 EDF at different pump power where the input signal is fixed to (a) -30 dBm (b) -4 dBm.

A flat gain bandwidth is also broader with the Zr-based EDFA, which is probably due to the smaller energy transfer from shorter wavelength to a longer wavelength at 1565nm region. This is due to the up-conversion process which suppresses the amplification at extended L-band region. This effect is less in the Bi-EDFA due to incorporation of Lanthanum ions. The noise figure characteristic is strongly dependant on the cavity loss as well as the gain of the EDFA. Therefore the noise figure reduces as the operating wavelength increases due to cavity loss that is lower at the longer wavelength. The noise figure is also lower with the Zr-EDFA which exhibits a higher gain.

Figure 10 shows the measured gain and noise figure of both amplifiers at input signal power of 0 dBm. In the experiment, the Zr-EDF and Bi-EDF are set at the optimized lengths of 2 m and 0.5 m respectively and the pump power is fixed at 160 mW. As shown in the figure, the gain varies from 12 dB to 14 dB within wavelength region from 1530 nm to 1575 nm for the Zr-EDFA. The gain is about 2 dB higher compared to the Bi-EDFA at flat gain region, which is due to the longer length used. On the other hand, the noise figure is maintained below 9.2 dB and 9.6 dB within the flat gain region for Zr-EDFA and Bi-EDFA, respectively. At wavelength region above 1590 nm, however, the gain of Bi-EDFA is relatively higher due to the suppression of excited state absorption at this wavelength region. For the extended study, we also investigated the performance of both EDFAs at longer gain medium or higher Erbium ion concentration.

Figure 11 shows the measured gain and noise figure of Zr-EDFA and Bi-EDFA with the doped fiber length of 3 m and 0.5 m respectively, as the input signal and 1480 nm pump power is set at -30dBm and 120mW, respectively.

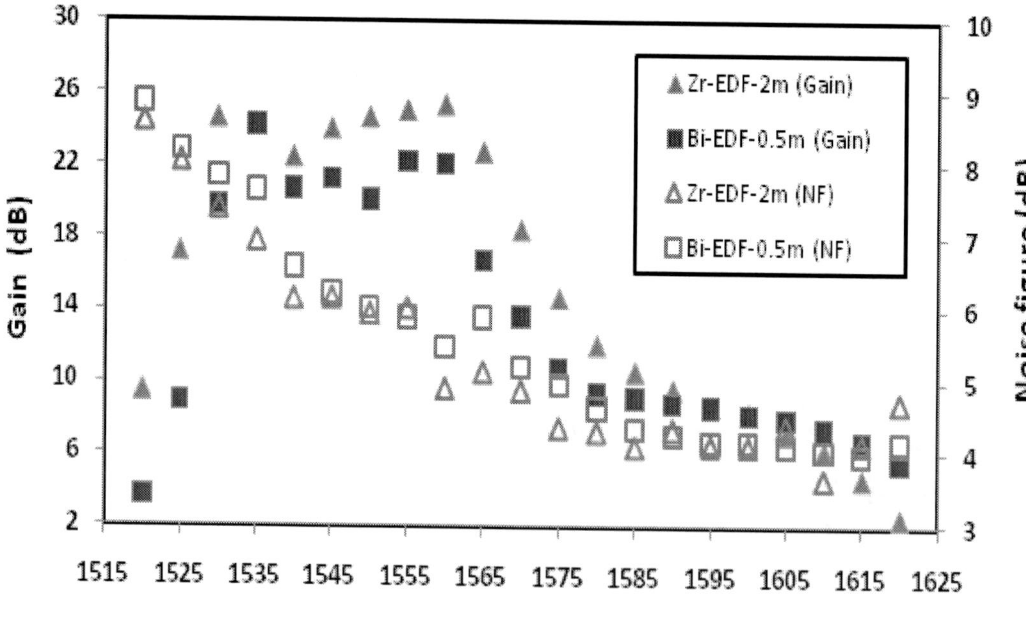

Figure 9. Comparison of gain and noise figure spectra between Zr-EDFA and Bi-EDFA at input signal power of -30 dBm.

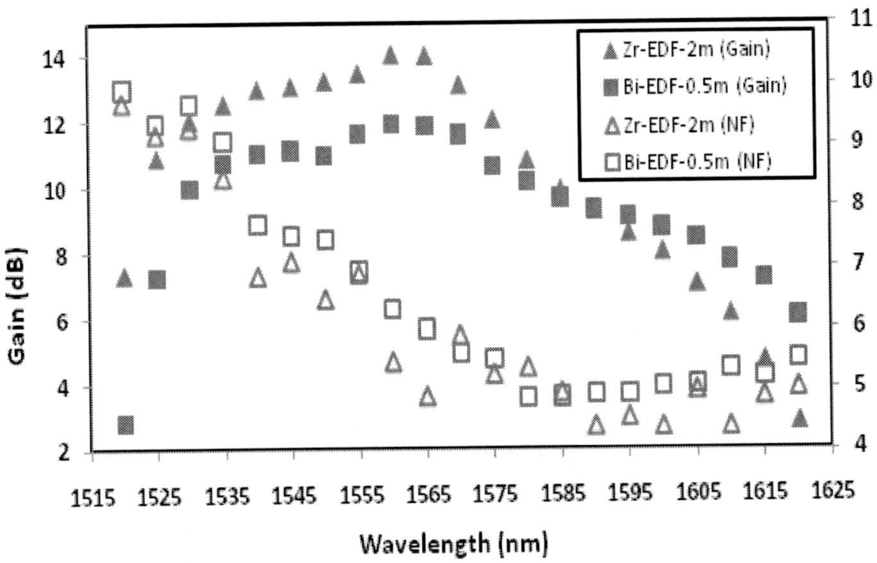

Figure 10. Comparison of gain and noise figure spectra between Zr-EDFA and Bi-EDFA at input signal power of 0 dBm.

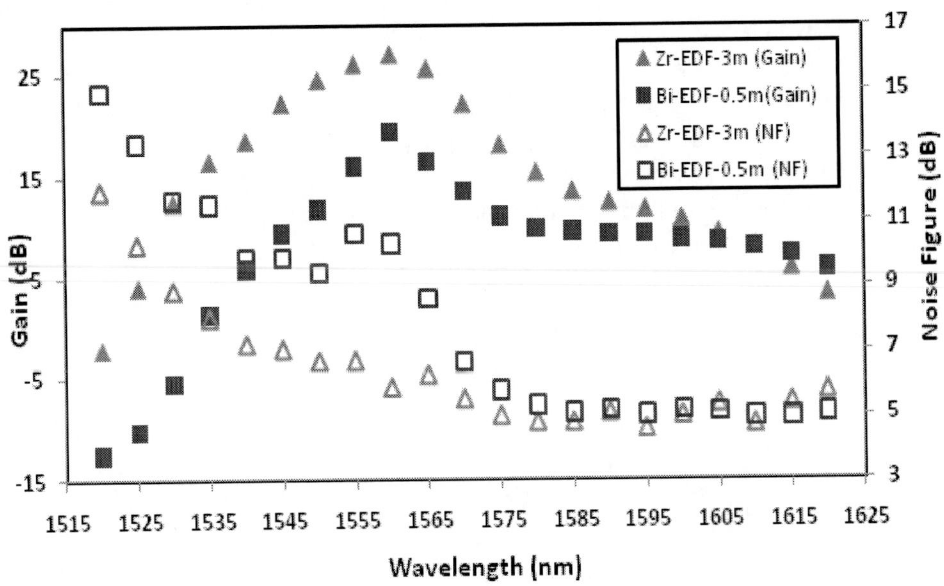

Figure 11. Gain and noise figure spectra for both Zr-EDFA and Bi-EDFA at input signal power of -30 dBm when both EDFs are optimized for L-band operation.

In this experiment, the Erbium ion concentration of Zr-EDF and Bi-EDF is set at 2800 ppm and 6500 ppm, respectively and the doped fiber length was optimized for L-band operation.

It was found that the measured gain of the Zr-EDFA is maintained at above 11 dB within wavelength region from 1530 nm to 1600 nm while the gain of the Bi-EDFA was above 8 dB within wavelength region from 1545 nm to 1610 nm. The maximum gains of 27 dB and 19.5 dB are observed for Zr-EDFA and Bi-EDFA, respectively. On the other hand, the noise figures of 8.8 dB and 11.5 dB are obtained for Zr-EDFA and Bi-EDFA respectively, at operating wavelength of 1535 nm. It was also found that the noise figure reduces as the operating wavelength reduces for both amplifiers. This result shows that the increase of Erbium ion concentration reduces the attainable gain and the performance of the EDFA is relatively better by the use of Zr-EDF as the gain medium.

6. Flat-Gain and Wide-Band Zr-EDFA

The current trend of the internet and data traffic growth has created an ever-increasing demand for transmission bandwidth. The bandwidth can be increased by fully exploiting the low-loss band region of silica fibers using a wide-band amplifier. The wide-band amplification is usually achieved using a hybrid amplifier that combines several amplifiers with different gain bandwidths. Normally, the wide-band amplifier is constructed by connecting two amplifiers in parallel [12-13]. In this configuration, the input signal is first de-multiplexed into different band by the wavelength division multiplexing (WDM) coupler, amplified by amplifiers that are suitable for the corresponding wavelength band, and finally multiplexed again with a WDM coupler. This type of hybrid amplifier has the advantage of extensibility, in which one amplifier can initially work independently while another amplifier can be added into the system according to the demand for expansion. The unusable wavelength region that exists between each gain band originated from the guard band of the WDM coupler is the main disadvantage of the hybrid amplifier. The noise figure of this amplifier also increases due to the insertion loss of the WDM coupler, which is located at the input end of each amplifier. To cope with this problem, a new amplifier which uses a single gain medium for wide-band amplification is required.

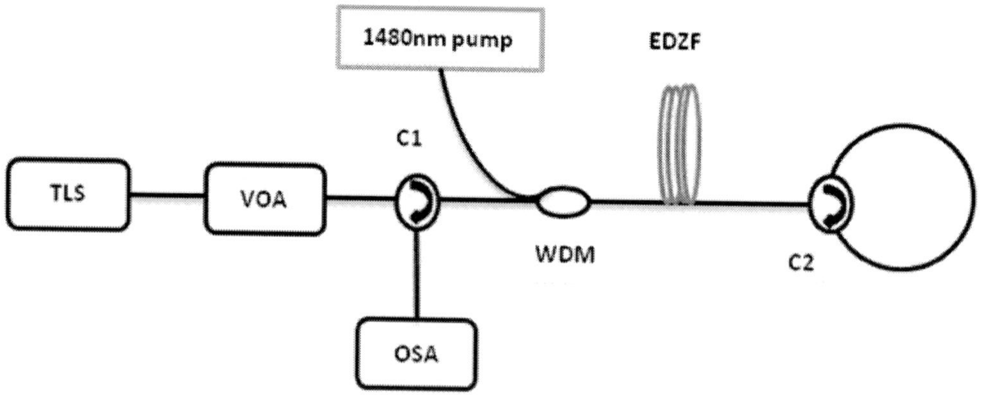

Figure 12. Configuration of the proposed double-pass Zr-EDFA.

In this section, a double-pass Zr-EDFA is proposed to achieve a wide-band and flat-gain operations. This amplifier utilizes a piece of Zr-EDF, which was fabricated through solution doping technique along with MCVD process as described in the previous section, as a gain medium. Figure 12 shows the proposed double-pass Zr-EDFA using a short piece of Zr-EDF as a gain medium and an optical circulator as a reflector. The Zr-EDF is forward pumped by a 1480nm laser diode via a 1480/1550nm WDM coupler. 3 ports optical circulator is used to route the forward ASE and the input signal back into the amplifier's system by joining port 3 to port 1. The peak absorption of the Zr-EDF was measured to be 14.5 dB/m at 980nm region, which is equivalent to the erbium ions concentration of 2800 ppm wt. The fiber length is varied from 2 m to 5.6 m in the experiment. The performance of the proposed double-pass Zr-EDFA is characterized using a TLS in conjunction with an OSA. The amplifier performance is then compared to the conventional single-pass Zr-EDFA, which was obtained by removing the optical circulator C2 and measuring the amplified signal at the output end of Zr-EDF.

Figure 13 shows the gain and noise figure spectra of the single pass and double-pass Zr-EDFAs when the input signal power is fixed at -30dBm and signal wavelength is varied from 1520nm to 1620nm. In the experiment, the Zr-EDF length is fixed at 2 m and the 1480 nm pump power is fixed at 120 mW. As shown in the figure, the gain is so much higher in the double-pass Zr-EDFA compared to that of the single-pass due to the increase of effective length of the gain medium. For instance, at input wavelength of 1560 nm, the maximum gains of 40.8 dB and 25.4 dB are achieved with the double-pass and single-pass Zr-EDFA, respectively. The gain enhancement of 15.3 dB in the double-pass Zr-EDFA is due to the double propagation of the signal in the gain medium, which increases the effective length of the amplifier and thus the gain. On the other hand, the noise figure is higher in the double-pass configuration compared to that of the single-pass especially at a shorter wavelength region, as shown in Figure 13. For instance, the noise figure penalty of about 1.4 dB was observed at input signal wavelength of 1560 nm. This is attributed to the increased backward propagating ASE power at the input part of the amplifier, which reduces the population inversion. The reduced population inversion increases the noise figure at the input part of the amplifier, which in turn increases the overall noise figure of the amplifier.

Figure 14 shows the gain and noise figure of the single pass and double-pass Zr-EDFA at the input signal and pump powers of 0 dBm and 120 mW, respectively. As shown in the figure, the high input signal gains are improved with the double-pass configuration especially at L-band region due to the same reason as explained earlier. The peak gains of 14.1 dB and 16.5 dB are obtained at 1560 nm and 1565 nm for single-pass and double pass configurations, respectively. The gain bandwidth is observed to shift to a longer wavelength by allowing the test signal to double propagate in the gain medium as shown in the Figure 14. This is attributed to the increased effective length in the proposed double-pass system, which improves the energy transfer from a shorter wavelength to a longer wavelength region and thus the gain enhancement is larger in this region. The maximum gain enhancement of 4.8 dB is observed at 1590nm.

Besides shifting a gain bandwidth to a slightly longer wavelength, the proposed double-pass configuration also flattens the gain spectrum of the Zr-EDFA by having a better gain enhancement at a longer wavelength. As shown in Figure 14, the gain spectrum of the proposed double-pass amplifier is flat with the variation of approximately 2.5 dB throughout the wavelength range from 1530 nm to 1590 nm. However, the noise figure spectrum is slightly higher with the proposed amplifier compared to that of single-pass configuration as explained earlier. The highest noise figure of 15.5 dB is obtained at 1530 nm for the double-pass while 9.7 dB is obtained at 1520 nm for the single-pass configuration. For the proposed double-pass configuration, the noise figure is maintained below 7 dB at wavelength region from 1565 to1610 nm.

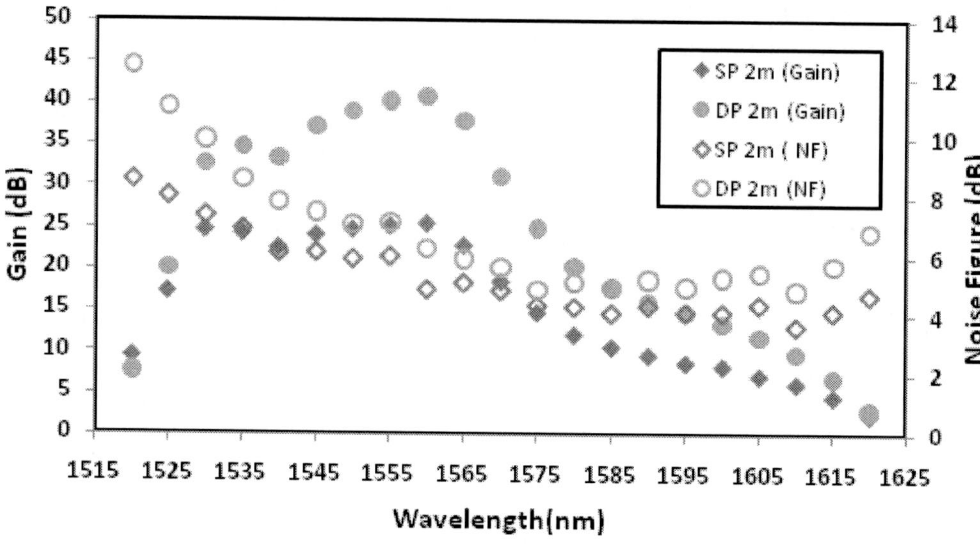

Figure 13. The gain and noise figure spectra for the single pass and double pass Zr-EDFAs at input signal power of -30 dBm.

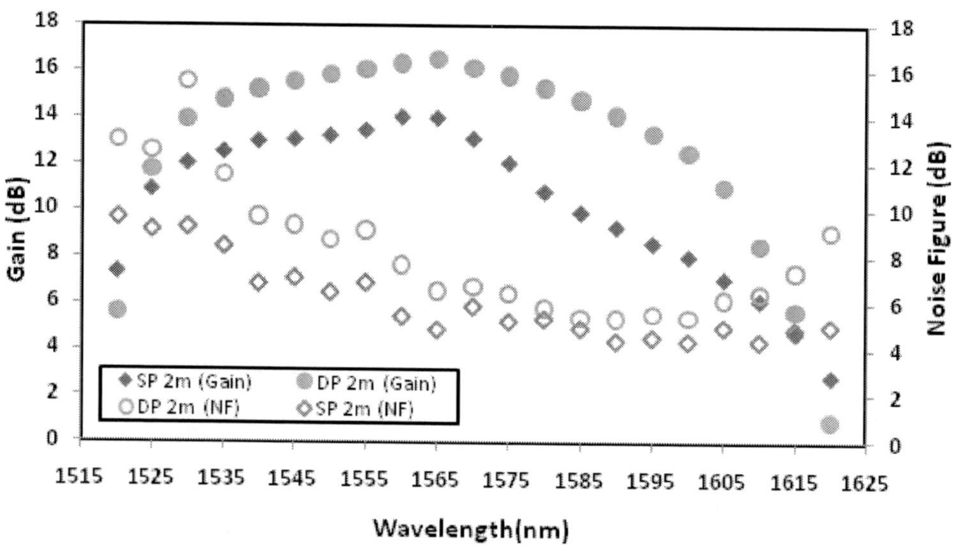

Figure 14. The gain and noise figure spectra for the single pass and double pass Zr-EDFAs at input signal power of 0 dBm.

Figire 15 shows the gain and noise figure spectrum of the proposed double-pass Zr-EDFA at various length of Zr-EDF. In the experiment, the input signal wavelength is fixed at 0 dBm and the length of Zr-EDF is varied from 2 m to 5.6 m. As shown in the figure, the optimum gain spectrum is obtained at Zr-EDF length of 2 m. As the EDZF length increases, the initial gain spectrum shifts to a longer wavelength due to the re-absorption process in the gain medium. The shorter wavelength (C-band) light was absorbed to emit a longer wavelength (L-band) light. Due to the saturation effect, the gain level reduces as the Zr-EDF length increases. The flattest gain spectrum with gain variation of less than 1 dB within a wavelength region from 1555 nm to 1610 nm was achieved with 5.6 m long Zr-EDF due to this saturation effect. However, the lowest noise figure is achieved at Zr-EDF length of 2 m due to the optimum length which provides an efficient population inversion. These results show that the proposed double-pass Zr-EDFA is capable of providing a wide-band amplification as well as flat-gain operation in both the C- and L-band regions using only a single gain medium.

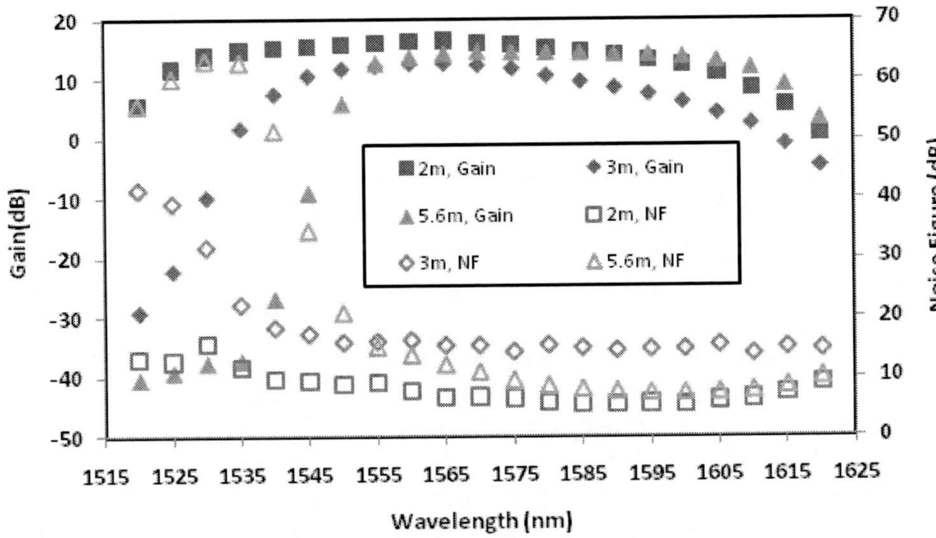

Figure 15. The gain and noise figure spectra of the proposed double pass EDZFA for different length of gain medium with the input signal power of 0dBm.

CONCLUSION

A Zircornia-based erbium-doped fiber amplifier (Zr-EDFA) is proposed and comprehensively reviewed as a compact optical amplifier with a short length of gain medium. The Zirconia fibers are drawn from Er_2O_3-doped preform, fabricated through deposition of porous layer by the MCVD process in conjunction with a solution doping technique. With a combination of both Zr and Al, we could achieve an erbium doping concentration as high as 4320 ppm in the glass host without any phase separations of rare-earths. At high input signal of -4 dBm, a flat-gain at average value of 8.6 dB is obtained with a gain variation of less than 4.4 dB within a wavelength region from 1535 nm to 1605 nm using 3 m of EDF. The corresponding noise figure is maintained below 9.6 dB at this wavelength region.

It is also found that Zr-EDFA can achieve comparable gain and noise figure characteristics as the conventional Bi-EDFA for both C-band and L-band operations. A flat gain of 13 dB was achieved with gain variation of less than 2 dB within the wavelength region of 1530 - 1575 nm at input signal of 0 dBm by using 2 m of Zr-EDF and 120 mW pump power. The corresponding noise figures are less than 9.2 at this wavelength region. Compared to the Bi-EDF, the Zr-EDF is compatible and easier to be spliced with the standard

SMF. The double-pass Zr-EDFA has been demonstrated to provide a wide-band amplification as well as flat-gain operation in both the C- and L-band regions using only a single gain medium.

Compared to a single-pass operation, the double-pass Zr-EDFA shows a better gain performance. At high input signal power of 0 dBm and optimum Zr-EDF length of 2 m, a flat gain of around 16 dB is achieved by the proposed double-pass amplifier with the gain variation of approximately 2.5 dB throughout the wavelength range from 1530 nm to 1590 nm. However, a noise figure is slightly higher in the double-pass amplifier compared to that of the single-pass due to inefficient population inversion at the input part of the amplifier.

References

[1] X. S. Cheng, R. Parvizi, H. Ahmad, S. W. Harun, " Wide-Band Bismuth-Based Erbium-Doped Fiber Amplifier With a Flat-Gain Characteristic," *IEEE Photonics Journal,* vol. 1, pp. 259 – 264 (2009).

[2] S. Jiang, B-C. Hwang, T. Luo, K. Seneschal, F. Smektala, S. Honkanen, J. Lucas, and N. Peyghambarian, "Net gain of 15.5 dB from a 5.1 cm-long Er3+ -doped phosphate glass fiber", *Optical Fiber Communications Conference Proceedings,* 2000, PD5-1.

[3] Y. Ohishi, A. Mori, M. Yamada, H. Ono, Y. Nishida, and K. Oikawa, "Gain characteristics of telluride-based erbium-doped fiber amplifiers for 1.5-mum broadband amplification," *Opt. Lett.*, vol. 23, no. 4, pp. 274–246, 1998.

[4] A. Cucinotta, F. Poli, and S. Selleri, "Design of erbium doped triangular photonic-crystal-fiber-based amplifiers," *Photonics Technology Letters*, *IEEE,* Vol.16, Issue 9, September 2004, pp. 2027 – 2029.

[5] E. Snoeks, P. G. Kik, and A. Polman, "Concentration quenching in erbium-implanted alkali–silicate glasses," *Opt. Mater.*, vol. 5, pp.159–167 (1996).

[6] D. M. Gill, L. McCaughan, and J. C. Wright, "Spectroscopic site determinations in erbium-doped lithium niobate," *Phys. Rev. B*, vol. 53, pp.2334–2344 (1996).

[7] S. W. Harun, R. Parvizi, X. S. Cheng, A. Parvizi, S. D. Emami, H. Arof and H. Ahmad, " Experimental and theoretical studies on a double-pass C-band bismuth-based erbium-doped fiber amplifier," *Optics and Laser Technol.,* vol. 42, pp. 790-793 (2010).

[8] N. Sugimoto, "Erbium doped fiber and highly non-linear fiber based on bismuth oxide glasses," *Journal of Non-Crystalline Solids*, Vol. 354, No. 12-13, pp. 1205-1210 (2008).

[9] A. Dhar, M. C. Paul, M. Pal, A. K. Mondal, S. Sen, H. S. Maiti, and R. Sen, "Characterization of porous core layer for controlling rare earth incorporation in optical fiber," *Opt. Express*, vol. 14, pp. 9006-9015 (2006).

[10] B. Rayner, R. Therrien, and G. Lucovsky, "The structure of plasma-deposited and annealed pseudo-binary ZrO2-SiO2 alloys," *Mater. Res. Soc. Symp. Proc*. 611, C1.3.1 (2000).

[11] P. F. James, "liquid-phase separation in glass-forming systems," J. *Mater. Sci.*, vol. 10, pp. 1802 (1975).

[12] Y. Miyamoto, H. Masuda, A. Hirano, S. Kuwahara, Y. Kisaka, H. Kawakami, M. Tomizawa, Y. Tada, S. Aozasa "S-band WDM coherent transmission of 40×43-Gbit/s CS-RZ DPSK signals over 400 km DSF using hybrid GS-TDFAs/Raman amplifiers" *Electronics Letters* vol. 38, no. 24, pp. 1569 – 1570 (2002).

[13] T. Sakamoto, S. Aozasa, M. Yamada, M. Shimizu "Hybrid fiber amplifiers consisting of cascaded TDFA and EDFA for WDM signals" *IEEE Journal of Lightwave Technology,* vol. 24, no. 6, pp. 2287 – 2295 (2006).

In: Optical Amplifiers
Editor: Galina Nemova

ISBN 978-1-61209-835-7

Chapter 5

OPTICAL FIBER AMPLIFIERS

Ali Reza Bahrampour[1], Laleh Rahimi,[2,*] and Ali Asghar Askari[2]

[1]Department of Physics, Sharif University of Technology, Tehran, Iran
[2]Department of Physics, Shahid Bahonar University, Kerman, Iran

ABSTRACT

Wideband optical fiber amplifiers are key components in optical fiber communication networks. Among various technologies of light amplification, rare earth doped fiber amplifiers (REDFAs) and fiber Raman amplifiers (FRAs) are of prime importance in wideband optical fiber links. Among REDFAs, silica-based erbium doped fiber amplifier (EDFA) is the most common one which is used for supporting dense wavelength division multiplexing (DWDM) systems. EDFAs work in the C-band (1530-1565 nm). Optical fibers based on different materials and various doping components of rare earth ions (RE^{3+}-ions) can provide gain in different ranges from O to U-band. In this chapter, some recent reports on the S (1460-1530 nm), C and L-band (1565-1625 nm) REDFAs are reviewed briefly. Also, the inhomogeneity and high concentration effects in EDFAs are considered. Almost every long-distance optical fiber transmission system uses FRAs. Gain characteristics of these amplifiers such as gain amplitude, bandwidth, and gain ripple are strongly dependent to the relative state of polarizations (SOPs) of pump and signal as well as the pumping configuration. These effects are discussed in this chapter. Recently, semiconductor nano-particles (quantum dots) have attracted attention to construct new gain media for optical fiber amplifiers. High gain and extra large bandwidth are the most important expected advantageous of this type of amplifiers. Authors have assigned the final section of this chapter to quantum dot doped fiber amplifiers (QDFAs)

Keywords: Rare earth doped fiber amplifier, quantum dot doped fiber amplifier

---

[*] E-mail: rahimi.laleh@gmail.com

1. INTRODUCTION

The advent of low-loss optical fibers triggered a revolution in communication field. Although very low-loss optical fibers (~ 0.2 dB/km in a conventional band) are now widely available, even this low attenuation is important in long distances. Optical fiber amplifiers are key devices that make optical fiber telecommunication possible. Historically, REDFAs are the first invented type of such devices. REDFAs were not born until 1960s when E. Snitzer invented the first neodymium doped fiber amplifier [1]. Although the potential of this device was to act as an amplifier for fiber optic communication systems which was pointed out by Snitzer, this technology did not have any other development until the 1980s. By the improvement of optical fiber manufacturing techniques such as modified chemical vapor deposition (MCVD) as well as the invention of laser diodes (LDs), REDFAs again became attractive for researchers. At present there are numerous reports on the various types of these optical amplifiers. In addition to neodymium (Nd^{3+}), other rare earth elements such as erbium (Er^{3+}), thulium (Tm^{3+}), ytterbium (Yb^{3+}) and praseodymium (Pr^{3+}) are utilized as the gain medium of an optical fiber amplifier. The various types of REDFAs (NDFAs, EDFAs, TDFAs, YDFAs, PDFAs, etc.) are very different in their main characteristics.

Various types of REDFAs are available in a very wide frequency band [2]. The next section of this chapter focuses on REDFAs in communication bands (S, C, and L bands) REDFAs will be reviewed. The minimum-loss window of standard silica-based optical fibers locates in the conventional band (C-band: 1530-1565 nm). Thus, the C-band is already the main transmission window of wavelength division multiplexing (WDM) systems. In the last two decades, the rapid growth of internet and data traffic has led to a great demand for extending the amplification bandwidth of REDFAs [3-5]. Since the optical fibers attenuation remains low in two neighbors of the C-band (S and L-bands), these two bands are obviously the best choices for adding to the present communication window. A brief review on S, C, and L-band REDFAs is presented in the first three parts of section 2. In addition to the RE^{3+}-ion type, the host material plays an important role in determining the gain performance of a REDFA. Besides silica glass, other host glasses doped with appropriate RE^{3+}-ions have been proposed for using in various spectral bands. All of these hosts are amorphous materials which cause inhomogeneous effects such as spectral hole burning (SHB) in REDFAs. Nevertheless, more reported works on REDFAs are based on homogeneous modeling. One of the main reasons is the complexity of governing equations system in an inhomogeneous model. Bahrampour and Mahjoei [6] have proposed a novel and effective method, which is called moment method (M-method), for simplifying the inhomogeneous analysis of REDFAs. The second part of section 2 of this chapter is devoted to the basis of this method. Ion concentration is another important parameter for REDFAs. The gain performance of a REDFA is very different in high and low concentration regimes. In the high concentration case, the energy transfer may occur between neighboring ions via several mechanisms. In the third part of section 2, the theoretical analysis of high concentration REDFAs is explained.

FRAs are one of the key technologies to obtain broadband optical amplifications [7]. FRAs present several advantages such as wide bandwidth, low noise, simple configuration and low gain ripple when compared with other amplifiers such as REDFAs. However, low efficiency and relative intensity noise transfer from pump to signal are two main challenges in this type of optical fiber amplifier [8]. The major goal of section 3 is to review some recent works for modeling the FRAs in different configurations. The theory of fiber Raman amplification is reviewed shortly in the first part of section 3. Polarization plays an important role in FRAs. Second part of section 3 is devoted to the explanation of polarization dependent gain (PDG). The gain performance of these amplifiers is strongly dependent to the pumping configuration. Different pumping schemes such as WDM, continuous spectrum pumping and time-division-multiplexing (TDM) with some new models such as Lagrange multiplier optimization method, M-method, and perturbation method will be presented in the third part of section 3.

In recent years, attention to the special optical properties of quantum dots (QDs) [9] (especially lead-based ones) had led to the starting of some efforts for utilizing these materials as new gain media for optical fiber amplifiers. Theoretical studies on QDFAs predict an important role for QDFAs among the further communication systems. In the fourth section of this chapter, the homogeneous and inhomogeneous models for QDFAs are presented.

2. Rare Earth Doped Fiber Amplifiers

The 3+ ions of lanthanide rare earth elements (RE^{3+}-ions) are very good candidates for using as the active medium of optical fiber amplifiers and lasers. The luminescence spectra of these ions locate in the near infrared region and most of them are highly soluble in various basic glasses of optical fibers. In addition, in a RE^{3+}-ion the 4f level electrons are shielded by the outer complete 5s and 5p levels. This property causes that the external environment can induce a little change in the 4f energy levels of RE^{3+}-ions. This means that the optical properties of RE^{3+}-ions are relatively independent of their host glass. Although the host glass influence on the energy levels is small, it is a determinative parameter for the gain performance of REDFAs. Figure 1 shows the emission cross section of Er^{3+}-ions in four different host glasses.

As seen in Figure 1, the width and amplitude of the emission cross section (also the absorption cross section) of a RE^{3+}-ion is dependent to its host. If different types of RE^{3+}-ions are being doped in appropriate host glasses, they can approximately cover the entire wavelength range from 455-2900 nm [2]. Figure 2 shows the loss spectrum of a typical single mode silica fiber (SMF) and the amplification bands of some common REDFAs. In Figure 2, spectral bands have been named according to ITU-T definitions.

As seen in Figure 2, an optical fiber loss is low in S, C and L-band. For this reason these bands have a large importance for WDM communication systems.

2.1. S-Band REDFAs

Both erbium and thulium ions (Er^{3+} and Tm^{3+}) can provide the light amplification in the S-band region [11-18]. Silica-based fibers are utilized as the host glass of both S-band EDFAs and TDFAs. In [11] Watekar et al. have examined a silica-based Tm^{3+} doped fiber amplifier (Si-TDFA) with 10 m length and 7×10^{25} ions/m^{-3} ions concentration. The measured gain and noise figure (NF) were 11.3 and 5 dB, respectively, for this amplifier in the 1470-1502 nm wavelength range. This gain has achieved for signal channels with approximately -16 dBm/ch input powers upon pumping the Si-TDFA by a 1064 nm fiber laser with 2000 mW output power. This very high required pump power is because of the low quantum efficiency of Si-TDFAs [19]. For this reason, other host glasses such as fluoride are usually preferred for S-band TDFAs [13, 19].The gain of a fluoride-based TDFA (Fl-TDFA) is about three times larger than a Si-TDFA with the same pumping power [20].

The gain spectrum of a conventional TDFA locates in the lower end of the S-band (S^--band: 1460-1480 nm), whereas the upper end of the S-band (S^+-band: 1480-1530 nm) is more usable for optical communication systems because it locates in the neighborhood of the C-band. High Tm^{3+}-ions concentration doping [12] and dual wavelength pumping [14] are two reported techniques for simultaneously shifting the gain spectrum of a Fl-TDFA to longer wavelengths (red shift) and increasing its gain value.

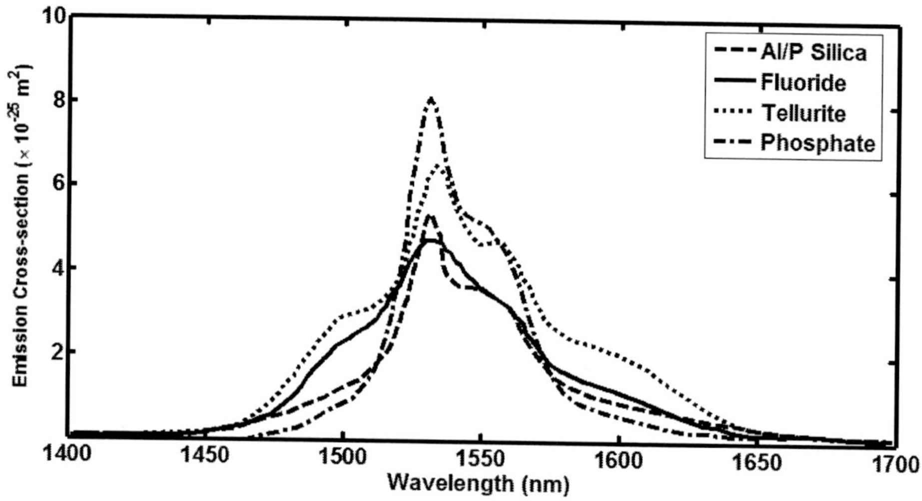

Figure 1. Emission cross-section of Er^{3+}-ions in different host glasses [2, 10].

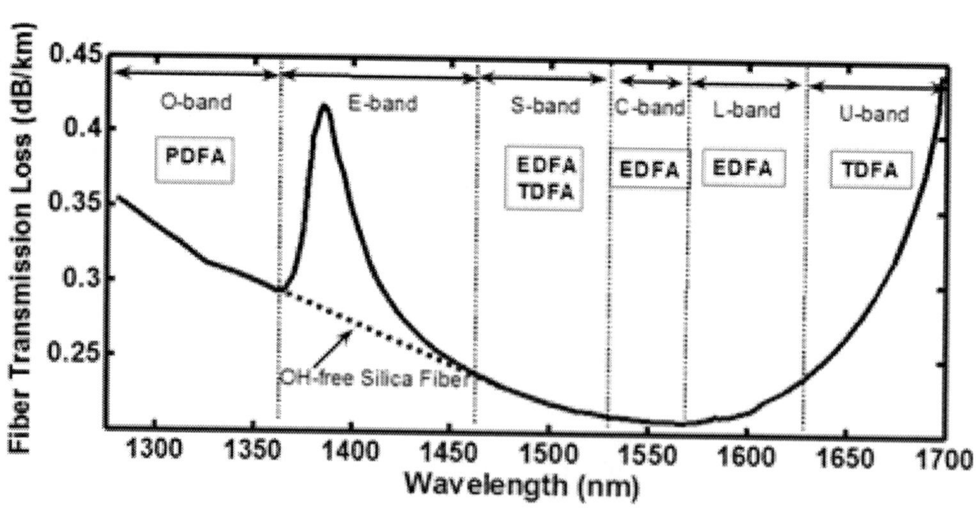

Figure 2. Loss spectrum of a typical single mode silica fiber and spectral ranges covered by standard REDFAs.

In [12], Aozasa et al. have proposed a cascaded configuration consists of a single and a double pass high concentration TDFAs with 3 and 5 m lengths, respectively. Aozasa et al. employed fluoro-zirconate (ZBLAN) based TDFAs with 6000 ppm ions concentration and pumped them with five 1400 nm LD The total pumping power was 580 mW. They measured a gain of over 26 dB and NF lower than 7 dB in the 1480-1510 nm wavelength range for signal channels with -16 dBm/ch input power. Tanabe and Tamaoka [14] have used dual pumping method for shifting the gain spectrum of a fluoride-based TDFA with 2000 ppm ions concentration and 20 m length. In [14], two main and auxiliary pumps have been used at 1051 and 1560 nm (1600 and 1640 nm) with 150 and 5 mW powers, respectively. Both pumps travel in the forward direction along the fiber. For such amplifier a gain of over 17 dB and NF lower than 5 dB have been measured in the 1450-1491 nm wavelength region. Reported results show that the largest red shift (~ 10 nm) is achieved by utilizing the 1560 nm auxiliary pump. The signal input power is -30 dBm in [14].

EDFAs are another type of S-band amplifiers [15-18]. As shown in Figure 1, the emission peak of Er^{3+}-ions is located in the C-band. For this reason, S-band EDFAs have high C-band amplified spontaneous emission noise (C-band ASE) which decreases their gain efficiency. In addition, the gain spectrum of an S-band EDFA is very non-uniform such that the gain coefficient around 1490 nm is much smaller than that around 1530 nm. Using C-band optical filters [15, 16] and W-shape index profile EDFs [17] (alone or combined with distributed filtering effect given by bending losses [18]) are some of reported solutions for the mentioned problems of S-band EDFAs. In [16] a cascaded configuration of identical S-band EDFAs has been designed such that a C-band filter has been located between two successive amplifier stages. Optical filters suppress the C-band ASE and flatten the gain spectrum simultaneously. The characteristics of this multi-stage configuration have been measured for various amplifier stage numbers, pumping powers, input signal powers and temperatures. For example for a multi-stage configuration with five stage number, a gain of over 21 dB and NF less than 6.7 dB have been reported in the 1491-1518 nm wavelength range. Each amplifier stage contains an S-band EDFA with 4 m length and 1000 ppm ions concentration and a pumping source at 980nm with 96 mW output power. In [16] the gain spectrum and NF have been measured by scanning desired bandwidth (1491-1518 nm) with a -30 dBm signal in the presence of five other signals with -12 dBm/ch input powers. Against TDFAs, the gain spectra of S-band EDFAs are located in the S^+-band.

The silica and non-silica- based EDFAs can fully cover the C and L-bands. The gain spectrum of a fluoride or telluride-based C-band EDFA is broader than that of silica-based ones (see Figure 1). Nevertheless, because of considerable properties such as simple manufacturing methods, low cost, great durability, high gain, low NF, high pump conversion efficiency, and compatibility to the existing optical fiber lines, silica-based C-band EDFAs are already key components of optical fiber communication networks. Of course, these amplifiers have some disadvantages, too. For example the gain spectrum of a C-band EDFA is naturally non-uniform. This reduces the usable gain bandwidth of these amplifiers. Various EDFAs gain equalization/flattening techniques have already been proposed which can be classified as intrinsic and extrinsic methods.

Intrinsic gain equalization techniques are based on producing a change in the structure of EDFs by co-doping some impurities like certain amounts of Al_2O_3 and P_2O_5 [21], replacing silica with other host glasses such as fluoride, telluride, and bismuth [22,23] or using special fiber designs [24,25].

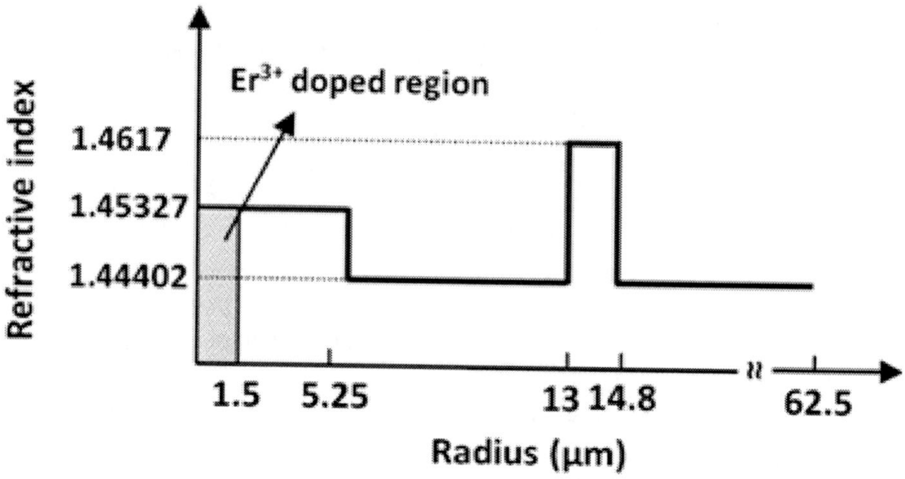

Figure 3. Schematic diagram of intrinsically gain flattened C-band EDFA reported in [25].

In [22], Ohara et al., have achieved a 3 dB bandwidth of 75 nm (1530-1605 nm) from a bismuth-based EDFA (Bi-EDFA) with 154 cm length and 3090 ppm Er^{3+}-ions concentration. The Bi-EDFA has been pumped bi-directionally by two similar pump sources at 1480 nm with 180 mW output powers. In [22] the total input power of signals (11 channels) was 0 dBm. Although Bi-based fibers are sliceable to silica-based fibers, it is difficult to achieve mass fabrication due to the difficulty of their manufacturing.

In [25], Nagaraju et al., have suggested a co-axial dual-core silica fiber as an appropriate host for an intrinsically gain flattened C-band EDFA. Figure 3 shows schematically the refractive index profile of their reported EDFA.

Only the central portion of inner core is doped with Er^{3+}-ions. The fiber parameters (refractive indexes and the radius of various layers) are selected such that the fundamental modes of two co-axial cores are being phase matched at a wavelength near 1530 nm (the emission peak of Er^{3+}-ions locates at this wavelength, see Figure 1). Thereby, a large fraction of signal channels with shorter wavelengths than 1530 nm couple to the outer core while signals with larger wavelengths than 1530 nm propagate through the inner core which is doped with Er^{3+}-ions. In other words, the geometric overlap between the Er^{3+}-ion doped region and transverse profile of signal intensity depends on its wavelength. Nagaraju et al., [25] have reported an average gain of over 28 dB and NF less than 6 dB in the 1532-1558 nm wavelength range. The proposed EDFA has 12 m length and 1.7×10^{25} ions/m^3 Er^{3+}-ions concentration. The pump and signal powers are 400 mW and -20 dB/ch,, respectively. The reported gain excursion is below ±2.2 dB in the amplifier bandwidth. The difficulty of fabrication and coupling pump and signal channels from the standard silica fiber of transmission line to EDFA and vice versa are two important problems of these types of EDFAs.

In extrinsic gain flattening methods, appropriate active or passive optical filters are connected in series with the EDFAs. These filters act as lossy devices for large amplified signals to reduce their output power to the level of those of other signal channels. Various types of gain flattening filters (GFFs) such as, long period fiber Bragg gratings (LPGs) [26], fiber Bragg gratings (FBGs) [27], Mach-Zehnder optical filter (MZFs) [28], acousto-optic tunable filters (AOFs) [29], fiber loop mirror filter (FLMs) [30], thin film interference filters

(TFIs) [31], and side-polished fiber (SPFs) based filters [32, 33] have been demonstrated for flattening the non-uniform gain spectrum of EDFAs. Generally, an ideal GFF should be reliable, low-cost, easily tunable, independent of a special technology (for example a special type of fibers), flexible (easily adaptive to the gain profile changes), temperature independent, low-noise and low loss inserting. A GFF is designed such that its resonance wavelength (notch position), peak loss at resonant (notch depth) and filtering bandwidth (notch width) be appropriate for flattening the gain spectrum of a desired EDFA.

In [32, 33], Varshney et al., have proposed a side-polished fiber (SPF) half-coupled to a multimode overlay waveguide (MMOW). It has been shown that a SPF-MMOW can be utilized as a GFF for EDFAs. Two designs for SPF-MMOW filters by uniform and tapered MMOWs are reported in [32] and [33], respectively. Figure 4 shows a schematic diagram of a typical uniform SPF-MMOW.

Coupling between the SPF mode and one of the MMOW modes causes the energy transfer from SPF to MMOW. Thus the SPF-MMOW filter acts as a wavelength selective lossy device. Periodic notches are observed in the transmission spectrum of this filter. The position of notches and spacing between them depends on the geometric parameters of the SPF and MMOW. Varshney et al. [32] have introduced a normal mode analysis to study the characteristics of a SPF-MMOW filter. It has shown that the filter characteristics (notch position, depth and width) depend on distance between the fiber core and MMOW (S_0), radius of curvature of the fiber, effective interaction length (L_{eff}), MMOW thickness (d), its refractive index and the position of MMOW relative to the SPF in tapered MMOW type [33]. Low insertion loss, low back reflections and good mechanical reliability are the main reported advantages of SPF-MMOW filters.

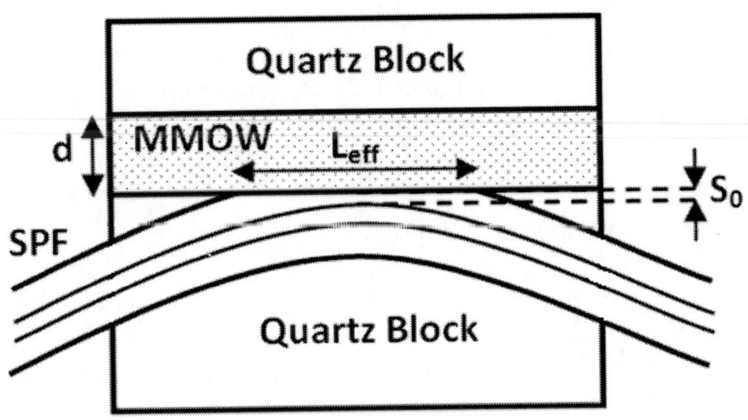

Figure 4. Schematic diagram of a uniform SPF-MMOW.

2.3. L-Band REDFAs

EDFAs are also good candidates for amplification in the L-band region. However like to S-band EDFAs, L-band EDFAs have a lower gain and higher C-band ASE noise in comparison with C-band ones. There are many papers on the gain enhancement and NF improvement in L-band EDFAs [34-39]. For example, Chio et al. [34] have replaced the conventional pumping wavelength, 1480 nm, with a C-band one. It has been shown that utilizing a 1545 nm pump wave increases the gain value to 2.25 times. No NF degradation has been reported in comparison to 1480 nm pumping case.

Dual-pass configuration is another proposed solution for the low gain of L-band EDFAs [35]. In this technique, the effective length of the L-EDFA is increased by locating it in a loop such that signals pass the amplifier twice. But in such configuration the generated ASE also propagates through the amplifier twice. Thus, using dual-pass configurations can enhance the gain of L-band EDFAs whereas degrades the NF. In [35], Harun et al. have proposed a new dual pass configuration which consists of three EDFAs such that two of them have been located in a loop and another one acts as a pre-amplifier for them. Since the ASE noise is low in the pre-amplifier in fact signal channels pass through all EDFAs while the ASE only passes through two of them. The pre-amplifier used in [35] is an EDFA with 400 ppm ions concentration and 4 m length. This EDFA is pumped forwardly with a 30.4 mW at 980 nm LD. Both of two other amplifiers have 400 ppm ions concentration and 50 m length. Their pump wavelengths are also 980 nm but with 15.6 and 50.6 mW output powers. Harun et al., has compared this new dual pass configuration with a conventional dual pass one. Although for both configurations a 33.5 dB gain has been measured in the 1568-1600 nm range, but a considerable difference between their NFs has been reported. The NF of new configuration is between 5.9 to 6.6 dB lower than conventional configuration in this range. Also it has been shown that NF of a single pass configuration is lower than both dual pass configurations, but its gain value is around 20 dB in this wavelength region. Signal input power is -30dBm in [35]. Yi et al., [36] have used a FBG for reducing the NF of a dual pass L-band EDFA.

Using the generated unwanted C-and ASE as a second pump for L-band EDFA [37], utilizing other host glasses instead of silica like fluorophosphate glass [38] and suppressing the ASE with the help of an injected C-band signal [39] are other examples of reported L-band EDFAs gain enhancement methods. The gain spectra of the L-band EDFAs locate in the L^--band (1565-1600 nm) and L^+-band (1600-1625 nm) can be covered by silica and fluoride-based TDFAs [19, 40].

2.4. Inhomogeneous Modeling of EDFAs

As mentioned before, silica-based EDFAs are the main S, C and L-band REDFAs. Inhomogeneous broadening is a very important effect which must be considered in EDFAs. Nevertheless, more reported works on EDFAs are based on homogeneous analysis. The main reasons of this are the complexity of the governing equation systems in an inhomogeneous model, requirement to quantify the homogeneous absorption and emission spectra and the effectiveness of the homogeneous models in most situations. An accurate study on saturation behavior of an EDFA and an explanation of some important effects such as SHB can be done only in an inhomogeneous analysis [21]. The inhomogeneous broadening effects of L-band EDFAs are more significant than those of C-band EDFAs [41]. Thus, the inhomogeneous model is more important in the L-band EDFAs analysis.

Energy levels of RE^{3+}-ions are composed of relatively well separated multiples that each of them is made of several individual sublevels. The energy diagram of an Er^{3+}-ion doped in silica glass is reported in many text-books [21]. Figure 5 shows schematically only three energy levels which are responsible for amplification process in EDFAs.

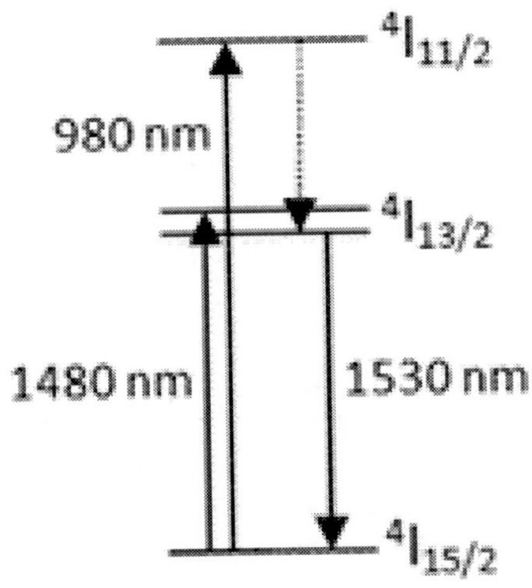

Figure 5. Energy diagram of Er^{3+} ion in silica. Dashed arrow is corresponding to a non-radiative transition.

The pumping of an Er^{3+}-ion may occur in two configurations $^4I_{15/2} \rightarrow {}^4I_{11/2}$ or $^4I_{15/2} \rightarrow {}^4I_{13/2}$. In the first case the ion is exited from ground state $^4I_{15/2}$ to $^4I_{11/2}$ level by absorbing a pump photon. This level has a short lifetime (~7μs [21]) and the exited ion quickly cascades back down to the lower $^4I_{13/2}$ level via a non-radiative transition. Then a radiative transition occurs from $^4I_{13/2}$ level to the ground state. The long lifetime of $^4I_{13/2}$ level (~10 ms [21]) makes it possible to achieve stimulated emission and as a result amplification process in an EDFA. For this reason, $^4I_{13/2}$ level is sometimes referred to as meta-stable level. The population of $^4I_{11/2}$ level is, in most situations, negligible in comparison with that of $^4I_{13/2}$ level because of its too shorter lifetime. So $^4I_{11/2}$ level can be ignored and a two level system is usually appropriate for studying the EDFAs. The most common pump wavelengths of EDFAs are 980 and 1480 nm for the first and second pumping cases, respectively. In two-level modeling, the rate equation for the ground state population is as follows [21]:

$$\frac{\partial n_1}{\partial t} = -n_1(r,z,t)\sum_{k=1}^{m}\frac{P_k(z,t)}{h\nu_k}i_k(r)\sigma_a(\nu_k) + n_2(r,z,t)\sum_{k=1}^{m}\frac{P_k(z,t)}{h\nu_k}i_k(r)\sigma_e(\nu_k) + \frac{1}{\tau_2}n_2(r,z,t) \qquad (2\text{-}1)$$

where n_1 (n_2) is the number of Er^{3+}-ions in the ground (exited) state per unit volume, m is the total number of propagating waves (pumps + signals), P_k is the power of kth wave at ν_k frequency, h is the Planck constant, i_k is the normalized transverse mode of kth wave [42], $\sigma_a(\nu_k)$ ($\sigma_e(\nu_k)$) is the absorption (emission) cross-section of Er^{3+}-ions at ν_k frequency [21] and τ_2 is the exited state life time. The first term of the right hand side of equation (2-1) shows loss in the ground state population due to thc absorption effect, the second is corresponding to the stimulated emission of exited ions and the spontaneous emission process is shown in the third term. The rate equation of the exited state population can be obtained from continuity relation: $n_1(r,z,t) + n_1(r,z,t) = n_0$, where n_0 is the volume density of doped ions. Obviously, the right hand side of the continuity relation is time independent. This model is sometimes called homogeneous model that the reason of this name will become clear in the following discussion.

As seen in equation (2-1), the absorption and emission cross-sections are identical for all doped Er^{3+}-ions while in amorphous materials such as fused silica, the total electric field changes from site to site. Hence in general there is a different Stark splitting for each doped ion. This means that different RE^{3+}-ions response differently to a monochromatic light beam. Obviously, each light field at frequency interacts more strongly with that group of ions which have the same transition frequency. The absorption and emission cross-section spectra of each group are named the homogeneous spectra. Any homogeneous spectrum has a natural broad which arises from the life time and de-phasing time of the exited state. The homogeneous broadening depends to the ion types, host glass and temperature. Generally, measured absorption and emission cross-section spectra (inhomogeneous spectra) are related to the distribution function of ions central transition frequencies (DFF) and the homogeneous spectra via the convolution integral [43]:

$$\sigma_{a,e}(\nu) = \int_0^\infty g(\nu')\, \sigma_{a,e}^h(\nu - \nu')\, d\nu', \tag{2-2}$$

where $\sigma_{a,e}^h(\nu - \nu')$ is the homogeneous absorption/emission cross-section corresponding to a specific subgroup of Er^{3+}-ions with central transition frequency ν' at frequency ν and $g(\nu')$ is the DFF which satisfies $\int_0^\infty g(\nu)\, d\nu = 1$ relation. DFF is, in most situations, a Gaussian function [21, 43, 6]. This function is wider for a more inhomogeneous medium. The center and width of the DFF depend on ions type and host glass, respectively. For example, it is wider for Al-silica fibers in comparison with Ge-silica fibers [21].

Figure 6 shows homogeneous spectra of a typical Al co-doped EDFA calculated by de-convolution method in comparison with inhomogeneous (measured) spectra [6]. The DFF has been assumed a Gaussian function with 1/e width 11.5 nm.

In the inhomogeneous model, the population rate equations are rewritten as follows [6, 44]:

$$\begin{aligned}\frac{\partial n_1(\nu)}{\partial t} &= -n_1(\nu,r,z,t)\sum_{k=1}^{m}\frac{P_k(z,t)}{h\nu_k} i_k(r)\sigma_a^h(\nu,\nu_k) + n_2(\nu,r,z,t)\sum_{k=1}^{m}\frac{P_k(z,t)}{h\nu_k} i_k(r)\sigma_e^h(\nu,\nu_k) \\ &+ \frac{1}{\tau_2} n_2(\nu,r,z,t) + F\left[n_{1qe}(\nu,r,z) + n_1(\nu,r,z,t)\right] \qquad \nu \in R\end{aligned} \tag{2-3}$$

$$\begin{aligned}\frac{\partial n_2(\nu)}{\partial t} &= n_1(\nu,r,z,t)\sum_{k=1}^{m}\frac{P_k(z,t)}{h\nu_k} i_k(r)\sigma_a^h(\nu,\nu_k) - n_2(\nu,r,z,t)\sum_{k=1}^{m}\frac{P_k(z,t)}{h\nu_k} i_k(r)\sigma_e^h(\nu,\nu_k) \\ &- \frac{1}{\tau_2} n_2(\nu,r,z,t) + F\left[n_{2qe}(\nu,r,z) + n_2(\nu,r,z,t)\right] \qquad \nu \in R\end{aligned} \tag{2-4}$$

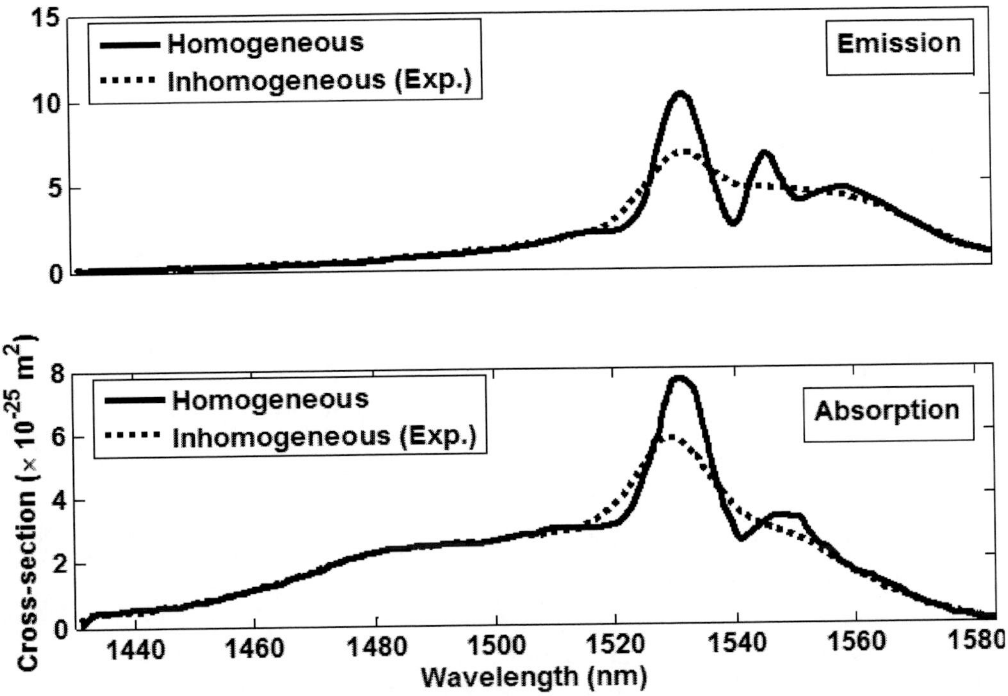

Figure 6. Calculated homogeneous (a) emission (b) absorption spectra (solid curves) of a typical Al co-doped EDF by deconvolution method, in comparison with inhomogeneous spectra (dashed curves) [6].

here $n_1(v)$ ($n_2(v)$) is the ground (exited) state population of ions with central transient frequency v, $\sigma_a^h(v,v_k)$ ($\sigma_e^h(v,v_k)$) is the absorption (emission) cross-section of this group of ions (homogeneous spectra) at v_k frequency, F is the cross-relaxation rate and n_{1qe} (n_{2qe}) is the quasi-equilibrium population of the ground (exited) sate. The continuity relation, here $n_1(v,r,z,t)+n_2(v,r,z,t)=n_0\,g(v)$, can be used for eliminating equation (2-3). In this case equation (2-4) is rewritten as follows:

$$\frac{\partial n_2(v)}{\partial t}=n_0\,g(v)\sum_{k=1}^{m}\frac{P_k(z,t)}{hv_k}i_k(r)\sigma_a^h(v,v_k)-n_2(v)\sum_{k=1}^{m}\frac{P_k(z,t)}{hv_k}i_k(r)\sigma_k^h(v)$$
$$-\frac{1}{\tau_2}n_2(v)+F\left[n_{2qe}(v)+n_2(v)\right] \qquad v\in R \qquad (2\text{-}5)$$

where $\sigma_k^h(v)=\sigma_a^h(v,v_k)+\sigma_e^h(v,v_k)$.

When a strong signal enters an inhomogeneous medium interacts only with that group of particles which have non-zero homogeneous spectra at the signal frequency. So the signal saturates this group and a hole appears in the gain spectrum of amplifier (SHB effect). The location of hole is determined by the saturating signal frequency and its width is proportional to the homogeneous emission cross-section width. Almost in all inhomogeneous media the energy transfer from unsaturated groups to saturated one is possible. In other words, after switching off the saturating signal the produced hole starts to be filled and spreads across the whole inhomogeneous gain spectrum. This phenomenon is named cross-relaxation or spectral diffusion effect. Cross-relaxation factor shows the speed of this energy transfer and has the same dimension as the Einstein parameter (s^{-1}). For large values of F, the hole deep is small and the medium behaviors is more similar to that of a homogeneous medium. Therefore the quasi-equilibrium population of each energy level is equal to the population of that level in the steady-state ($\partial/\partial t = 0$) situation and in the absence of signals. Thus, the quasi-equilibrium population of the exited level is given by:

$$n_{2qe}(v) = \left(n_0\, g(v) \sum_{k=1}^{m_p} \frac{P_k(z)}{h v_k} i_k(r) \sigma_a^h(v, v_k) \right) \Bigg/ \left(\frac{1}{\tau_2} + \sum_{k=1}^{m_p} \frac{P_k(z)}{h v_k} i_k(r) \sigma_k^h(v) \right) \qquad v \in R \tag{2-6}$$

where m_p is the total pumps number.

By employing the slowly varying envelop approximation (SVEA), the propagating equation of kth wave is as follows:

$$u_k \left(\frac{\partial P_k(z,t)}{\partial z} + \frac{u_k}{V_k} \frac{\partial P_k(z,t)}{\partial t} \right) = \left(P_k(z,t) + 2h v_k \Delta v_k \right) \int_0^{\infty} dr\, 2\pi r\, i_k(r) \int_0^{\infty} dv\, n_2(v) \sigma_e^h(v, v_k)$$
$$- P_k(z,t) \left[\int_0^{\infty} dr\, 2\pi r\, i_k(r) \int_0^{\infty} dv\, n_1(v) \sigma_a^h(v, v_k) + l_k \right] \tag{2-7}$$
$$k = 1,2,\ldots,m$$

where V_k is the velocity of kth wave, u_k is a constant factor which gets ± 1 values for forward and backward propagating cases respectively, Δv_k is the effective noise bandwidth and l_k is the intrinsic fiber loss at v_k frequency. By using the continuity relation equation (2-7) is written as follows:

$$u_k \left(\frac{\partial P_k(z,t)}{\partial z} + \frac{u_k}{V_k} \frac{\partial P_k(z,t)}{\partial t} \right) = 2h v_k \Delta v_k \int_0^{\infty} dr\, 2\pi r\, i_k(r) \int_0^{\infty} dv\, n_2(v) \sigma_e^h(v, v_k)$$
$$+ P_k(z,t) \left[\int_0^{\infty} dr\, 2\pi r\, i_k(r) \int_0^{\infty} dv\, n_2(v) \sigma_k^h(v) - n_0 \int_0^{\infty} dr\, 2\pi r\, i_k(r) \int_0^{\infty} dv\, g(v) \sigma_a^h(v, v_k) - l_k \right] \tag{2-8}$$
$$k = 1,2,\ldots,m$$

A desired, EDFA can be studied as an inhomogeneous medium by solving the presented governing equation system (equations (2-5), (2-6) and (2-8)). Since v is a real number in equations (2-5) and (2-6), we have an uncountable system of coupled partial differential equations (USPDE) (equations (2-5) and (2-8)) which must be solved. Obviously, the problem is unsolvable in this form and first of all an appropriate technique is needed for reducing the dimension of the governing equation system.

In [44], Bahrampour et al., have used a standard technique to convert the USPDE of the inhomogeneous model of EDFAs to a finite system of partial differential equations (FSPDE). In the standard method the Er^{3+}-ions are assumed to can be classified into N homogeneous groups. In other words the convolution integral (equation (2-2)) is approximated with a limited summation as follows:

$$\sigma_{a,e}(\nu) = \sum_{j=1}^{N} g_j\, \delta\nu_j\, \sigma_{a,e}^{h}(\nu_j, \nu) \tag{2-9}$$

where $g_j\,\delta\nu_j$ is the percentage of jth group which consists all Er^{3+}-ions with central transition frequencies located in a $\delta\nu_j$ bandwidth around ν_j frequency. Subdivisions ($\delta\nu_j$ s) are not selected equal to each other. They are more concentrated around the frequencies of present waves where SHB occurs and there are rapid temporal variations. In this approximation the continuity relation is in the $n_{1j} + n_{2j} = g_j\,\delta\nu_j$ form and the governing equations ((2-5), (2-6) and (2-8)) are as follows:

$$\frac{\partial n_{2j}}{\partial t} = n_0\, g_j\, \delta\nu_j \sum_{k=1}^{m} \frac{P_k(z,t)}{h\nu_k} i_k(r)\sigma_a^h(\nu_j,\nu_k) - n_{2j}\sum_{k=1}^{m}\frac{P_k(z,t)}{h\nu_k} i_k(r)\sigma_k^h(\nu_j) - \frac{1}{\tau_2}n_{2j} + F\left[n_{2qej} + n_{2j}\right] \tag{2-10}$$

$$j = 1,2,\ldots,N$$

$$n_{2qej} = \left(n_0\, g_j\, \delta\nu_j \sum_{k=1}^{m_p} \frac{P_k(z)}{h\nu_k} i_k(r)\sigma_a^h(\nu_j,\nu_k)\right) \Bigg/ \left(\frac{1}{\tau_2} + \sum_{k=1}^{m_p}\frac{P_k(z)}{h\nu_k} i_k(r)\sigma_k^h(\nu_j)\right) \qquad j = 1,2,\ldots,N \tag{2-11}$$

$$u_k\left(\frac{\partial P_k(z,t)}{\partial z} + \frac{u_k}{V_k}\frac{\partial P_k(z,t)}{\partial t}\right) = 2h\nu_k\Delta\nu_k \int_0^\infty dr\, 2\pi r\, i_k(r) \sum_{j=1}^{N} n_{2j}\,\sigma_e^h(\nu_j,\nu_k)$$
$$+ P_k(z,t)\left[\int_0^\infty dr\, 2\pi r\, i_k(r)\sum_{j=1}^{N} n_{2j}\,\sigma_k^h(\nu_j) - n_0\int_0^\infty dr\, 2\pi r\, i_k(r)\sum_{j=1}^{N} g_j\,\delta\nu_j\,\sigma_a^h(\nu_j,\nu_k) - l_k\right]$$
$$k = 1,2,\ldots,m \tag{2-12}$$

By using of this simple method the governed USPDE is reduced to a set of $N + m$ PDEs (equations (2-10) and (2-12)) which can be solved easily by the fourth order Runge-Kutta method. By using the boundary values: $n_{2j}(r,z,0) = 0$, $P_k(z,0) = 0$ and $P_k(0,t) = P_{k,input}$ for forward propagating and $P_k(L,t) = P_{k,input}$ for backward propagating waves that L is EDF length.

EDFAs generally operate near saturation. In this case the total output power of the amplifier is nearly constant and independent of the input power. The total input power may change due to add/drop operations or dynamic changes in the input power of signal channels.

This means that the output power of each surviving channel is dependent to the number and the input power of other channels. Gain clamping techniques are proposed methods for fixing the gain spectrum of an EDFA. All gain clamping techniques are based on stabilizing the population inversion of doped Er^{3+}-ions with the help of an oscillating laser. Generally, there are two kinds of all optically gain clamping (AOGC) techniques. In both methods, a portion of ASE, which is generated during the amplification process, is used to form an oscillating laser in the amplifier. Changes in the input power of amplifier transfer to the oscillating laser such that the total population inversion and as a result output power of surviving signal channels remain constant. Usually the laser cavity is made by a fiber feedback loop [45] or by using fiber Bragg gratings (FBGs) [46, 47]. A single oscillating laser can lock the population inversion of a homogeneous gain medium and is sufficient for fixing the gain spectrum while for the gain clamping in an inhomogeneous active medium several oscillating lasers are needed, each for stabilizing the output power of a specific group of signal channels [44, 48, 49].

Thus, propagating equations of oscillating lasers must also add to the governing equations system and in equation (2-12) m is the total number of pumps, signal channels and oscillating lasers. There is a boundary condition for oscillating lasers instead of boundary value: $P_k(0,t) = f_k\, P_k(L,t)$ where f_k is the feedback gain for kth oscillating laser. In an EDFA with optical feedback loop the temporal variation of signal channels and oscillating lasers is comparable with their spectral variation, while in open loop situation, the system can usually be considered in the steady-state situation.

In [44], the inhomogeneous model has been used for analyzing a dual feedback loop EDFA as an inhomogeneous medium. A schematic diagram of desired system is shown in Figure 7.

In Figure 7 configuration, signal channels enter the EDFA after passing a 3dB coupler (C_1), an isolator (ISO_1) and a WDM coupler. Pump wave is launched into the amplifier by the WDM coupler. The energy transfer from pump to signal channels occur in EDFA section and then amplified signals and generated ASE are sent to the second isolator (ISO_2) and 3dB coupler (C_2). A potion amount of ASE is coupled into the fiber loop via the 3dB-C2 and signal channels are routed into the output end of configuration. In the fiber loop, the first wavelength selective coupler (WSC_1) divides ASE into two parts which join again together through WSC_2 after passing a variable optical attenuator (VOA_1 and VOA_2) and an optical bans pass filter ($OBPF_1$ and $OBPF_2$ centered at λ_1 and λ_2 wavelengths). These two filtered waves are fed to the input end of configuration by C_1, amplified by EDFA and repeat this process again and again. Thereby two oscillating lasers with λ_1 and λ_2 wavelengths are made from the ASE of amplifier for clamping its gain.

Figure 8 shows obtained results from the inhomogeneous model for studying the transient response of a signal output power at 1550 nm to adding/dropping of another signal in a dual feedback EDFA [44]. Two oscillation lasers are located at 1531 and 1541 nm wavelengths.

As seen in Figure 8, the output power of a signal is more sensitive to adding/dropping of closer wavelengths. This is expected for an inhomogeneous gain medium. Temporal variations of two oscillating lasers have also been calculated in [44]. The reported results are presented in Figure 9.

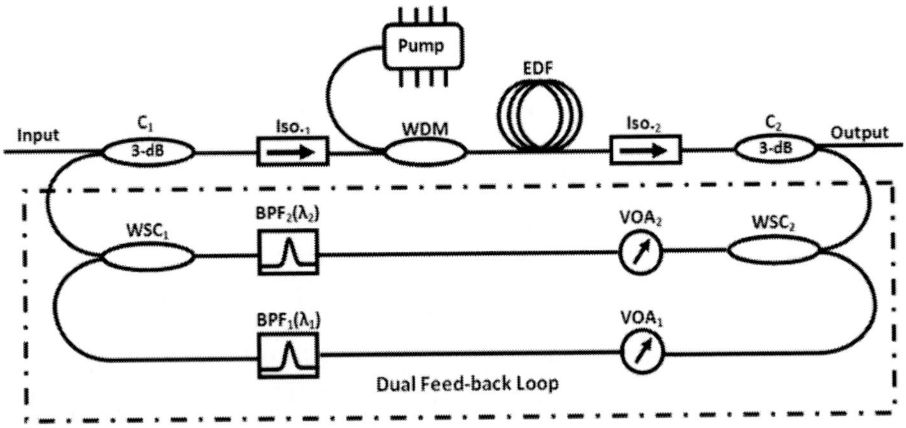

Figure 7. Schematic diagram of dual feedback loop EDFA.

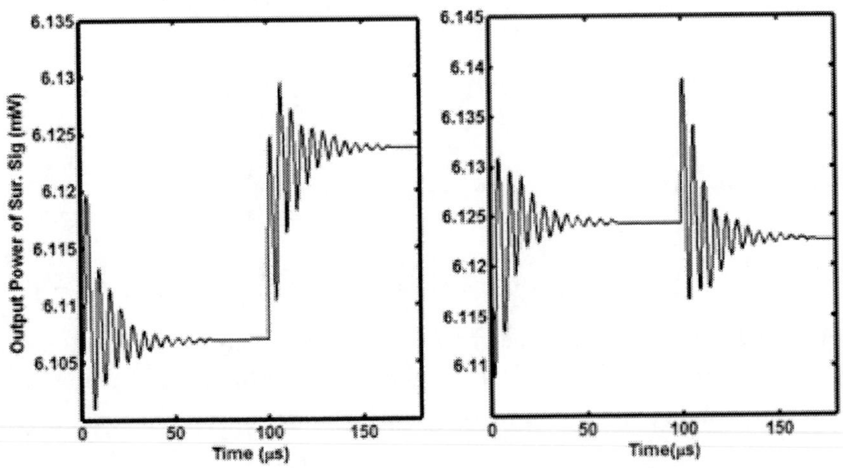

Figure 8. Transient response of the output power of the surviving signal at 1550 nm to adding/dropping of another signal at 1552 (left) 1540 (right) with two oscillation lasers at 1531 and 1541 nm wavelengths [44].

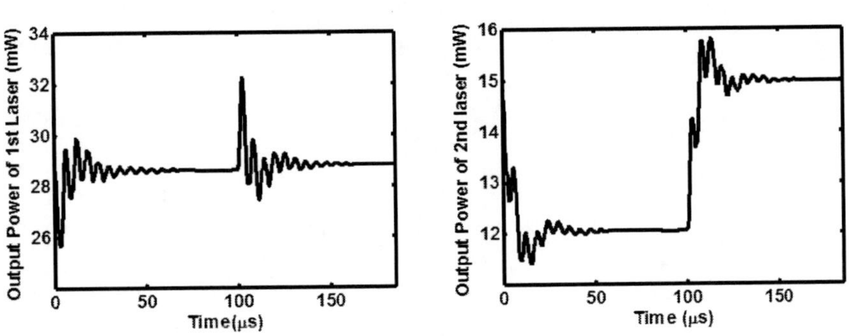

Figure 9. Temporal variation of two oscillating lasers at 1531 nm (left) 1541 nm (right) when a 1540 nm signal is added and dropped at t=0 and t=100μs, respectively [44].

As shown in Figure 9, adding/dropping of a specific signal has more influence on that oscillating laser which is closer it. Presented results in Figure 9 verify that more than one oscillating laser are needed for clamping the gain spectrum of an inhomogeneous active medium. Also, Figure 9 shows that add/drop process causes a relaxation oscillation (RO) in the signal output power. These ROs are the result of energy oscillation between the exited state of Er^{3+}-ions and the oscillating laser. RO of the exited state population is transferred to the output power of signal channels [50].

In [6], Bahrampour and Mahjoei has presented another novel method, is named the M-method, for reducing the governing USPDE of an inhomogeneous EDFA to a FSPDE. The effectiveness of the M-method for solving a USPDE depends on the possibility of a special function definition which is named moment function (M-function). The jth order M-function must be defined with respect to the uncountable functions (here $n_2(v)$ in equation (2-5)) such that it converges to zero by increasing its order. Fortunately, an appropriate definition for the M-function is possible in inhomogeneously analysis of EDFAs. jth order M-functions can be defined as follows:

$$Q_{i_1,i_2,\ldots,i_j} = \int_0^\infty n_2(v)\sigma_{i_1}^h(v)\sigma_{i_2}^h(v)\ldots\sigma_{i_j}^h(v)\,dv \tag{2-13}$$

$$q_{i_1,i_2,\ldots,i_j} = \int_0^\infty n_2(v)\sigma_e^h(v,v_{i_1})\sigma_{i_2}^h(v)\sigma_{i_3}^h(v)\ldots\sigma_{i_j}^h(v)\,dv \tag{2-14}$$

Where $i_f = 1,2,..,m$ for $f = 1,2,..,j$ and as before $\sigma_{i_f}^h(v)$ is $\sigma_a^h(v,v_{i_f})+\sigma_e^h(v,v_{i_f})$. Low order of absorption and emission spectra (10^{-25}) causes that the M-functions be rapidly decreasing functions of their order. The differential equation of the M-functions can easily be obtained by multiplying two sides of equation (2-5) by $\sigma_{i_1}^h(v)\sigma_{i_2}^h(v)\ldots\sigma_{i_j}^h(v)$ and $\sigma_e^h(v,v_{i_1})\sigma_{i_2}^h(v)\sigma_{i_3}^h(v)\ldots\sigma_{i_j}^h(v)$ and integrating over frequency. Then the system of governing equations will be obtained as follows:

$$\begin{aligned}\frac{\partial Q_{i_1,i_2,\ldots,i_j}}{\partial t} &= n_0\sum_{k=1}^{m}\frac{P_k(z,t)}{hv_k}i_k(r)\eta_{i_1,i_2,\ldots,i_j,k} - \sum_{k=1}^{m}\frac{P_k(z,t)}{hv_k}i_k(r)Q_{i_1,i_2,\ldots,i_j,k}\\ &-\frac{1}{\tau_2}Q_{i_1,i_2,\ldots,i_j} + F\left[\delta_{i_1,i_2,\ldots,i_j} + Q_{i_1,i_2,\ldots,i_j}\right]\end{aligned} \tag{2-15}$$

$$\begin{aligned}\frac{\partial q_{i_1,i_2,\ldots,i_j}}{\partial t} &= n_0\sum_{k=1}^{m}\frac{P_k(z,t)}{hv_k}i_k(r)\alpha_{i_1,i_2,\ldots,i_j,k} - \sum_{k=1}^{m}\frac{P_k(z,t)}{hv_k}i_k(r)q_{i_1,i_2,\ldots,i_j,k}\\ &-\frac{1}{\tau_2}q_{i_1,i_2,\ldots,i_j} + F\left[\beta_{i_1,i_2,\ldots,i_j} + q_{i_1,i_2,\ldots,i_j}\right]\end{aligned} \tag{2-16}$$

$$u_k\left(\frac{\partial P_k(z,t)}{\partial z}+\frac{u_k}{V_k}\frac{\partial P_k(z,t)}{\partial t}\right)=2h\nu_k\Delta\nu_k\int_0^\infty dr\,2\pi r\,i_k(r)q_k(r,z,t)$$
$$+P_k(z,t)\left[\int_0^\infty dr\,2\pi r\,i_k(r)Q_k(r,z,t)-n_0\eta_k\int_0^\infty dr\,2\pi r\,i_k(r)-l_k\right] \quad (2\text{-}17)$$
$$k=1,2,\ldots,m$$

where,

$$\eta_{i_1,i_2,\ldots,i_j,k}=\int_0^\infty g(\nu)\sigma_a^h(\nu,\nu_{i_1})\sigma_{i_1}^h(\nu)\sigma_{i_2}^h(\nu)\ldots\sigma_{i_j}^h(\nu)\,d\nu,$$

$$\delta_{i_1,i_2,\ldots,i_j}=\int_0^\infty n_{2qe}(\nu)\sigma_{i_1}^h(\nu)\sigma_{i_2}^h(\nu)\ldots\sigma_{i_j}^h(\nu)\,d\nu$$
$$=\tau_2\sum_{k=1}^{m_p}\frac{P_k(z)}{h\nu_k}i_k(r)\left(n_0\eta_{i_1,i_2,\ldots,i_j,k}-\delta_{i_1,i_2,\ldots,i_j,k}\right),$$

and

$$\alpha_{i_1,i_2,\ldots,i_j,k}=\int_0^\infty g(\nu)\sigma_a^h(\nu,\nu_k)\sigma_e^h(\nu,\nu_{i_1})\sigma_{i_1}^h(\nu)\sigma_{i_2}^h(\nu)\ldots\sigma_{i_j}^h(\nu)\,d\nu$$

$$\beta_{i_1,i_2,\ldots,i_j}=\int_0^\infty n_{2qe}(\nu)\sigma_e^h(\nu,\nu_{i_1})\sigma_{i_1}^h(\nu)\sigma_{i_2}^h(\nu)\ldots\sigma_{i_j}^h(\nu)\,d\nu$$
$$=\tau_2\sum_{k=1}^{m_p}\frac{P_k(z)}{h\nu_k}i_k(r)\left(n_0\alpha_{i_1,i_2,\ldots,i_j,k}-\beta_{i_1,i_2,\ldots,i_j,k}\right)$$

In the first approximation all functions with two or more multiplied cross-sections are assumed to be equal to zero. Therefore, in the first approximation, the system of governing equations reduced to:

$$\frac{\partial Q_k}{\partial t}=\left(F-\frac{1}{\tau_2}\right)Q_k \qquad k=1,2,\ldots,m \qquad (2\text{-}18)$$

$$\frac{\partial q_k}{\partial t}=\left(F-\frac{1}{\tau_2}\right)q_k \qquad k=1,2,\ldots,m \qquad (2\text{-}19)$$

$$u_k\left(\frac{\partial P_k(z,t)}{\partial z}+\frac{u_k}{V_k}\frac{\partial P_k(z,t)}{\partial t}\right)=2h\nu_k\Delta\nu_k\int_0^\infty dr\,2\pi r\,i_k(r)q_k(r,z,t)+$$
$$P_k(z,t)\left[\int_0^\infty dr\,2\pi r\,i_k(r)Q_k(r,z,t)-n_0\,\eta_k\int_0^\infty dr\,2\pi r\,i_k(r)-l_k\right]$$
$$k=1,2,\ldots,m \tag{2-20}$$

So, in the first approximation, the USPDE is reduced to a set of $3m$ PDEs which can be solved easily. In the second approximation, the multiplying of three cross-sections is assumed to be zero and a set of 5m PDEs are solved. This process is continued until no difference is observed between power functions ($P_k(z,t)$ s for $k=1,2,\ldots,m$) obtained in two successive approximations.

In [6], it has reported that j gives accurate results. So, the USPDE of an inhomogeneous EDFA could be reduced to a set of 5m PDEs by utilizing M-method. Although the M-method is a little complex but it is faster than the standard method. In [6] the M-method has been used for studying the transient response of an inhomogeneous EDFA with a 980 nm pump, one oscillating laser and eight signal channels. Figure 10 shows temporal response of the surviving signal at 1550 nm to adding/dropping of other seven channels for two values of oscillating laser wavelength.

Similar to Figure 9, it can also be seen in Figure 10 that an oscillating laser which is far from the surviving channel cannot fix its output power during the add/drop process.

This means that the several oscillating lasers are needed for gain-clamping in an inhomogeneous EDFA. In addition, comparison of Figure 9 and Figure 10 shows that the amplitude of transient responses is smaller in dual feedback case. The reason for this comes from beating between the frequencies of ROs corresponding to two oscillating lasers.

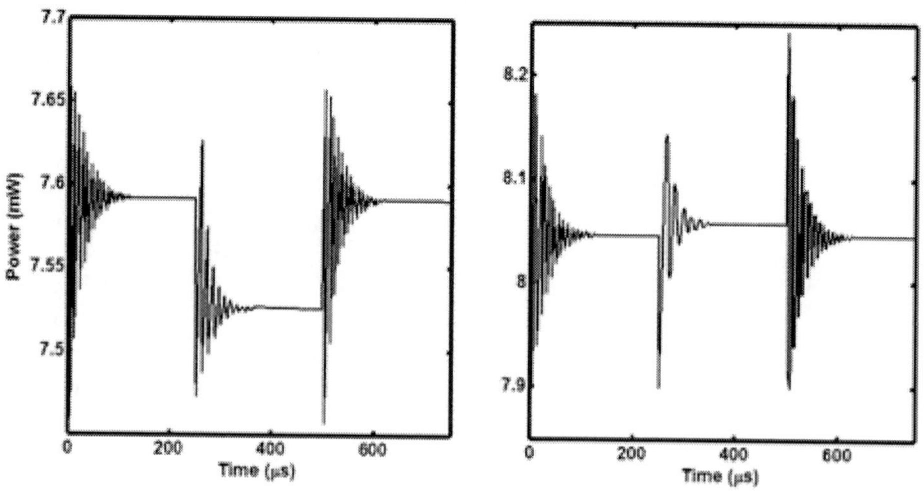

Figure 10. Transient response of a signal at 1550 nm to adding/dropping of other seven signal channels in the presence of a (a) 1540 nm (b) 1555 nm oscillating laser [6].

2.5. Er^{3+}-ions Interaction Effects

All previous discussions are valid only for low concentration EDFAs. In low concentrations, each Er^{3+}-ion acts independently of others while if the ions concentration is high enough, an energy transfer may occur between neighboring ions. Besides spontaneous and stimulated emission processes, this energy transfer acts as a new pathway for depleting Er^{3+}-ions exited to the $^4I_{13/2}$ level. This corresponds to a reduction in the overall lifetime of the exited state and thus the population inversion of gain medium [21]. This means that above an optimum concentration the optical gain of EDFA beings to decrease. For this reason, ions interaction effects sometimes named as concentration quenching effects, too. Therefore, ion-ion interaction is an important problem in high concentration EDFAs [2, 51] and in erbium doped waveguides which have a few centimeters length [52].

The simplest form of ions interaction effect is energy migration (see Figure 11a). This effect happens when an exited Er^{3+}-ion transfer its energy to a nearby ion in the ground sate. The energy transfer may occur without any loss in the total energy of two ions (resonant energy transfer) or be a lossy process (non-resonant energy transfer) if one of the ions couples to a defect. The energy transfer can also occur from an exited ion to another neighboring exited ion. This concentration quenching effect is named stepwise up-conversion or simply up-conversion (see Figure 11b). Generally, up-conversion is attributed to a process in which the energy is being given to an exited ion and promotes it to a higher energy state. So, the exited state absorption (ESA) is also an up-conversion process (see Figure 11c). ESA is not among the ions interaction effects and is independent of ions concentration. This process is important in high intensity pumping EDFAs and leads to the decrease gain.

Here we explain the influence of the up-conversion effect (Figure 11b) on the gain performance of a high concentration EDFA. Although it is possible that an up-converted ion has radiative transitions from $^4I_{9/2}$ and $^4I_{11/2}$ levels to the ground state, these transitions can be ignored because of their low probabilities [52].

The $^4I_{9/2}$ state has a very short lifetime and up-conversion ion cascades quickly back down to the $^4I_{13/2}$ level by two non-radiative transitions $^4I_{9/2}\rightarrow{}^4I_{11/2}\rightarrow{}^4I_{13/2}$ (see Figure 11b). Thus, the two-level approximation can also be used for analysis high concentration EDFAs with up-conversion effect. As before, two levels are $^4I_{13/2}$ and $^4I_{15/2}$ levels which are named 1 and 2 levels. As shown in Figure 11b, the up-conversion process can occur only when two nearby Er^{3+}-ions are in the exited state simultaneously. As a result the rate of depletion in the exited state population caused by up-conversion effect, is proportional to the square of the population of this level (n_2^2). So, in the homogeneous model the ground state population rate equation (equations (2-1)) can be rewritten as follows for high concentration EDFAs:

$$\frac{\partial n_1}{\partial t} = -n_1(r,z,t)\sum_{k=1}^{m}\frac{P_k(z,t)}{h\nu_k}i_k(r)\sigma_a(\nu_k) + n_2(r,z,t)\sum_{k=1}^{m}\frac{P_k(z,t)}{h\nu_k}i_k(r)\sigma_e(\nu_k) + \frac{1}{\tau_2}n_2(r,z,t) + C_{up}(n_2(r,z,t))^2 \quad (2\text{-}21)$$

where $C_{up}n_2^2$ is corresponding to the up-conversion effect and C_{up} is the up-conversion coefficient. The up-conversion lifetime is defined as $\tau_{up} = 1/n_0 C_{up}$ where n_0 is the total concentration of Er^{3+}-ions. C_{up} is a concentration independent and host-dependent parameter.

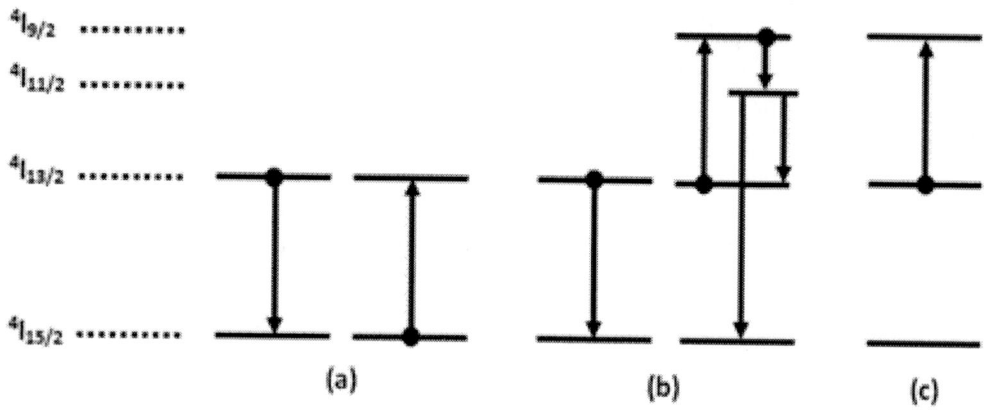

Figure 11. Schematic diagrams of three ion-ion energy transfer mechanisms, (a) exited state absorption process (b) Resonant energy transfer (c) stepwise up-conversion.

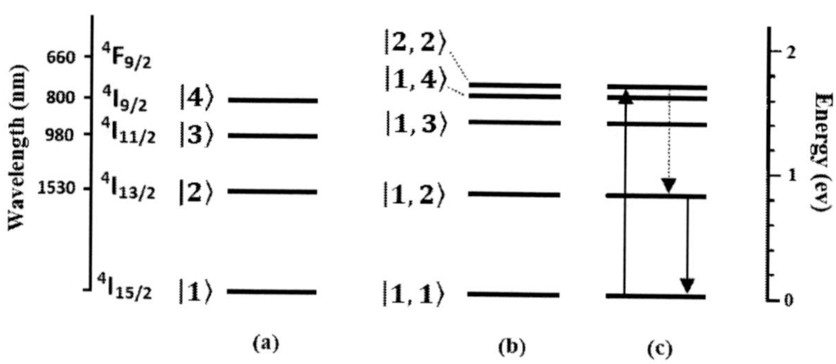

Figure 12. Schematic energy levels diagram of (a) an isolated Er^{3+}-ion (b) an Er^{3+}-ion pair and (c) PIQ mechanism.

Cooperative up-conversion is another concentration quenching effect. This process occurs when two or several ions (pair or cluster) are so strongly coupled that interact with light as a unique system. The cooperative up-conversion effect related to the ion pairs is named pair induced quenching (PIQ) effect. Before discussing the PIQ mechanism, it is important to explain the energy diagram of an ion pair. It has been shown that the screening effect of the d_{10} electron on the active ^{4}f electrons prevents the strong interaction between neighboring ions [44]. Therefore, the interaction term is negligible in the Hamiltonian of an ion pair system. In this case the ion pair eigen state can be written as the tensor product of two insolated ions states ($|\alpha,\beta\rangle = |\alpha\rangle \otimes |\beta\rangle$) and the eigen value of this state is the summation of the energies of those states ($E = E_\alpha + E_\beta$). Figures 12a and 12b show the energy diagram of an Er^{3+}-ion pair and an isolated Er^{3+}-ion, respectively. Also the mechanism of the PIQ effect is shown in Figure 12c.

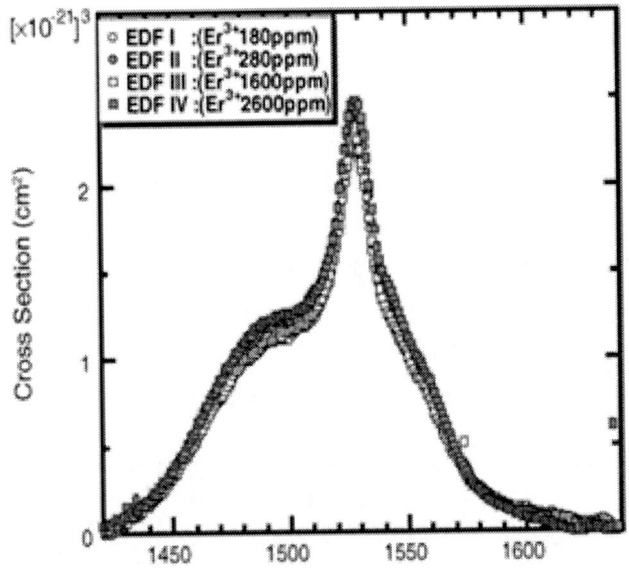

Figure 13. Absorption cross-sections of four EDFAs with different Er^{3+}-ions concentration [53].

As seen in Figure 12c, in the PIQ process the ion pair absorbs two pump photons and is exited from ground state $|1,1\rangle$ to level $|2,2\rangle$. This level is very close to $|1,4\rangle$ level and there is a strong coupling between them. The result is a quick non-radiative transition from $|2,2\rangle$ to $|1,4\rangle$ which has a short lifetime and the pair transits quickly to the metastable level $|1,2\rangle$ via a non-radiative transition by using $|1,3\rangle$ level as an intermediate step. Finally the system returns to the ground state $|1,1\rangle$ by radiating a photon.

Obviously, the high concentration level is different for various host glasses. For example, it has been shown that the portion of ion pairs is considerable in Ge-co doped silica fibers with concentration higher than 100 ppm while this level increases to 1000 ppm for Al-co doped silica fibers [21].

In high concentration EDFAs, it is usually assumed that the Er^{3+}-ions population can be broken up into two groups: isolated ions and ion pairs. In a good approximation, an ion pair can also be considered as a two-level system (with $|1,1\rangle$ and $|1,2\rangle$ levels). In this case there are two continuity relations for single ions and ion pairs, $n_1^s + n_2^s = (1-2x)n_0$ and $n_1^p + n_2^p = xn_0$ that x is the fraction of ion pairs relative to the total Er^{3+}-ions density n_0. By using the continuity relations, the population rate equations can be written as follows in the homogeneous model:

$$\frac{\partial n_2^s}{\partial t} = (1-2x)n_0 \sum_{k=1}^{m} \frac{P_k(z,t)}{h\nu_k} i_k(r)\sigma_a(\nu_k) - n_2^s \sum_{k=1}^{m} \frac{P_k(z,t)}{h\nu_k} i_k(r)\sigma_k - \frac{1}{\tau_{2s}} n_2^s + C_{up}(n_2^s)^2, \tag{2-22}$$

$$\frac{\partial n_2^p}{\partial t} = x n_0 \sum_{k=1}^{m} \frac{P_k(z,t)}{h\nu_k} i_k(r) w_p \sigma_a(\nu_k) - n_2^p \sum_{k=1}^{m} \frac{P_k(z,t)}{h\nu_k} i_k(r)\sigma_k - \frac{1}{\tau_{2p}} n_2^p \tag{2-23}$$

In equation (2-23) w_p is a constant parameter which is 1 for signal and 2 for pump waves [21]. Because of this fact that each ion pair absorbs two pump photons to be exited to its 2th level. The other parameters have been defined previously. Although the lifetime of an ion pair exited state depends on the Er^{3+}-ions concentration, but τ_{2s} and τ_{2p} can assumed to be equal for mid concentrations (not very high) [21].

Also, experimental results show that the absorption and emission cross-sections are independent from Er^{3+}-ions concentration in the C and L-band regions [53]. For this reason absorption and emission cross sections need no superscript in equations (2-22) and (2-23).

Figure 13 shows the absorption cross-sections of four EDFAs with different Er^{3+}-ions concentration measured by Ono and Tanabe [53].

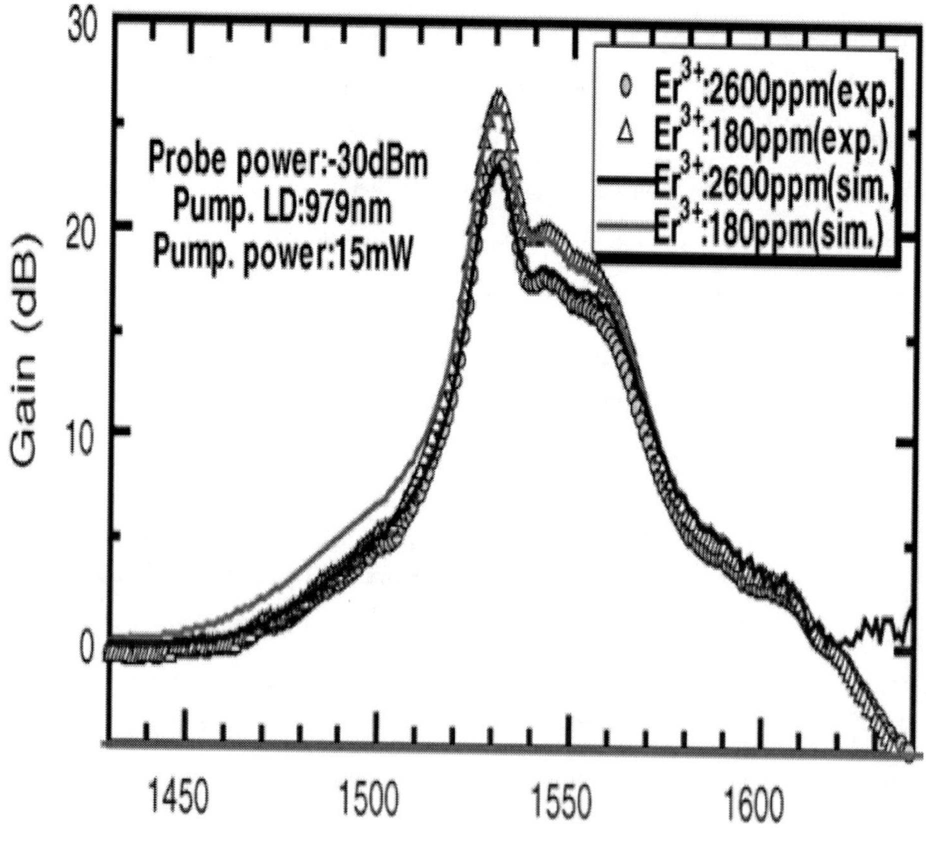

Figure 14. Gain spectrum of two EDFAs with different concentrations [53].

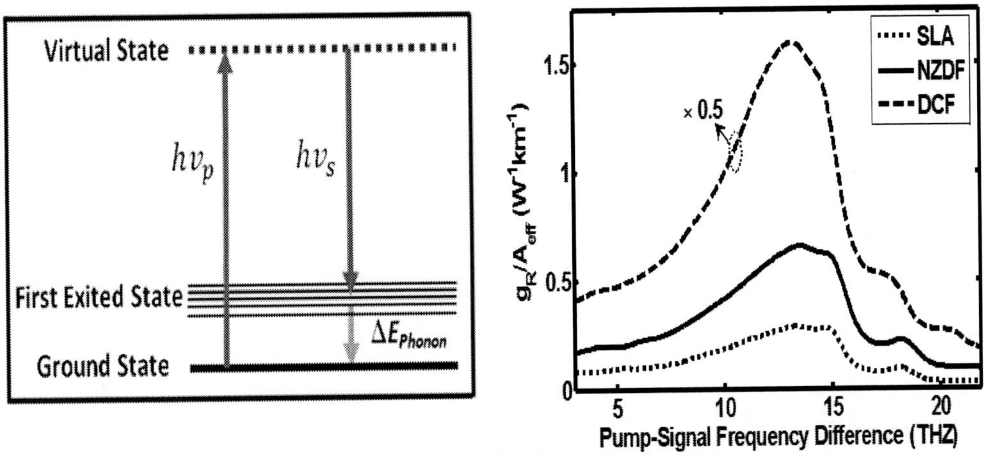

Figure 15. Left: Quantum-mechanical description of SRS. Right: Raman gain spectrum for dispersion compensated fiber (DCF) with A_{eff}=15μm^2, non-zero dispersion fiber (NZDF) with 55μm^2 and supper large area (SLA) fiber with 105μm^2 pumped at 1450 nm ([54]).

Figure 14 shows measured the gain spectra of two EDFAs with different concentrations [53]. As shown in this figure, the up-conversion and PIQ effects lead to the gain degradation. It has been shown that even a %2 ion pair fraction can leads to 1 dB degradation in the gain [53].

In [44] and [49], Bahrampour et al. have used the inhomogeneous model for analyzing an optically gain stabilized high concentration EDFA. Details of their works for low concentration EDFAs ([6] and [44]) have been presented in the previous section. The basis of the inhomogeneous model for high and low concentration states is identical and the only difference is in the addition of the rate equations of a new group of active particles i.e. ion pairs. As mentioned in the previous section, in the inhomogeneous model the governing equations form an USPDE and two different techniques have been proposed for reducing this USPDE to a FSPDE: M-method and standard technique. In [44] the standard technique and in [49] the M-method have been utilized for studding a high concentration EDFAs with a dual feedback loop as an inhomogeneous medium. Calculations show that in high concentration of Er^{3+}-ions, depending on the pump rate, the ROs in oscillating lasers output powers may convert to nT-periodic or even chaotic behaviors [44, 49].

3. FIBER RAMAN AMPLIFIERS

3.1. Theory of Raman Amplification

Stimulated Raman scattering (SRS) is the main basis of FRAs. Quantum-mechanical description of this phenomenon is depicted in Figure 15. A laser pump at frequency ν_p excites the silica molecules to a virtual state. Virtual states are states of atom-field system and have a very short lifetime. A signal photon at frequency ν_s can depopulate the virtual state and hence generate a Stokes photon. As this figure shows the SRS can occur when the frequency of signal is down-shifted from the pump to the extent of Stokes shift (i.e., the difference between ground and first exited states). Unlike real states, the location of virtual states is not fixed and depends on atomic energy states and the states of quantized field. This is an interesting property, because by properly choosing a pump frequency, the desired signal can be amplified. It also means that, by using a multi-pump configuration, a broadband of signal channels can be amplified which will be discussed later. The Stokes shift of silica fibers is approximately

13.2 THz in the communication window. Raman gain spectrum for 3 kinds of optical fibers is also shown in Figure 1.

This figure shows a relatively broad profile (~40 THz) due to amorphous nature of silica glass. This means that with one pump a relatively broad band of data signal channels can be amplified. In addition, asymmetrically nature of Raman gain can be improved by multi-pump configuration. It is important to note that despite EDFA, no special fiber is needed to obtain Raman amplification and the fiber line acts as an amplifier itself. However, depending on fiber effective area and Germania (GeO_2) concentration doping, Raman gain efficiency is different from fiber to fiber (Figure 15).

According to isotropic nature of SRS, different pumping configurations such as forward, backward or bi-directional can be implemented in FRAs. Neglecting noise effects, the evolution of a signal channel $P_{s,i}$ and pump $P_{p,k}$ along the fiber axis z of a FRA with N_p pumps and N_s signal channels can be described by the following coupled differential equations [55]

$$\left(\frac{\partial}{\partial z}+\frac{1}{V_s}\frac{\partial}{\partial t}\right)P_{s,i}(z,t)=P_{s,i}(z,t)\left[-\alpha_{s,i}+\sum_{\substack{i=1\\ j\neq i}}^{N_s} g_{i,j}\,P_{s,i}(z,t)+\sum_{k=1}^{N_p} g_{i,k}\,P_{p,k}(z,t)\right] \tag{3-1}$$

$$\left(\pm\frac{\partial}{\partial z}+\frac{1}{V_p}\frac{\partial}{\partial t}\right)P_{p,k}(z,t)=P_{p,i}(z,t)\left[-\alpha_{p,i}+\sum_{i=1}^{N_s} g_{k,i}\,P_{s,i}(z,t)+\sum_{\substack{j=1\\ j\neq k}}^{N_p} g_{k,j}\,P_{p,j}(z,t)\right], \tag{3-2}$$

where $V_{s(p)}$ and $\alpha_{s(p)}$ are signal (pump) group velocity and fiber attenuation, respectively. The composite symbol ± means the propagation directions $\pm z$ of the pump and Raman gain $g_{i,j}$ is defined as $g_R(v_i-v_j)/\Gamma A_{eff}$ for $v_i<v_j$ and $-v_i\,g_R(v_i-v_j)/v_j\,\Gamma A_{eff}$ for. $v_i>v_j$ .where g_R and A_{eff} correspond to Raman gain coefficient and fiber effective area respectively and parameter Γ denotes the polarization dependent of Raman gain. In these equations, variations of group velocities in signal and pump bands are neglected.

In addition to effective area and Germania concentration, relative SOP of pump and signal is another important factor in Raman gain, therefore the following section is devoted to this parameter and its effect on FRA performance.

3.2. Polarization Effects in FRAs

Ideally, any single mode fiber supports two degenerate, orthogonally-polarized uncoupled modes. Cylindrical symmetry in real optical fibers breaks due to effects such as intrinsic stresses, imperfections during the fabrication process and stochastic deformations due to unpredictable changes in the environmental conditions. This leads to slightly difference in propagation constants for each polarization and hence breaking the polarization degeneracy. This property is known as birefringence and is responsible for polarization variation of propagating field in an optical fiber. In a special case, fiber symmetry is broken with a well-controlled large amount of birefringence. Such fibers can preserve the SOP of propagating field; so, these types of fibers are known as polarization maintaining fibers (PMFs) [56-58].

In conventional optical fibers, birefringence is not constant along fiber and changes randomly due to mentioned effects. In fact, these random variations of birefringence also exist in PMF, but their amplitude is negligible compared to strong birefringence due to intentionally induced stresses. Random birefringence variations in real optical fibers leads to random changes in polarization of any lunched light during propagation along them. This phenomenon is referred to as polarization mode dispersion (PMD) and is a limiting factor in ultra long-haul high bit rate communication systems such as distributed Raman fiber amplifiers (DFRAs) [59-61]. Stimulated Raman scattering (SRS) is maximum for parallel SOPs of pump and signal and minimum for orthogonal condition. Experimental measurements show that energy transfer from pump to signal for parallel situation two orders of magnitude larger than the orthogonal condition near the Raman gain peak [54, 62].

In the presence of PMD, the SOPs of pump and signal in FRAs vary randomly. This leads to gain fluctuation and system impairment. Contrary to the negligibly small PDG of an EDFA, this effect is one of the drawbacks of Raman amplifiers based on silica fibers. Raman gain fluctuations due to PDG have been observed in several experiments [63-65]. Various theoretical models for describing PDG have been presented by several authors [66-72]. Based on vectorial model [72], it has been shown that in the limit of large PMD, average gain G_{avg} in a FRA with length L can be obtained as follows

$$G_{avg} = \exp\left[\frac{1}{2}(1+3\mu)g\, P_{in} L_{eff} - \alpha_s L\right], \tag{3-3}$$

where P_{in} is the input pump power and effective length can be defined as $L_{eff} = \left[1-\exp(-\alpha_p L)\right]/\alpha_p$ where μ is the ratio of the Raman gain for orthogonally and parallel SOPs. This equation shows that for large amount of PMD, Raman gain coefficient g, reduces to $(1+3\mu)g/2 \approx g/2$.

For small PMD coefficient, fiber PMD does not affect the relative SOPs of pump and signal. In this regime, fluctuation level of signal is very small and average gain depends on relative SOPs of pump and signal at the input end of the fiber amplifier. For large PMD coefficients, the initial relative SOPs of pump and signal is lost very rapidly and both parallel and orthogonal situations converge. Gain is polarization insensitive but its value is much smaller than previous situation (parallel SOP). In this region, because of high level of birefringence fluctuations, signal fluctuations also decreases. For PMD coefficient between these two regions, standard deviation of signal fluctuations is very large. In order to mitigate the detrimental effects of PDG in this region, pump depolarizing technique has been proposed by some authors [71-74]. By using depolarized pumps, PDG can be reduced at the expense of a reduced efficiency.

Figure 16 shows that how PDG changes for various pump degree of polarizations (DOPs) [75]. Raman gain, signal wavelength, pump wavelengths, pump power and direction are 13.3 dB, 1560 nm, 1460 nm, 200 mW and backward, respectively. PMD of any fiber is a constant value and as this figure shows, by controlling the pump DOP, small values of PDG can be obtained. A depolarized pump can be obtained by orthogonal polarization multiplexing [74], Mach-Zehnder interferometer configuration [71], fiber Lyot depolarizer [73] and crystal type depolarizer [76].

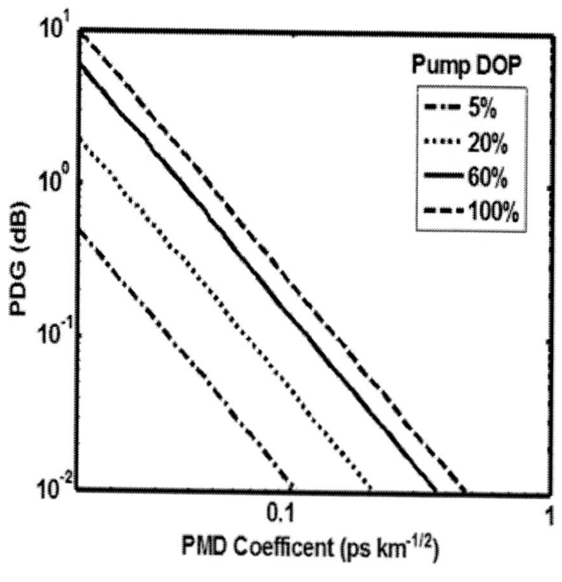

Figure 16. Raman gain as a function of PMD parameter for various pumps DOP [75].

Until this point, PDG caused by pump-to-signal Raman interaction has been investigated. Li et al., [77] experimentally demonstrated PDG enhancement due to signal-to-signal Raman interactions (SSRI) in a WDM system with 40 channels for two types of DCFs with different PMDs. Figure 17 shows that for fiber with small differential group delay (DGD), PDG increases between 0.1 to 0.5 dB, due to signal to signal interaction, depending on signal wavelength. As this figure shows, for large PMD, PDG due to SSRI is negligible. In order to suppress the PDG, they have used a polarization scrambler to improve the system performance [77].

As mentioned before, PDG can be greatly reduced by using depolarized pumps. But in WDM systems, in addition to SSRI, PDG can be further increased through signal induced pump depletion (SIPD) effect [78]. Signal channels can re-polarize the pump significantly in fiber with very low PMD. This leads to PDG enhancement (Figure 18). From this figure it is found that, SIPD effect is noticeable for large input signal powers especially in small PMD regime.

FRAs based on spun fibers have been proposed [79] to suppress polarization impairment. In this technique PMD and PDG can be reduced simultaneously by application of a two section fiber. First section has no spin and the second one is periodically spun. By optimizing system parameters PDG=0.1 dB and PMD coefficient =0.0072 ps. $km^{-1/2}$ have been reported [80].

3.3. Pumping Schemes in Fiber Raman Amplifiers

Unlike EDFAs in which gain bandwidth is determined by transition energies of ions, in FRAs pumping scheme is a determinative factor. This means that, any signal can be amplified by selecting a suitable pump. In addition, gain bandwidth could be broadened by employing an appropriate pumping scheme. In the following, WDM pumping, broad-band continuous spectrum pumping and time division multiplexing (TDM) pumping configurations as three types of pumping schemes in FRAs will be investigated and some methods such as M-method and Lagrange multiplier method will be employed for modeling and optimizing the performance of these types of optical amplifiers.

3.3.1. WDM Pumping Scheme

In WDM configuration, several laser diodes at different wavelengths in continuous mode (CW) are employed as a pump unit. Based on WDM pumping, Emory and Namiki [7] proposed a broadband high-power pumping unit including 12 fiber Bragg grating stabilized laser diodes and 11 Mach-Zehnder interferometer multiplexer Integrated on single substrate. To reduce the pump to pump interaction effect, a non-symmetrical pumping, in which short wavelengths have greater lunched power than long wavelength, has been proposed. By employing this configuration with an output power of 2.2 W in backward scheme, a Raman amplifier with bandwidth of 100 nm in the range 1520-1620 nm and gain ripple 0.5 dB in 25 km of SMF, RDF and DSF have been reported.

The strong Raman interaction between pump-to-pump, signal-to-signal and pump-to-signal make FRA a nonlinear system. In addition, several factors such as polarization, fiber length and fiber loss make optimization of multi-wavelength FRA somewhat complicated and time-consuming procedure. Perlin and Winful [81, 82] proposed an efficient method for a backward-multi-pump distributed FRA in order to broaden and flatten the Raman gain profile.

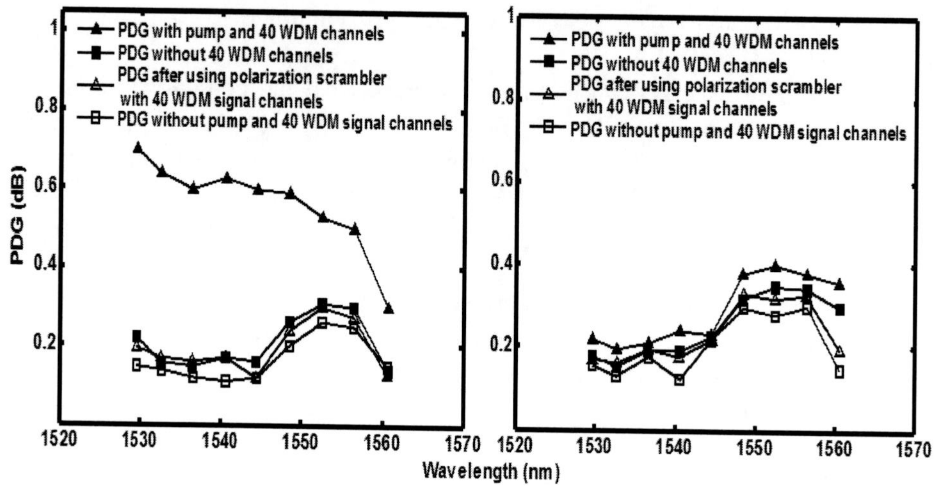

Figure 17. PDG for a WDM system with 40 channel signals with input power 0 dBm/channel and two pumps with powers 350 mW at 1425 nm and 450 mW at 1453 nm and ON-OFF gain 12 dB. Left: DCF with DGD 0.36 ps. Right: DCF with DGD 1 ps [77].

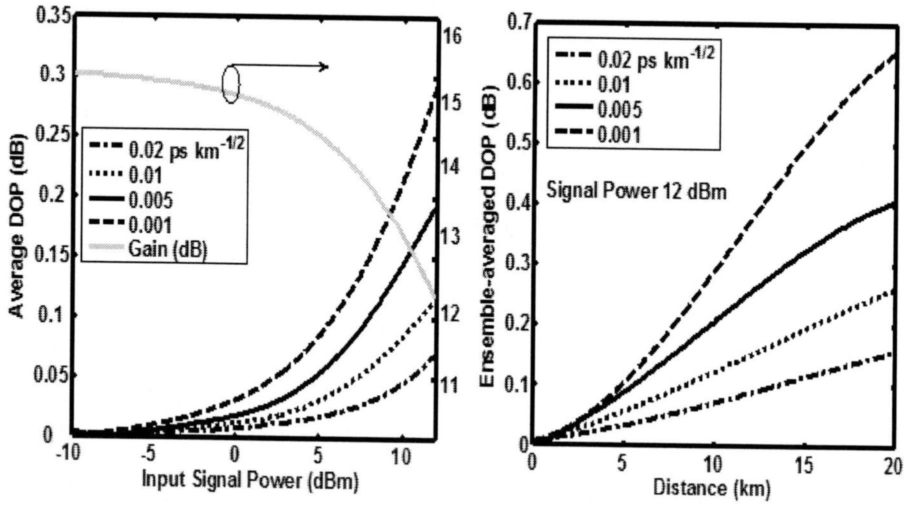

Figure 18. DOP versus input signal power (Left) and distance (Right) for different values of PMDs. Pump power 480 mW at 1488 nm [78].

From equation (3-1) in the steady state condition ($\partial/\partial t = 0$) Raman gain can be obtained as follows

$$G_{dB} = 4.343\left(-\alpha_s L + T_{SRS} + C_{p,s} I\right), \tag{3-4}$$

where for N_s signal channels and N_p pump wavelengths, T_{SRS}, $C_{p,s}$ and I matrices, correspond to signal-to-signal Raman gain tilt ($N_s \times 1$), Raman gain coefficient ($N_s \times N_p$)

and pump power integral ($N_p \times 1$), $I = \int_0^L P_p(z)\,dz$, respectively. Optimal pump configuration has been obtained by separating the optimization procedure based on Genetic algorithm from the solution of the propagation equation through a self consistent procedure. Relative gain flatness smaller than 1% has been reported for 80-100 nm bandwidth, using more than 12 pumps.

In order to avoid time-consuming calculations to obtain optimal input pumps from pump power integrals, Cui et al., [83] have reported an efficient linear method. Using perturbation method, they have shown that, changes in input pump power $\Delta P_p(z=L)$ and pump-power integral ΔI , are related via an approximately linear equation $\Delta I = M\,\Delta P_p(L)$. Where M is a $N_p \times N_p$ matrix and can be calculated through a neural network procedure. Replacing solution of propagation equation in [81, 82] with this procedure, optimal input pump powers are obtained without time-consuming calculations.

Even though optimizing both wavelength and power of the Raman pump unit leads to small values of gain ripple, some problems may occur during realization. Multiplexer couplers for two closely spaced wavelengths and laser pumps at desired wavelengths are two existing challenges. To overcome these problems Hu et al., [84] have proposed a flat-gain FRA based on equally spaced pumps using optimization procedure of [81]. It has been shown through this method that, using equally spaced pumps with either a fixed or optimized central wavelength, gain flatness on the same order as [81] can be obtained.

All optimization methods based on traditional genetic algorithm [81-84], have an intrinsic weakness. Their optimum solution may be trapped in a local optimal point of search space. To overcome these difficulties, hybrid genetic algorithm based on techniques such as clustering, sharing, crowing and adaptive probability have been employed for multi wavelength FRAs [85, 86]. Zhou et al., [87] have proposed an optimization method based on neural network algorithm to obtaining optimal values for pump powers.

Variational method is a powerful technique with strong mathematical basis for optimizing a cost function in the presence of constraints in the form of differential and/or algebraic equations. In a FRA system, Raman gain ripple can be considered as the cost function and upper limit of total pump power and lower bond of the average gain are constraints of the minimization problem [88]. For a backward pumping scheme the boundary values are given at $z=0$ for signal powers and $z=L$ for pump powers where for pump powers we have $P_{p,i}(z) \geq 0$. For applying this constraint on the pump equation (Eq. 25 in steady state condition and counter-pumping scheme), it is assumed that $P_{p,i}(z) = Q_i^2(z)$ where $Q_i(z)$ is a real number. So; the governing equation for pump can be rewritten as follows

$$\frac{dQ_i(z)}{dz} = \frac{Q_i(z)}{2}\left[\alpha_{p,i} - \sum_{j=1}^{N_s} g_{i,j} P_{s,j}(z) - \sum_{j=1}^{N_p} g_{i,j} Q_j^2(z)\right] \tag{3-5}$$

It is reasonable to have an upper bond for total pump power $P_{p,\max}$, (i.e., $\sum_{i=1}^{N_p} Q_i^2(L) \leq P_{p,\max}$). This inequality can be rewritten in the form of equality

$\sum_{i=1}^{N_p} Q_i^2(L) - P_{p,\max} = D^2$, where D is a real number. In addition, we assume that the average gain $\overline{G}$ of FRA is greater than a lower bond G_0 where average gain is defined as

$$\overline{G} = \frac{1}{N_s}\sum_{j=1}^{N_s}\frac{P_{s,j}(L)}{P_{s,j}(0)} \tag{3-6}$$

Furthermore, the inequality for gain can be converted to equality using real number E (i.e., $\overline{G} - G_0 = E^2$). Cost function F can be defined as

$$F\left(P_{s,1}(L), P_{s,2}(L), \ldots, P_{s,N_s}(L)\right) = \sum_{k=1}^{N_s}\left(G_k - \overline{G}\right)^2 . \tag{3-7}$$

Now the problem is to find an appropriate set of pumps (for wavelength and power) at the input end of the fiber line, which minimizes the cost function subject to the mentioned constraints. Using Lagrange multiplier method, generalized cost function J is written as

$$\begin{aligned} J = F + \omega\left(P_{p,\max} - \sum_{i=1}^{N_p} Q_i^2(L) - D^2\right) + \varphi\left(\overline{G} - G_0 - E^2\right) \\ + \sum_{i=1}^{N_s}\int_0^L \gamma_i(z)\left[\frac{dP_{s,i}(z)}{dz} + P_{s,i}(z)\left(\alpha_{s,i} - \sum_{j=1}^{N_s} g_{i,j}P_{s,j}(z) - \sum_{j=1}^{N_p} g_{i,j}Q_j^2(z)\right)\right]dz \\ + \sum_{i=1}^{N_p}\int_0^L \beta_i(z)\left[\frac{dQ_i(z)}{dz} - \frac{\alpha_{p,i}Q_i(z)}{2} + \frac{Q_i(z)}{2}\left(\sum_{j=1}^{N_s} g_{i,j}P_{s,j}(z) + \sum_{j=1}^{N_p} g_{i,j}Q_j^2(z)\right)\right]dz \end{aligned} , \tag{3-8}$$

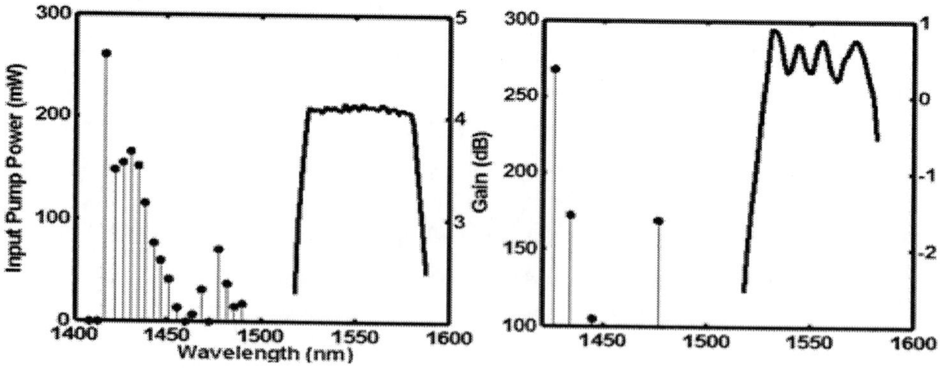

Figure 19. Left: calculated pump powers for a 100 km SMF-28, FRA with 20 pumps which are distributed uniformly over 1408-1490 nm band and 50 signals in the range 1528-1578 nm with -10dBm/channel. Average gain and gain ripple are 4 and 0.096 dB, respectively. Right: calculated pump powers and wavelengths for previous setup with 4 pumps. Average gain and gain ripple are 0.5 and 0.72 dB, respectively [88].

where ω, φ, γ and β are Lagrange multipliers correspond to total pump power equality, minimum gain equality, signal and pump differential equations constraints, respectively. Therefore, the optimization problem reduces to classical variational problem $\delta J = 0$. After some mathematical manipulation, from $\delta J = 0$ a brief summary of results are listed as follows (more mathematical details is available in [88])

a. N_s differential equations governing on signals as Eq. (24 in steady state).
b. N_p differential equations governing on pumps as Eq. (25 in steady state).
c. N_p differential equations governing on Lagrange multiplier β.
d. N_s differential equations governing on Lagrange multiplier γ.
e. $2N_p + N_s$ boundary condition (it is assumed that the N_s input signal powers at $z = 0$ are available).
f. Two switching equations ($D\omega = 0$ and $E\varphi = 0$).
g. A set of algebraic equations for pump wavelengths optimization.

The optimization is converted to a two-point boundary problem. Optimal powers and wavelengths of pump unit can be obtained by solving coupled differential-algebraic set of equations for each binary vector which was defined in *f*. This method has been employed to minimize the gain ripple of a 100 km SMF-28 fiber with 20 pumps which is uniformly distributed over 1408-1490 nm band and 50 signals in the range 1528-1578 nm. Average net gain has been fixed on 4 dB. Results of optimization have been shown in Figure 19. In addition, optimization for a 4 pumps unit for powers and wavelengths are shown in this figure.

Based on multiple pump gain flattening, Jiang et al., [89] have proposed improved particle swarm optimization (PSO) and shooting algorithm. In their work, three improvements have been performed. At first, they have introduced a correction factor in shooting algorithm for preventing bad initial guess. Then, adjustable step size has been proposed to modify Newton-Raphson method. Finally PSO performance has been enhanced by introducing velocity acceptability probability definition. By this optimization method, the pump wavelengths and powers of a 4 counter pumping Raman amplifier with an on-off gain of 13.3dB, bandwidth of 80 nm and gain ripple of ±0.5 dB has been optimized.

3.3.2. Continuous Spectrum Pumping Scheme

As mentioned before, gain flatness of FRA can be improved by choosing appropriate wavelengths and powers for pumping unit using various optimization algorithms. Broad-band pumping is an alternative method to have a broad-band FRA with minimum gain ripple. The Raman gain profile produced by a broad-band pump is the convolution of pump spectrum and Raman gain curve. High degree of gain flatness can be obtained by selecting appropriate pump line-shapes. The governing equation of FRA with broad-band counter-pumping scheme is a set of uncountable coupled differential equations as follows [90]

$$\frac{dP_s(\nu_k,z)}{dz}=P_s(\nu_k,z)\left[-\alpha(\nu_k)+\sum_{k'=1}^{N_s}g(\nu_k,\nu_{k'})P_s(\nu_{k'},z)+\int_0^{\infty}P_p(\nu',z)g(\nu_k,\nu')d\nu'\right] \quad k=1,2,\ldots,N_s \tag{3-9}$$

$$\frac{dP_p(\nu,z)}{dz}=P_p(\nu,z)\left[\alpha(\nu)-\sum_{k'=1}^{N_s}g(\nu,\nu_{k'})P_s(\nu_{k'},z)-\int_0^{\infty}P_p(\nu',z)g(\nu,\nu')d\nu'\right]; \quad \nu\in R \tag{3-10}$$

where $P_s(\nu_k,z)$ and $P_p(\nu,z)$ are signal power at frequency ν_k and pump power spectrum, respectively. Full numerical methods for solving such set of equations have been reported by [91, 92].

M-method, which is previously introduced for modeling the inhomogeneous EDFA, can be employed to reduce the system of uncountable governing equations into a countable set of equations [90]. The (m,n,l)th order moment $q(m,n,l)$ is defined as

$$q(m,n,l)=\left\{q^{(m,n)}_{k_1,k_2,\ldots,k_l}(z)\,|\,k_1,k_2,\ldots,k_l=0,1,\ldots,N_p\right\} \quad m,n=0,1,\ldots \quad l=1,2,\ldots \tag{3-11}$$

Each element is given by

$$q^{(m,n)}_{k_1,k_2,\ldots,k_l}(z)=\int_0^{\infty}\alpha^m(\nu')g(k_1,\nu')g(k_2,\nu')\ldots g(k_l,\nu')P_p(\nu',z)Q^n(\nu',z)d\nu' \tag{3-12}$$

where, $Q(\nu,z)=\int_0^{\infty}P_p(\nu',z)g(\nu,\nu')d\nu'$. Rewriting the governing equations (Eq. 3-9 and Eq. 3-10) with respect to M-functions lead to

$$\frac{dP_s(\nu_k,z)}{dz}=P_s(\nu_k,z)\left[-\alpha(\nu_k)+\sum_{k'=1}^{N_s}g(\nu_k,\nu_{k'})P_s(\nu_{k'},z)+q_k^{(0,0)}\right] \quad k=1,2,\ldots,N_s \tag{3-13}$$

$$\frac{dP_p(\nu,z)}{dz}=P_p(\nu,z)\left[\alpha(\nu)-\sum_{k'=1}^{N_s}g(\nu,\nu_{k'})P_s(\nu_{k'},z)-Q(\nu,z)\right] \quad \nu\in R. \tag{3-14}$$

Boundary conditions are defined at $z=0$ for signal powers (i.e., $P_s(\nu_k,0)=P_{0,k}$ $k=1,2,\ldots,N_s$) and at $z=L$ for pump power spectrum (i.e., $P_p(\nu_k,L)=P_p(\nu)$). Differential equations of moment elements is obtained by multiplying both side of equation (3-14) by $\alpha^m(\nu')g(\nu_{k1},\nu')g(\nu_{k2},\nu')\ldots g(\nu_{kl},\nu')Q^n(\nu',z)$ and integrating over all frequencies

$$\frac{d}{dz}q^{(m,n)}_{k_1,k_2,\ldots,k_l}(z) = q^{(m+1,n)}_{k_1,k_2,\ldots,k_l}(z) - \sum_{k'} P_s(v_k, z) q^{(m,n)}_{k_1,k_2,\ldots,k_l,k'}(z) - q^{(m,n+1)}_{k_1,k_2,\ldots,k_l}(z) \qquad (3\text{-}15)$$

Since the magnitudes of $g(v,v')$, $Q(v,v')$ and $\alpha(v)$ are less than unity, the M-functions are decreasing functions with respect to moment indices (m,n,l) and rapidly converge to zero as indices increases. Signal power variation along the fiber line for each frequency ($k=1,2,\ldots,N_S$) is obtained by solving coupled equations (3-13) and (3-15), which are explicitly independent of pump power, for various moment indices. Therefore the system of uncountable differential equations has been converted to a set of countable equations using M-method. In addition, boundary condition for moment elements at $z = L$ is given as

$$q^{(m,n)}_{k_1,k_2,\ldots,k_l}(L) = \int_0^\infty \alpha^m(v') g(k_1,v') g(k_2,v') \ldots g(k_l,v') P_p(v') Q^n(v',L) dv' \qquad (3\text{-}16)$$

A FRA consists of a broad-band pump with Gaussian spectrum, central wavelength at 1443.2 nm, FWHM 15 nm, output power 388.35 mW and 15 signal channels with input power 0 dBm/channel and channel spacing 100 GHz have been considered for numerical simulations. Evolution of a signal channel (195 THZ) along the fiber line and Raman gain for various moment indices are shown in Figure 20. For comparison, signals output powers based on full numerical method and moment method are calculated and presented in this figure. As seen in Figure 20, Raman gain for large values of moment indices converge to the results of full numerical method.

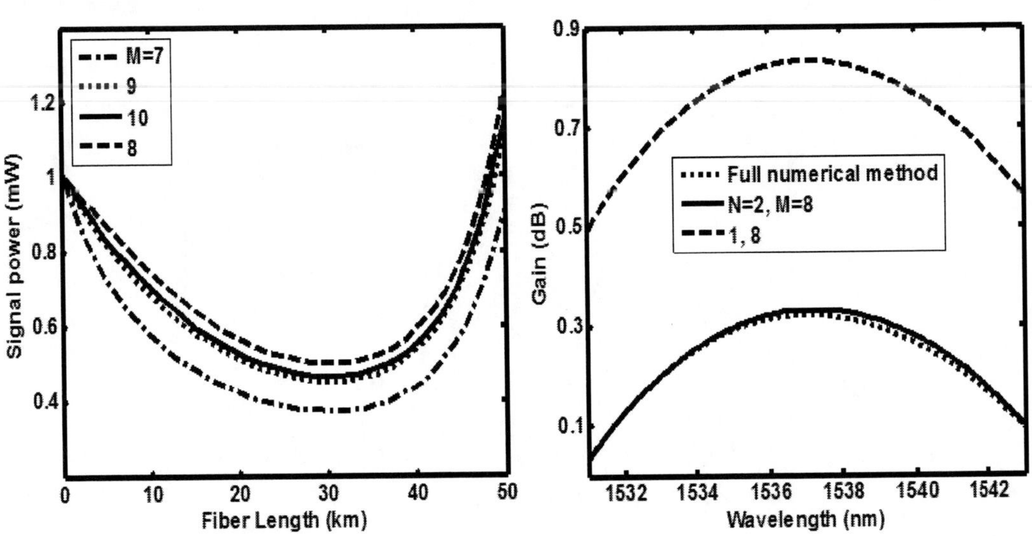

Figure 20. Left: evolution of signal power along fiber line for various moment indices $M=m$. Right: signal output power versus wavelength for $M=8$ and various moment indices $N=n+l$, [90].

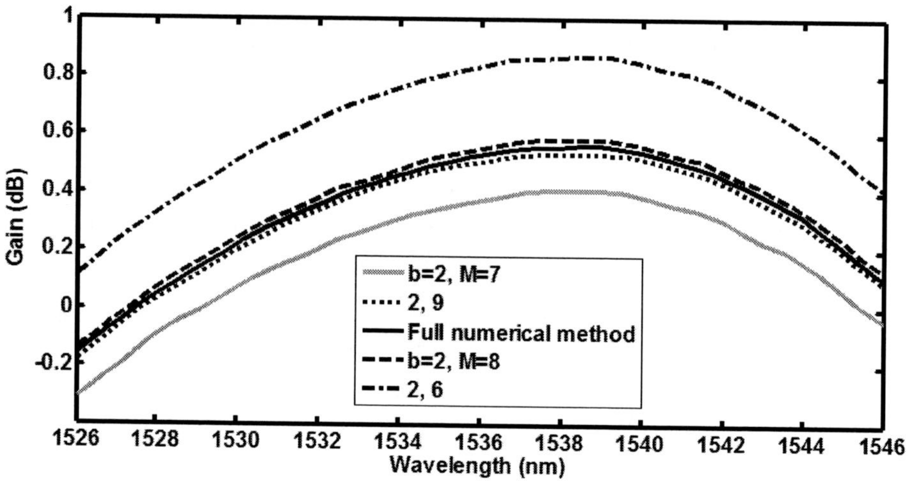

Figure 21. Raman gain calculation using M-method for various moment indices and full-numerical method. 10 signals from 1525.6 to 1546.3 nm with lunch power 0dB/channel along 25 Km SMF-28 have been considered [93].

By some modification, semi-analytical M-method has been employed in continuous bi-directional pumping scheme [93]. For this purpose, a bi-directional pumping scheme consists of two wide-band pumps with Lorentzian line-shapes and output powers of 168 and 158 mW have been considered. FWHM and central wavelength for both of pumps are 25 nm and 1443.13 nm, respectively. Gain ripples calculation using M-method for various moment indices and full numerical method has been compared in Figure 21. As this figure shows, by increasing moment indices, both M-method and Full-numerical method converge to same results.

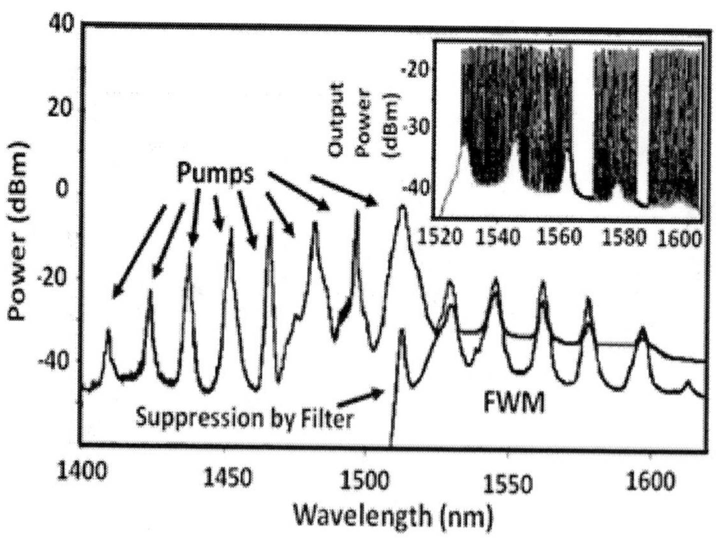

Figure 22. Effects of FWM in a WDM pumping scheme [94].

3.3.3. TDM Pumping Scheme

WDM and continuous-spectrum pumping schemes are two efficient methods for broadening and flattening the Raman gain spectrum. However, some problems occur during employment of these configurations. Four wave mixing (FWM) and pump-to-pump energy transfer are two well-known drawbacks. FWM between pumps, especially in dispersion shifted fiber lines, can induces significant noises through Rayleigh backscattering in signal band and degrades optical signal to noise ratio. Figure 22 shows, how FWM affects the signal channels and corrupts them. Moreover, Raman interaction between pumps lead to energy transfer from short pump wavelengths to longer ones and hence increasing noise figure (NF) for short wavelength signal channels.

These problems can be solved when interaction length between various pumps is reduced. This can be achieved by an alternative approach of pumping scheme known as time-division-multiplexing (TDM). In a TDM configuration, an array of high power laser diodes at different wavelengths sequentially switched on and off to produce a train of pulses with various amplitudes and durations (Figure 23).

The switching rate must be large enough to suppress temporal gain variation. For example for a system with fiber attenuation of 0.23 dB/km, pump repetition rate 500 kHz and duty cycle 25%, gain ripple is approximately 0.8% of on-off gain which is essentially negligible [95]. Switching speed requirement and temporal gain variation have been investigated theoretically and experimentally in [96, 97]. In addition, several numerical methods for both steady state and time-dependent regimes have been presented in [98, 99]. In the following, a time dependent solution for a backward FRA with TDM pumping configuration, based on Picard iterative method is presented [100]. By introducing $\xi = z - V_s t$ and $\eta = z + V_p t$, equations (3-1) and (3-2) is rewritten as

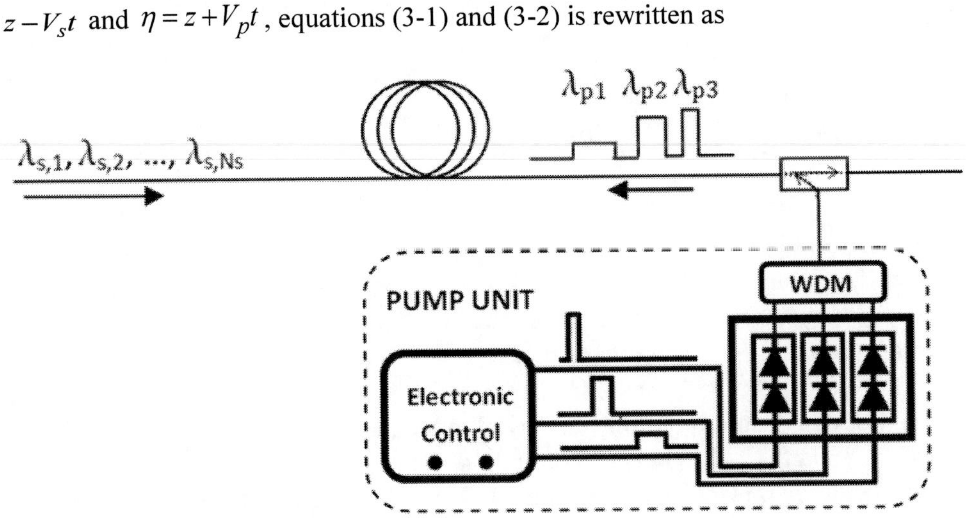

Figure 23. Generalized schematic of TDM configuration.

$$\frac{1}{P_{f,i}(\xi,\eta)}\frac{\partial P_{f,i}(\xi,z)}{\partial \eta} = -\alpha_i^f + \sum_{j=1}^{N_s} g_{i,j}^f P_{f,j}(\xi,\eta) + \sum_{k=1}^{N_p} g_{i,k}^f P_{b,k}(\xi,\eta); \ i = 1,2,\ldots,N_s \tag{3-17}$$

$$\frac{1}{P_{b,k}(\xi,\eta)}\frac{\partial P_{b,k}(\xi,z)}{\partial \xi} = -\alpha_k^b + \sum_{j=1}^{N_s} g_{k,j}^b P_{f,j}(\xi,\eta);\; k = 1,2,\ldots,N_p, \tag{3-18}$$

where $P_{f,i}(\xi,\eta) = P_{s,i}(z,t)$, $P_{b,k}(\xi,\eta) = P_{p,k}(z,t)$, $\alpha_{i(k)}^{f(b)} = (V_{s(p)}\alpha_{s(p)})/(V_s + V_p)$ and $g_{i,j}^{f(b)} = V_{s(p)}/(V_s + V_p)$. It is assumed that both pump and signal waves have arbitrary shapes as $P_{s,i}(0,t) = s_i(t)$ and $P_{b,k}(L,t) = h_k(t)$ at the input ends of fiber line. Solutions of this set of partial equations are

$$P_{f,i}(\xi,z) = s_i(-\frac{\xi}{V_s})\exp\left[-\alpha_i^f\left(\eta + \frac{V_p}{V_s}\xi\right)\right]G_{f,i}(\xi,\eta) \tag{3-19}$$

$$P_{b,k}(\xi,z) = h_k\left(\frac{\eta - L}{V_p}\right)\exp\left[-\alpha_{p,k}L + \alpha_k^b\left(\xi + \frac{V_s}{V_p}\eta\right)\right]G_{b,k}(\xi,\eta) \tag{3-20}$$

where, the Raman effective gains corresponds to signal and pump $G_{f,i}(\xi,\eta)$ and $G_{b,k}(\xi,\eta)$ are given as

$$G_{f,i}(\xi,\eta) = \exp\left[\sum_{j=1}^{N_s} g_{i,j}^f \int_{\eta_0}^{\eta} P_{f,j}(\xi,\eta')d\eta' + \sum_{k=1}^{N_p} g_{i,k}^f \int_{\eta_0}^{\eta} P_{b,k}(\xi,\eta')d\eta'\right]; \qquad \eta_0 = -\frac{V_p}{V_s}\xi \tag{3-21}$$

$$G_{b,k}(\xi,\eta) = \exp\left[\sum_{j=1}^{N_s} g_{k,j}^b \int_{\xi_0}^{\xi} P_{f,j}(\xi',\eta')d\xi'\right]; \qquad \xi_0 = \frac{V_s}{V_s + V_p}L - \frac{V_s}{V_p}\eta \tag{3-22}$$

Equations (3-19) and (3-20) can be solved through an iterative procedure. For this purpose, at first step it is assumed that Raman effect can be treated as a perturbation. Therefore for the 0-th order both Raman effective gains are set to one and 0-th order of solutions are resulted. Replacing these solutions into equation (3-21) and (3-22), first order Raman effective gains are obtained. Higher order of solutions is achieved by repeating this procedure. For demonstration of this numerical model, a TDM-pumped FRA consists of four pumps at wavelengths 1433, 1447, 1470 and 1490 nm with powers 3.9, 3.5, 5.6 and 5.6 W have been considered, respectively. Figure 24 shows the temporal evolution of output signal power (at 1535 nm) for a 13 km dispersion compensated fiber in 3-th iteration (forth order is similar to third order).

It is seen that after approximately 170 μs a steady state is reached. In addition temporal gain fluctuation is negligible compared to average gain. Using this method together with Lagrange multipliers method, the performance of TDM-pumped FRA can be optimized for various numbers of pumps and results are shown in Figure 25. As this figure shows the higher the number of pumps the lower the gain ripple.

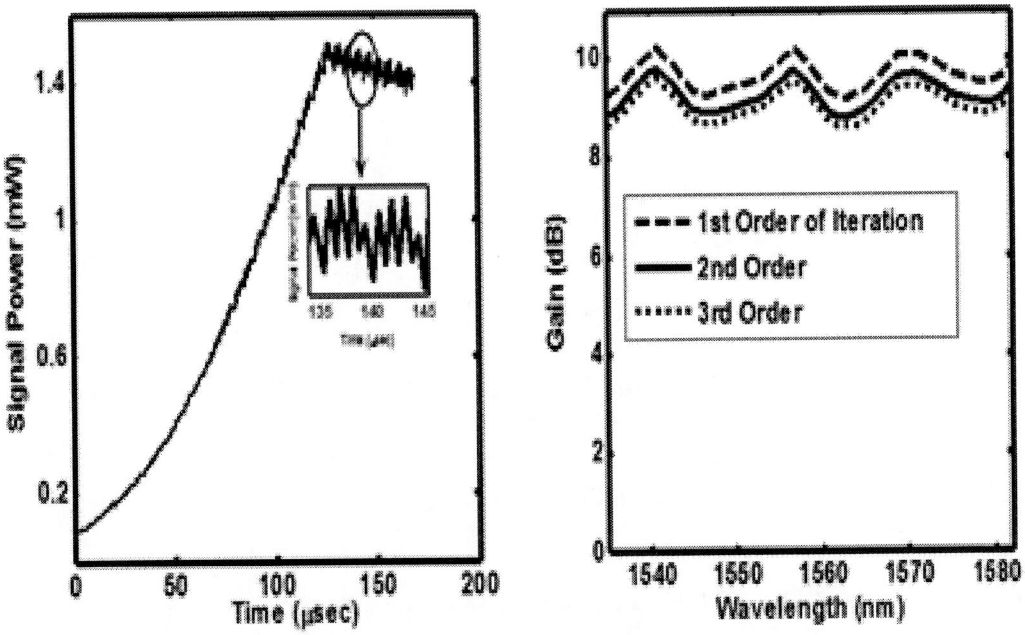

Figure 24. (Left) Temporal variation of output signal power (1535 nm) of a TDM-pumped FRA consists of four square-shapes pumps with duty cycle 25% and modulation frequency 200 kHz. (Right): Raman Net gain for 25 WDM signal channels with 2 nm channel spacing and lunched power -7dBm/channel [100].

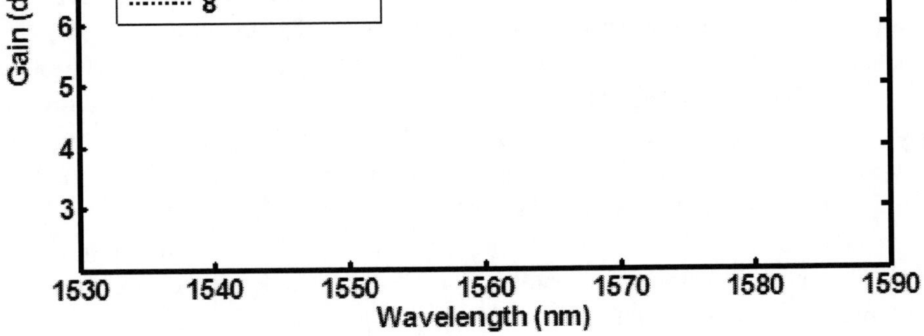

Figure 25. Net gain for a TDM-pumped FRA with 25 signal channels, 2 nm channel spacing and -7dBm/channel lunched power for various number of pumps with optimized wavelengths and powers. (All of pumps have same duty cycles without any time spacing and modulation.

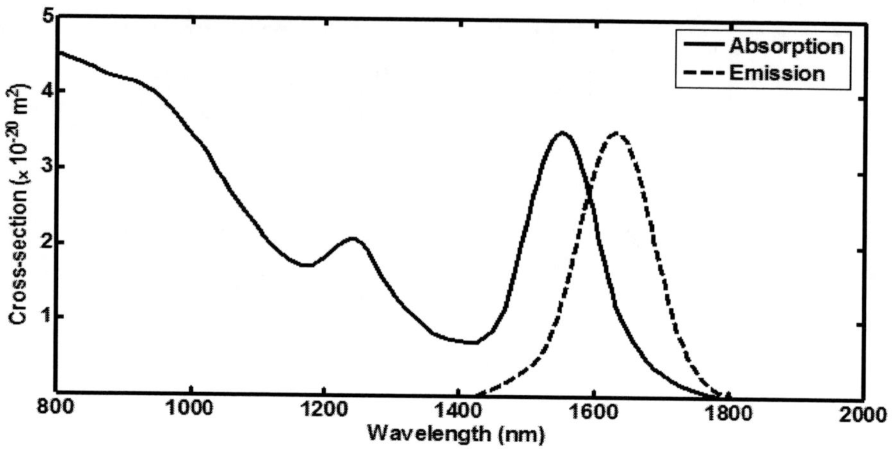

Figure 26. Absorption and emission cross section of a sample of PbSe QDs with 5.5 nm average diameter.

4. Quantum Dot Doped Fiber Amplifiers

The energy diagram of a semiconductor consists of several energy bands. The energy gap between the edges of two higher energy bands, valance, and conduction bands is a very important parameter which determines the optical and electrical properties of the semiconductor material. For macroscopic size semiconductors, the energy diagram and bandgap value are independent of the semiconductor size. But if the semiconductor size is reduced to the order of its exciton Bohr radius [9], then a pair of electron and hole feel the presence of boundaries and behave similar to a particle in the box quantum system [9]. This means that the energy bands of a semiconductor will be converted to semi-discrete energy levels if its size becomes less than its exciton Bohr radius [101]. A nonosize semiconductor is called a QD (or an artificial atom).

The energy diagram of QDs are strongly size dependent. In the simple spherical box model, the energy gap of a QD with radius R is simply proportional to $1/R^2$ [9, 101]. This means that two QDs with different sizes absorb and emit different frequencies even if they are made from the same semiconductor [102]. As the QD size increases, its absorption and emission spectra shift to larger wavelengths (red shift) [102]. Hence, the optical properties of QDs can be engineered during the synthesis procedure by their size controlling to suit a particular application. Several synthesis methods have already been reported which can be used to produce high quality (with narrow size distribution) QD samples [102, 103].

Among various types of QDs, lead salt ones (PbSe, PbS, and PbTe) have attracted a large attention because of their significant and unique properties. The large Bohr radius (46 nm for PbSe and 18 nm for PbS QDs) and emission in near infrared spectral region are the most important properties of lead salt QDs [102]. Figure 26 shows the measured emission and absorption cross sections of a sample of PbSe QDs with 5.5 nm average diameter.

These important properties lead to the large potential of Pb-based QDs for application in various photonic devices such as optical amplifiers [104-108], photodetectors [109], solar cells [110] and Q-switched infrared lasers [111].

A homogeneous model for PbSe QDFAs in steady-state regime has been presented by Cheng and Zhang [106, 107]. In [106, 107] it has been shown that a PbSe QD can be modeled as a three-level system. The reported lifetimes for the second and third energy levels are 0.3 μs and 4 ps,, respectively. Thus, the population of level 3 depopulates to the level 2 immediately and makes it possible to use a two-level model for PbSe QDFAs as done in [108]. Because of the existence of the QDs with different sizes, the medium is inhomogeneous. In [108], we have proposed an inhomogeneous model for analyzing the QDFAs. In this case the rate equation of the exited sate population is given by

$$\frac{\partial n_2(R)}{\partial t} = n_1(R,r,z,t)\sum_{k=1}^{m}\frac{P_k(z,t)}{h\nu_k}i_k(r)\sigma_a^h(\nu_k,R) - n_2(R,r,z,t)\sum_{k=1}^{m}\frac{P_k(z,t)}{h\nu_k}i_k(r)\sigma_e^h(\nu_k,R)$$
$$-\frac{1}{\tau_2}n_2(R,r,z,t) + F\left[n_{2qe}(R,r,z) + n_2(R,r,z,t)\right] \qquad (4\text{-}1)$$

where $n_1(R)$ ($n_2(R)$) is the number of QDs with R diameter in the ground (exited) energy level per unit volume. Other parameters are defined similar to the parameters of equation (2-4). The rate equation of $n_2(R)$ can be obtained easily by using the continuity relation: $n_1(R)+n_2(R)=g(R)n_q$ where n_q is the total concentration of doped QDs and $g(R)$ is their size distribution function which satisfies $\int_0^\infty g(R)\,dR=1$ relation. Homogeneous and inhomogeneous absorption and emission cross sections are related to each other by the averaging equation [108,112].

$$\sigma_{a,e}(\nu)=\int_0^\infty g(R)\,\sigma_{a,e}(\nu,R)\,dR \qquad (4\text{-}2)$$

$g(R)$ can be measured in lab TEM imaging technique and is nearly a Gaussian function [113]. Then homogeneous spectra can be calculated by the solution of the integral equation (4-2). In the absence of noise and by employing the SVEA, the propagation equation of kth wave with frequency ν_k can be written as follows

$$\frac{\partial P_k(z,t)}{\partial z}+\frac{u_k}{V_k}\frac{\partial P_k(z,t)}{\partial t}=$$
$$u_k P_k(z,t)\left[\int_0^\infty dr\,2\pi r\, i_k(r)\int_0^\infty dR\, n_2(R)\sigma_e^h(\nu_k,R)-\int_0^\infty dr\,2\pi r\, i_k(r)\int_0^\infty dR\, n_1(R)\sigma_a^h(\nu_k,R)-l_k\right] \qquad (4\text{-}3)$$
$$k=1,2,\ldots,m$$

Parameters of equation (4-3) have previously been defined in equation (2-7).

In the steady state situation and by using the continuity relation, the governing equations can be rewritten as follows:

$$n_2(R)=\left(F n_{2qe}(R)+n_q g(R)\sum_{k=1}^{m}\frac{P_k(z)}{h\nu_k}i_k(r)\sigma_a^h(\nu_k,R)\right)\Bigg/\left(F+\frac{1}{\tau_2}+\sum_{k=1}^{m}\frac{P_k(z)}{h\nu_k}i_k(r)\sigma_k^h(R)\right) \qquad (4\text{-}4)$$

$$\frac{dP_k(z)}{dz} = u_k P_k(z)\left[\int_0^{\infty} dr\, 2\pi r\, i_k(r)\int_0^{\infty} dR\, n_2(R)\sigma_k^h(R) - n_0\int_0^{\infty} dr\, 2\pi r\, i_k(r)\int_0^{\infty} dR\, g(R)\sigma_e^h(v_k,R) - l_k\right]$$

$$k = 1,2,\ldots,m \qquad (4\text{-}5)$$

where $\sigma_k^h(R) = \sigma_a^h(v_k,R) + \sigma_e^h(v_k,R)$. And the quasi equilibrium population of the exited state ($n_{2qe}(R)$) is defined as

$$n_{2qe}(R) = \left(n_q g(R)\sum_{k=1}^{m_p}\frac{P_k(z)}{h v_k} i_k(r)\sigma_a^h(v_k,R)\right) \Bigg/ \left(\frac{1}{\tau_2} + \sum_{k=1}^{m_p}\frac{P_k(z)}{h v_k} i_k(r)\sigma_k^h(R)\right) \qquad (4\text{-}6)$$

The system of nonlinear ordinary differential equations (equation (4-5)) can be solved easily by the fourth order Runge-Kutta method. In [108] the size distribution function has been defined as a Gaussian function centered at $\overline{R}$ and with standard deviation D (i.e., $g(R) = \frac{1}{D\sqrt{2\pi}}\exp\left(-\left(R-\overline{R}\right)^2/2D^2\right)$).

Figure 27 shows calculated gain spectra in [108] corresponding to various amounts of standard deviation. The QDs concentration is 2×10^{20} m^{-3} in [108].

The gain spectra of Figure 27 have been calculated for fifty signals ranging from 1570 to 1720 nm and -30 dBm/ch input powers. The cross relaxation coefficient, pump wavelength and power are 10^8 s^{-1}, 1540 nm and 300 mW, respectively. As expected, the gain value decreases while the gain bandwidth increases with increasing the standard deviation.

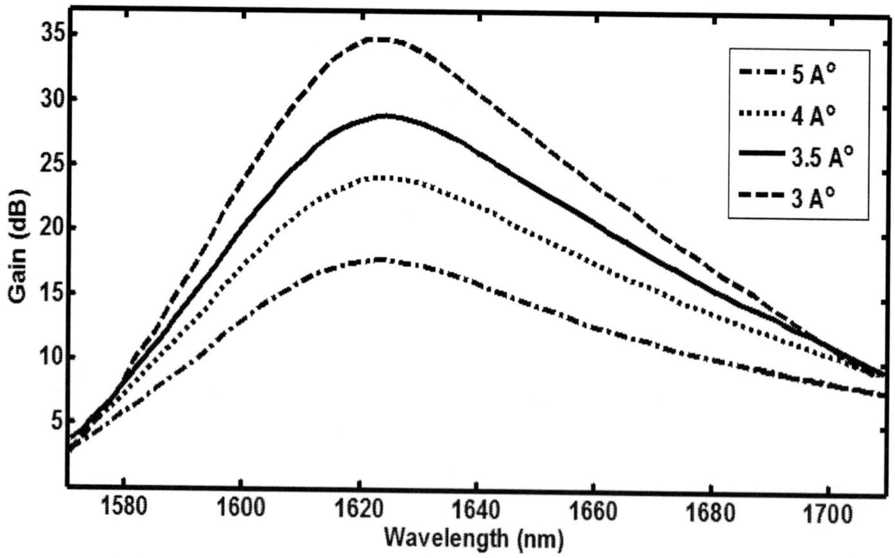

Figure 27. Gain spectrum of a typical PbSe QDFA for various standard deviations. $\overline{R}$=5.5 nm [108].

For studding the SHB effect in [108], a strong signal has been located at 1640 nm with 7 dBm input power and the wavelength range from 1580 to 1740 nm has been scanned with a -30 dBm probe signal. Obtained results are presented in Figure 28.

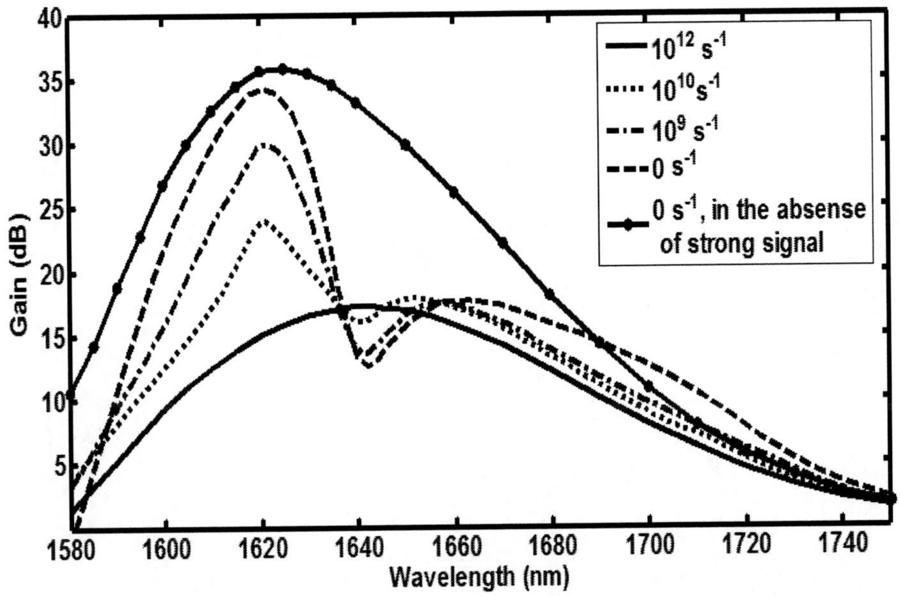

Figure 28. Hole burning effect in the gain spectrum of a PbSe QDFA. A strong signal with 7dBm power is fixed at 1640 nm and the wavelength region is scanned with a -30 dBm probe signal. ($\overline{R}$=5.5 nm and $D/\overline{R}$=0.05) [108].

As seen in Figure 28, by increasing the cross relaxation coefficient, the gain amplitude decreases while the hole is disappeared. This means that for large values of F, the inhomogeneous medium behaves like a homogeneous medium.

To see the saturation effects and energy transfer between the signals in QDFAs, the gain spectrum for seventy signal channels has been calculated for different signals input powers (Figure 29). The gain stabilization in QDFAs has also been studied in [108]. Figure 29 (right) shows obtained results for -20 dBm/ch signals input power.

As expected, the gain spectrum of a QDFA is strongly dependent to its size distribution function (see Figure 27). This property is too beneficial for flattening the gain spectrum of these amplifiers. In fact, an appropriate selection of the size distribution function can easily flatten the gain spectrum of the QDFA without any requirement to external equipments. Figure 30 shows the gain spectrum presented in [108] for a multi-size QDFA doped with three samples of QDs with ($\overline{R}_1$= 4.9 nm, D_1= 3.5 A°, n_{q1}=1.45×10^{20} m^{-3}), ($\overline{R}_2$= 5.5 nm, D_2 = 3 A°, n_{q2}=1.25×10^{20} m^{-3}) and ($\overline{R}_3$= 6 nm, D_3= 2.5 A°, n_{q3}=1.35×10^{20} m^{-3}) characteristics. In this case the SDF is defined by

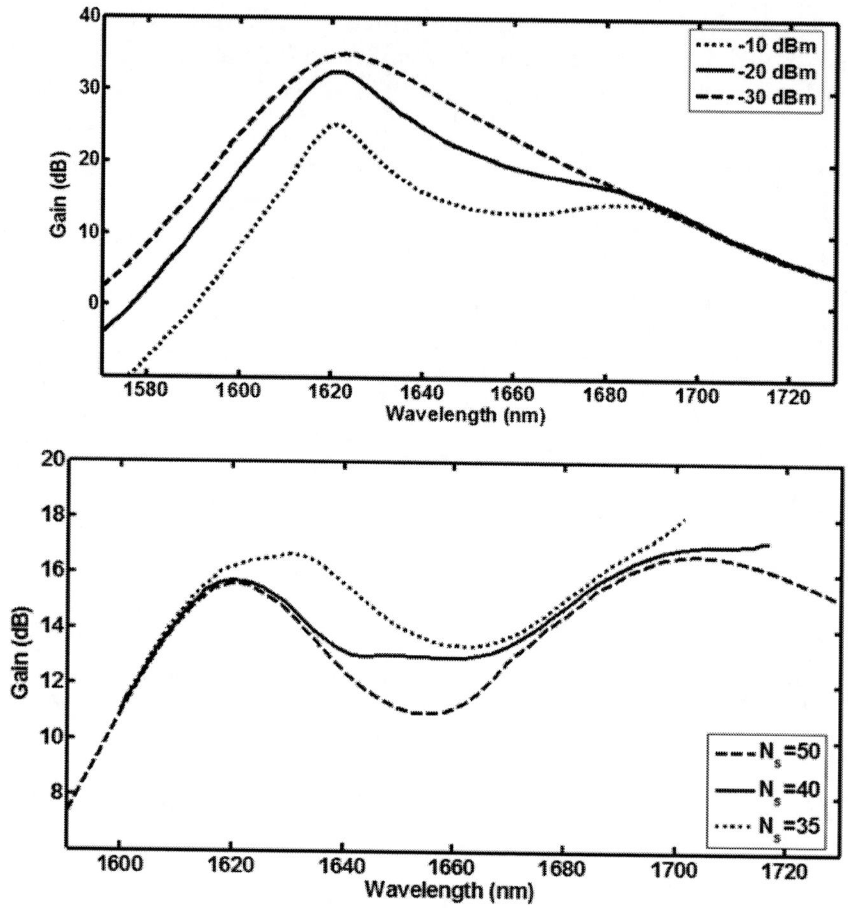

Figure 29. Gain spectrum of a typical PbSe QDFA right: with seventy signal channels for different signals input powers left: for various numbers of input channels. ($\overline{R}$ =5.5 nm and $D/\overline{R}$ =0.05) [108].

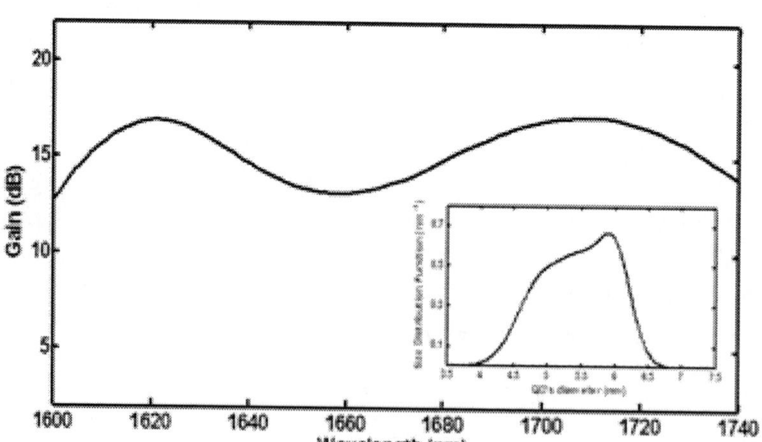

Figure 30. Gain spectrum of a typical multi-size PbSe QDFA. The inset graph shows the size distribution function of this amplifier [108].

$$g(R)=\sum_{i=1}^{3}\frac{w_i}{D_i\sqrt{2\pi}}\exp\left(-\left(R-\overline{R}_i\right)^2/2D_i^2\right) \quad (4\text{-}7))$$

where w_i is the partial fraction concentration of ith sample which is defined as $w_i = n_{qi}/n_q$ that $n_q = \sum_{i=1}^{3} n_{qi}$ is the total concentration of QDs.

These reported theoretical results [106-108] are very interesting and predict that Pb-based QDFAs can provide high gain and wide bandwidth for WDM and DWDM systems. Recently, some works have been reported on various glasses doped with Pb-based QDs [114-116]. These valuable efforts can make the QDFAs achievable in the near future.

CONCLUSION

EDFAs have been theoretically analyzed as the inhomogeneous media by employing the standard and M-methods in both low and high concentration doping regimes. The M-method is a powerful technique for reducing uncountable system of differential equations to a finite system. This method has also been applied for studying FRAs with broad band continues pumps. Reported results show that the M-method is a fast and accurate technique in comparison with the standard methods. Other pumping configurations in FRAs such as WDM and TDM pumping are studied using the Lagrange multiplier and perturbation methods, respectively. An inhomogeneous model for QDFAs has been presented, too. This model has been used for analyzing the gain spectrum, SHB, and saturation effects in QDFAs.

REFERENCES

[1] Koester, C. J.; Snitzer, E. *Appl. Optics* 1964, vol 3, 1182-1186.
[2] Miniscalco, W. J. in Rare-Earth-Doped Fiber Lasers and Amplifiers; Digonnet, M. J. F.; Ed.; Marcel Dekker, Inc; 2001, pp 17-112.
[3] Tanabe, S. *j. Alloys & Compounds* 2006, 675-679.
[4] Labio, L. D. et al. *Applied Optics* 2008, vol 47, no 10, 1581-1584.
[5] Cheng, X. S.; Parvizi, R.; Ahmad, H.; Harun, S. W. *IEEE Photonics Journal* 2009, vol 1, 259-264.
[6] Bahrampour, A. R.; Mahjoei, M. *J. Lightwave Technol.* 2001, vol 19, no 8, 1130-1139.
[7] Emori, Y.; Namiki, S. *Furukawa Review,* 2000, No 19, 59-62.
[8] Islam, M.N. *IEEE J. Selected Topics in Quantum Electron.* 2002, vol 8, no 3, 548-559.
[9] Woggon, U. Optical Properties of Semiconductor Quantum Dots; Springer-Verlag: Berlin, 1997.
[10] Jiang, C. et al. *IEEE J. Quantum Electron.* 2003, vol 39, 1266-1271.

[11] Watekar, P. R.; Ju, S.; Han, W.T. *IEEE Photon. Technol. Lett.* 2007, vol 19, no 19, 1478-1480.

[12] Aozasa, S.; Masuda, H.; Shimizu, M.; Yamada, M. *J. Lightwave Technol.* 2007, vol 25, no 8, 2108-2114.

[13] Wang, F.; Li, C. *IEEE Photon. Technol. Lett.* 2007, vol 19, no 21, 1771-1773.

[14] Tanabe, S.; Tamaoka, T. *Optical Materials* 2004, vol 25, 165-169.

[15] Rosolem, J.B.; Juriollo, A.A.; Romero, M.A. *Electronics Lett.* 2007, vol 43, no 22.

[16] Ono, H.; Yamada, M.; Shimizu, M. *J. Lightwave Technol.* 2003, vol 21, no 10, 2240-2246.

[17] Arbore, M. A.; Zhou, Y.; Keaton, G.; Kane, T. Optical Amplifiers and Their Applications (OAA) 2002, Vancouver, PD-4-1.

[18] Vincetti, L. et al. *J. Lightwave Technol.* 2008, vol 26, no 14, 2168-2174.

[19] Dussardier, B.; Blanc, W. Optical Fiber Communication (OFC) Conf 2007. paper OMN1.

[20] Aozasa, S. et al. *NTT Technical Review* 2004, vol 2, no 12, 44-50.

[21] BECKER, P. C.; Olsson, N. A.; Simpsom, j. R. Erbium-Doped Fiber Amplifiers Fundamentals and Technology; Academic Press, 1999.

[22] Ohara, S. et al. *Optical Fiber Technol.* 2004, vol 10, 283-295.

[23] Marjanovic, S. et al. *J. Non-Crystalline Solids* 2003, vol 322, 311-318.

[24] Nagaraju, B.; Jaiswal, P.; Varshney, R. K.; Pal, B. P. *ICOP conf.* 2009, Chandigarh, India.

[25] Nagaraju, B. et al. *Opt. Communs.* 2009, vol 282, 2335-2338.

[26] Ni, N. et al. *Opt. Communs.* 2007, vol 271, 377-381.

[27] Dung, J. C.; Chi, S.; Wen, S. *Electronics Lett.* 1998, vol 34, no 6, 555-556.

[28] Sahu, P. P. *Opt. Communs.* 2008, vol 281, 573-579.

[29] Kim, H. S. et al. *IEEE Photon. Technol. Lett.* 1998, vol 10, no 6, 790-792.

[30] Li, S.; Chiang, K. S.; Gambling, W. A. *IEEE Photon. Technol. Lett.* 2001, vol 13, no 9, 942-944.

[31] Verly, P. G. Appl. Opt. 2002, vol 41, 3092-3096.

[32] Varshney, R. K.; Singh, A.; Pande, K.; Pal, B. P. *Opt. Communs.* 2007, vol 271, 441-444.

[33] Varshney, R. K. *Optics Express* 2007, vol 15, no 21, 13519-13530.

[34] Choi, B. H.; Park, H. H.; Chu, M. J. *IEEE J. Quantum Electron.* 2003, vol 39, no 10, 1272-1280.

[35] Harun, S. W.; Tamchek, N.; Poopalan, P.; Ahmad, H. *IEEE Photon. Technol. Lett.* 2003, vol 15, no 8, 1055-1057.

[36] Yi, L. L. et al. *IEEE Photon. Technol. Lett.* 2004, vol 16, no 4, 1005-1007.

[37] Jin, Y. et al. *Opt. Communs.* 2006, vol 266, 390-392.

[38] Aozasa, S. et al. *Electronics Lett.* 2008, vol 44, no 4, 273-274.

[39] Altuncu, A.; Basgumus, A. *IEEE Photon. Technol. Lett.* 2005, vol 17, no 7, 1402-1404.

[40] Kakkar, C.; Monnom, G.; Thyagarajan, K.; Dussardier, B. *Opt. Communs.* 2006, vol 262, 193-199.

[41] Kakui, M.; Ishikawa, S. *IEICE Trans. Electron.* 2000, vol E83-C, no 6, 799-815.

[42] Ghatak, A.; Thyagarajan, K. Introduction to Fiber Optics; Cambridge University Press, 1997.

[43] Desurvire, E. in Rare-Earth-Doped Fiber Lasers and Amplifiers; Digonnet, M. J. F.; Ed.; Marcel Dekker, Inc; 2001; pp 531-582.

[44] Bahrampour, A. R.; Mahjoei, M.; Rasouli, A. *Opt. Communs.* 2006, vol 265, 283-300.

[45] Mahdi, M. A. et al. *Laser Phys. Lett.* 2008, 5, no 2, 126-129.

[46] Mahdi, M. A. et al. *Laser Phys. Lett.* 2008, 5, no 4, 296-299.

[47] Yi, L. et al. Optics Express 2006, vol 14, no 2, 570-574.

[48] Zhao, C. L. et al. *Opt. Communs* 2003, vol 225, 157-162.

[49] Bahrampour, A. R.; Keyvaninia, S.; Karvar, M. *Optical Fiber Technol.* 2008, 14, 54–62.

[50] Bahrampour, A. R.; Khosravi, M.; Farahi, R. M. *Scientia Iranica* 2000, vol 7, no 1, 32-40.

[51] Chernyak, V.; Qian, L. *IEEE J. Selected Topics in Quantum Electron.* 2002, vol 8, no 3, 569-574.

[52] van den Hoven, G. N. et al. *J. Appl. Phys.* 1996, 79, 1258-1266.

[53] Ono, S.; Tanabe, S. *j. Alloys & Compounds* 2006, 732-736.

[54] Bromage, J. *J. Lightwave Technol.* 2004, vol 22, 79-93.

[55] karasek, M.; Kanka, J.; Honzatko,P.; Peterka, P. *Int. J.Numer.Model.* 2004, vol 17, 165-167.

[56] Payne, D. N.; Barlow, A. J.; Hansen, J. J. R. *J. Quantum Electron.* 1982, vol 18, 477-487.

[57] Messerly, M. J.; Onstott, J. R.; Mikkelson, R. C. *J. Lightwave Technol.* 1991, vol 9, 817-820.

[58] Dyott, R. B. Elliptical Fiber Waveguides, *Artec House*, Boston, 1995.

[59] Poole, C. D.; Winters, J. H.; Nagel, J. A. *Opt. Lett.* 1991, vol 16, 372-374

[60] Lichtman, E. *J. Lightwave Technol.* 1995, vol 13, 898-905.

[61] Mahgerefteh, D.; Menyuk, C. R. *IEEE Photon. Technol. Lett.* 1999, vol 11, 340-342.

[62] Stolen, R. H. *Tech. Dig. Symp. Optical Fiber Measurements,* Gaithersburg, 2000, 139-142.

[63] Ebrahimi, P. et al. Laser and Electro-Optics Conf., OSA Trends in Optics and Photonics Series, Washington, DC, 2001, vol 56, 143-144.

[64] Popov, S.; Vanin, E.; Jacobsen, G. *Opt. Lett.* 2001, vol 27, 848-850.

[65] Emori, Y.; Namiki, S. *Proc. Opt. Amplifier Appl.* 2001, vol 60, 44-48.

[66] Azami, N. *Opt. Commun.*, 2004, vol 230, 181-184.

[67] Lin, Q.; Agrawal, G. P. *Optical Fiber Communication (OFC) Conf.*, Atlanta, GA, 2003, paper: TuC4.

[68] Son, E.; Lee, J.; Chung, Y. Optical *Fiber Communication (OFC) Conf.,* Atlanta, GA, 2003, paper: TuC5.

[69] Lin, Q.; Agrawal G. P. *Opt. Lett.* 2003, vol 28 (4), 227-229.

[70] Lin, Q.; Agrawal, G. P. *J. Opt. Soc. Am. B.* 2003, vol 20, 1616.

[71] Takada, K.; Okamoto, K.; Noda, J. *J. Lightwave Technol.* 1986, vol 4, 213-219.

[72] Lutz, D. R. *IEEE Photon. Technol. Lett.* 1993, vol 5, 463-465.

[73] Wang, J. S.; Costelloe, J. R.; Stolen, R. H. *IEEE Photon.Technol. Lett.* 1999, vol 11, 1449 -1451.
[74] Kee, H.H.; Fludger, C.R.S.; Handerek, V. *Optical Fiber Communications (OFC) conf.*, Washington, DC, 2002, 180-181.
[75] Tokura, T. et al. *J. Lightwave Technol.* 2006, vol 24, 3889-3896.
[76] Kazami, H. et al. *Furukawa Review* 2003, No 23, 44-47.
[77] Li, Zhaohui et al. *IEEE Photon.Technol. Lett.* 2005, vol 17, 558 -560.
[78] Zhou, X.; Magill, P.; Birk, M. *J. Lightwave Technol.* 2005, vol 23, 1056-1062.
[79] Sergeyev, S.; Popov, S.; Friberg, A. T. *Opt. Express* 2008, vol 16, 14380-14389.
[80] Sergeyev, S.; Popov, S.; Friberg, A. T. *European Quant. Electro. Conf.,* Munich, 2009.
[81] Perlin, V. E.; Winful, H. G. *J. lightwave. Technol.* 2002, vol 20, 409-416.
[82] Perlin, V. E.; Winful, H. G. *J. lightwave. Technol.* 2002, vol 20, 250-254.
[83] Cui, S.; Liu, J.; Ma, X. *IEEE Photon. Technol. Lett.* 2004, vol 16, 2451-2453.
[84] Hu, J.; Marks, B. S.; Menyuk, C. R. *J. lightwave. Technol.*2004, vol 22, 1519-1522.
[85] Liu, X.; Li, Y. *J. Opt. Commun.* 2004, vol 230, 425-431.
[86] Liu, X.; Li, Y. *Chinese Phys. Lett.* 2004, vol 21, 84-86.
[87] Zhou, J. et al. *J. lightwave. Technol.* 2006, vol 24, 2362-2367.
[88] Bahrampour, A. R.; Ghasempour, A.; Rahimi, L. *Opt. Commun.* 2008, vol 281, 1545-1557.
[89] Jiang, H.; Xie, K.; Wang, Y. *Opt. Express*, 2010, vol 18, 11033-11045.
[90] Bahrampour, A. R.; Teimourpour, M. H.; Rahimi, L. *Opt. Commun.* 2008, vol 281, 587-591.
[91] Kobtsev, S. M.; Pustovskikh, A. A. *CLEO/Europe-2003 conf, Munich,* 2003, p.630.
[92] Kobtsev, S. M.; Pustovskikh, A. A. *Laser Physics* 2004, vol 14, 1488-1492.
[93] Bahrampour, A. R.; pourmoghadas, A. *Opt. & Laser Technol.* 2010, vol 42, 332-335.
[94] Neuhauser, R. E.; Krummrich, P. M.; Bock, H.; Glingener, C. *Optical Fiber Communication*(OFC) *Conf.*, Anaheim 2001.
[95] Grant, A. R.; Mollenauer, L. F. in Raman Amplifiers for telecommunications 1, physical principles; Islam, M. N.; Ed.; Springer-Verlag, Inc; 2004, pp 61-89.
[96] Winzer, P. J.; Sherman, K.; Zirngibl, M. *Photon. Technol. Lett.* 2002, vol 14, no 6, 789-791.
[97] Winzer, P. J. et al. *IEEE J. Lightwave. Technol.* 2004, vol 22, no 2, 401-408.
[98] karasek, M.; kanka, J.; Honzatko, P.; Peterka, P. *Int. J. Numer. Model.* 2004, vol 17, 165-167.
[99] Jiang, W.; Chen, J.; Zhou, J. *Opt. Eng.* 2005, vol 45(9), p.095004.
[100] Bahrampour, A. R.; Farhadi, M.; Farman, F. *Optical fiber Technol.* 2009, vol 15, 353-362.
[101] Hollingsworth, J. A.; Klimov, V. I. Semiconductor and Metal Nanocrystals ; Klimov, V. I.; Ed.; Marcel Dekker, 2004.
[102] Murray, C. B. et al. *IBM J. RES. & DEV.* 2001, vol 45, no 1, 47-56.
[103] Yu, W. W.; Falkner, J.C.; Shih, B. S.; Colvin, V. L. *Chem. Mater.* 2004, vol 16, 3318-3322.
[104] Bakonyi, Z. et al. *IEEE J. Quant. Electron.* 2003, vol 39, no 11, 1409–1414.
[105] Herz, E. et al. *U.S. Patent.* 2008, no: 7, 362, 938 B1.
[106] Cheng, C.; Zhang, H. *Opt. Communs.* 2007, vol 277, 372–378.
[107] Cheng, C. *J. Lightwave Technol.* 2008, vol 26, no 11, 1404-1410.

[108] Bahrampour, A. R.; Rooholamini, H.; Rahimi, L.; Askari, A. A. *Opt. Communs*. 2009, vol 282, 4449-4454.

[109] McDonald, S. A. et al. *Nat. Mater.* 2005, vol 4, 138-142.

[110] Law, M. et al. *Nano Lett.* 2008, vol 8, 3904-3910.

[111] Brumer, M. et al. *Appl. Opt.* 2006, vol 45, no 28, 7488-7497.

[112] Ferreira, D. L.; Alves, J. L. A. *Nanotechnology* 2004, vol 15, 975–981.

[113] Moreels, I. et al. *Acs Nano* 2009, vol 3, no 10, 3023-3030.

[114] Lipovskii, A.; Kolobkova, E.; Petrikov,V. *Appl. Phys. Lett.* 1997, vol 71, no 23, 3406-3408.

[115] Melekhin, V. G. et al. *Glass Physics and Chemistry* 2008, vol 34, no 4, 351–355.

[116] Chang, J.; Liu, C.; Heo, J. *Nano-Cryst. Solids* 2009, vol 355, 1897-1899.

In: Optical Amplifiers
Editor: Galina Nemova

ISBN 978-1-61209-835-7
2012 Nova Science Publishers, Inc.

Chapter 6

Optimization of Hybrid Erbium-Doped Fiber Amplifier/Fiber Raman Amplifier

Alireza Mowla and Nosrat Granpayeh*

Faculty of Electrical and Computer Engineering,
K.N. Toosi University of Technology, Tehran, Iran.

Abstract

Erbium-doped fiber amplifiers, fiber Raman amplifiers, and hybrid erbium-doped fiber amplifiers/fiber Raman amplifiers are the most important optical components which are used nowadays in the conventional and wavelength division multiplexing systems and networks. In this chapter, these three kinds of optical fiber amplifiers are introduced and studied theoretically. We will present different configurations for these amplifiers, investigate the simulation methods of them and finally after their simulation, we derive the optimum parameters for each one of the configurations.

After invention of the fast computational machines or computers, simulation and optimization have become an essential part of the engineering design of devices and systems. In the past decades we have witnessed the striking developments in the simulation software and optimization methods, which provide us with accurate estimations and predictions that we need to industrialize efficient devices. The three above mentioned amplifiers have many parameters which will affect their performances. To obtain amplifiers with better characteristics, optimization processes on them are quite necessary. After introducing genetic algorithm, it has become one of the most popular evolutionary population-based optimization processes which have been used in many problems. Newer optimization methods, such as particle swarm optimization, have also proved their efficiencies. Particle swarm optimization is a fast and straightforward optimization method which is used in this chapter to optimized erbium-doped fiber amplifier, fiber Raman amplifier, and also hybrid erbium-doped fiber amplifier/fiber Raman amplifier. Its performance is compared to genetic algorithm during the chapter.

---

* E-mail: granpayeh@eetd.kntu.ac.ir; a.mowla@ee.kntu.ac.ir

Keywords: erbium-doped fiber amplifiers, fiber Raman amplifiers, hybrid erbium-doped fiber amplifiers/fiber Raman amplifiers.

1. Introduction

1.1. Erbium Doped Fiber Amplifiers

Erbium-doped fiber amplifiers (EDFAs) are the vital components of the dense wavelength division multiplexing (DWDM) optical communication systems. Current improvements of the DWDM systems and networks are entirely indebted to the EDFAs. Merits of EDFAs are high gain, low noise, broad bandwidth, high output power, and high efficiency of the pump power. Major criterions of the compact optical fiber amplifiers are gain, noise, bandwidth, and gain spectrum flatness. Higher gain and lower noise let the prolongation of the distance between two repeaters, wider bandwidth enables the DWDM networks to embrace more channels that brings about higher capacities, and flatter gain spectrum causes avoiding transmission impairments due to heterogeneous amplifications.

After presenting thorough models for EDFAs that solved propagation and rate equations for a two level laser homogeneous medium [1, 2], tremendous efforts have been made to find the efficient configurations that appease voracious appetite of DWDM systems for higher capacities.

Yeh *et al.* [3] suggested parallel structure of S- and C-band EDFAs with wide 96nm gain bandwidth of 1480–1576 nm, and Lu *et al.* [4] proposed a parallel combination of Cand L-band dual-core EDFAs which results in 105nm bandwidth over 1515–1620 nm. Tellurite-based EDFAs have been investigated to attain flat amplification bandwidth of 76nm [5]. It is highly desired to reach the same amplification results for EDFAs employed in dynamic DWDM networks, since, adding or dropping channels can cause changes in the gain performance of other existing channels. Therefore, gain-clamping techniques have been proposed. Ring-type optical laser-cavities have been utilized to clamp the gain [6–8]. In addition, gain flattening filters (GFFs), such as fiber Bragg grating (FBG) filters [9–11], dielectric filters, and twin core fiber filters [12] are used to equalize the amplifier gain spectrum. Also, all reflection mirrors (ARMs) or fiber reflection mirrors (FRMs) are used in double-pass structures to enhance the pump efficiency and the amplifier gain [13]. Efficient pumping schemes, such as using a 980nm and a 1480nm pump lasers bi-directionally [14] or employing pumping wavelength of 1540nm [15] and other optimal structures [16], have been presented to modify the gain and noise performances. Martin [17] introduced erbium transversal distribution influence. In consequence, Cheng *et al.* [18, 19] worked on EDFAs radial effects to earn the most appropriate characteristics.

However, all the aforementioned methods offer worthwhile designs, but they may not work in their best possible configurations, since, too many parameters affect the performance of the EDFAs and for the optimal designs, all of them must be optimized. Not unexpectedly, an optimization algorithm is quite helpful here, in order to find the best global solution. Beside traditional optimization methods, there are some evolutionary computational search algorithms, called global optimization methods (GOMs), that are practical tools to search global maximums and minimums. Genetic algorithm (GA) is one of the most popular GOMs that has been used in diverse fields, recently [22–24]. GA is established upon natural selection and biological behavior of genes [25]. In a nutshell, GA works by evaluating the chromosomes of a random selected population based on their fitness, selecting fittest members of the population, mating the qualified chromosomes to produce offspring, and finally applying mutation to avoid suboptimal solutions. Cheng *et al.* [18–21] applied this algorithm and offered an optimal design including radial effects and dual pumped structure. Wei *et al.* [26] implemented GA algorithm to optimize the multistage EDFAs.On the other hand, Zhang *et al.* [27], used the improved GA for gain flattening of an ultra wide-band EDFAs by means of stitched long-period fiber gratings (SLPFGs). Also, hybrid GA was used in designing ultra wide-band optical amplifiers [28]. Despite of the effectiveness of GA, it is quite time consuming and needs a large amount of computational works. Particle swarm optimization (PSO) algorithm is a new and effective method of optimization. This algorithm has been used in solving different engineering problems [29-32].

In this chapter, we propose the PSO and GA algorithm to optimize EDFAs [29]. First, a quick review of PSO and GA is presented. Then, a model of EDFA is given and the parameters of EDFA are optimized by PSO algorithm. In the following, the performance of PSO is compared with that of continuous GA, and it is shown that PSO surpasses GA in speed and simplicity.

1.2. Distributed Fiber Raman Amplifiers

Distributed fiber Raman amplifiers (DFRAs) are used in long-haul high-capacity wavelength-division-multiplexed (WDM) transmission systems. Some critical merits of multi-pump DFRAs, such as broad amplification spectra, low-noise figure (NF), small gain fluctuation, and optional gain band, recently attracted researchers' attention and ended the exclusivity of erbium-doped fiber amplifiers (EDFAs). To nail these amplifiers down, a comprehensive theoretical model that takes all the different aspects into account is needed [33–37]. This model must include nonlinear pump-to-pump and signal-to-signal interactions, amplified spontaneous emission (ASE), and Rayleigh backscattering. Since bidirectionally pumped DFRAs have lower optical signal-to-noise-ratio (OSNR) tilt, it is logical to model bidirectionally pumped DFRAs.

So we have to solve a two-point boundary value problem of ordinary differential equations and use the shooting algorithm, which is a common method for solving such problems. Here we need to find the best possible pump scheme that results in the flattest and widest gain spectrum. Numerous efforts have been made to find optimized multi-pumped configurations. In almost all investigations, the genetic algorithm (GA) has been used for optimization [28, 38, 39]. Although GA is the most popular searching algorithm, it requires a tremendous amount of computational processing to be actuated. PSO can easily reduce the computational time and often finds the global maximum or minimum more conveniently [29-32, 40-42]. Besides GA and PSO, some other optimization methods can be used for solving engineering optimization problems that might have some drawbacks. In this chapter PSO is discussed and new pumping schemes of DFRAs composed of six and ten backward (BW) and bidirectional (BiD) pumped DFRAs have been introduced and optimized by PSO.

1.3. Hybrid EDFA/FRA

Hybrid erbium-doped fiber amplifier/fiber Raman amplifiers (EDFA/FRAs) have been a beneficial and effective component for dense wavelength division multiplexing (DWDM) in modern optical communication systems and networks [43]. Merits of the hybrid EDFA/FRAs are a moderately low noise figure and a large bandwidth. In addition, it can be designed to maximize the fiber amplifier spans and minimize the nonlinearity impairments [44]. Based on the demands, various types of EDFAs and FRAs can be combined to build a hybrid EDFA/FRA [45]. Consequently, these types of fiber amplifiers have a wide range of freedom for optimum design.

An effective design should expand the bandwidth, decrease the noise figure, flatten the gain spectrum, and compensate the dispersion as much as possible [46-48]. In some practical designs additional modules, such as dispersion-compensating fibers and gain-flattening filters, are used to compensate the dispersion and flatten the gain spectrum, respectively. To avoid using gain-flattening filters, which raise the attenuation of the system and the overall cost, gain spectra of the EDFA and FRA can be added up in a hybrid EDFA/FRA to have a reasonably flat gain spectrum amplifier. A large number of variables affecting the amplifier performance need to be optimized to minimize spectral gain variation. In continue, we present and analyze two configurations of hybrid EDFA/FRA consisting of a single mode fiber (SMF), a dispersion-compensating fiber (DCF), a dispersion-shifted fiber (DSF) as a multi-wavelength pumped FRA with six or 10 backward pumps, and a C-band EDFA is optimized by PSO.

2. IMPLEMEMTATION OF PSO AND GA

PSO strategy is a global search method that can handle different optimization problems. Because of its simplicity and high capability for searching for the global maximum or minimum, PSO has been used in solving hard optimization problems with a high number of variables to be optimized. Development of PSO is fundamentally based on a hypothesis that social sharing of information, within particles of a society, brings about an evolutionary improvement.

In this method, an adequate number of individuals, which are called particles, are spread through a multidimensional search space. Each particle shows a single crossing point of all searching dimensions. We associate the change of particles' positions with iterations with movement and define velocities for each of the particles. Particles evaluate their positions with respect to a goal or fitness function at each iteration and according to neighborhood structures, they share their memories of "best positions," afterward, particles utilize the shared information to define new velocities and subsequently modify their positions. In other words, particles find the best solution according to their own previous experiences and also that of their neighbors. There are different neighborhood structures such as the star neighborhood structure for global best version and the ring neighborhood structure for local best version [40-42]. Each particle $S_i(k)$ is identified by three features, position $x_i(k)$, velocity $v_i(k)$, and fitness $F(x_i(k))$, where k is the number of current iteration. The steps of the global

PSO algorithm are presented in the following, assuming that the optimal answer is the global minimum [49]:

Step 1. Initialize the swarm $S(k)$. Distribute particles S_is randomly within the design space $x_i(0)=x_{min}+rand(x_{max}-x_{min})$, where x_{min} and x_{max} are the vectors of the lower and upper limits of the space, respectively. There is no certain rule for swarm size, just that it should not be too small or too large. Also, initial velocities, personal bests, and global bests are determined, normally to their worst possible values.

Step 2. Evaluate fitness function $F(x_i(k))$ for the initial swarm $S(k)$.

Step 3. If the current fitness value of each particle $S_i(k)$ is smaller than its personal best $Sbest_i$, set $Sbest_i=F(x_i(k))$ and $x_{Sbest\,i}=x_i(k)$.

Step 4. If the current minimum value of all particles is smaller than the global best $Gbest_i$, set $Gbest_i=F(x_i(k))_{min}$ and $x_{Gbest\,i}=x_i(k)$.

Step 5. Update the velocities. The effects of current movement, particle memory, and swarm memory are added together to form particle's velocity:

$$\upsilon_i(k)=\varphi\upsilon_i(k-1)+\rho_1 C_f(x_{Sbest_i}-x_i(k))+\rho_2 C_f(x_{Gbest_i}-x_i(k)) \quad (1)$$

$$\varphi=1-0.7\frac{1-k}{1-k_{\max}} \quad (2)$$

where φ is the inertia weight whose value will decease during the process because, at the beginning, its larger value facilitates a global search, but at the end, its smaller value leads the particles to be gathered in the global best; k is the number of the iteration; and C_f is the accelerator constant whose value is 2 in this paper. A reasonable value of C_f speeds up the optimization process. Finally, ρ_1 and ρ_2 are random values uniformly distributed over [0 1].

Step 6. Update the positions by adding the velocities to particles' previous positions after setting $k=k+1$. If new positions were out of the limits, particles should be fetched into the design space,

$$x_i(k)=x_i(k-1)+\upsilon_i(k) \quad (3)$$

Step 7. Check the stopping criterion to find out whether convergence takes place. If yes stop the process, and if not go to *step 2*.

Flow charts of PSO and GA are compared in figure 1. As shown, PSO and GA have some similarities. Both of them start with a population of random individuals and put some deterministic and probabilistic rules into practice to change the positions of individuals and to move them toward the best solution. On the other hand, PSO has parameters such as *Gbest* and *Sbest*, which act as its memories, where GA has no memory. In PSO, variations are based on learning from other members of society, but in GA they are based on genetic recombination [49].

To perform optimization, we consider the optimization algorithm as the main program and the amplifier's model as a subprogram. Each time that sub program is called for; our defined fitness function will be evaluated. The fitness function is defined based on our expectations to have a flat-gain spectrum and a broad bandwidth.

In this paper we define two different fitness functions. The first one, formulated in Eq. (4), is a summation of differences between input and output signal powers on all channels. And the second one, formulated in Eq. (5), is made of two terms: the first term is the maximum of gain values in decibels, and the second term is the summation of all signal gains multiplied by an arbitrary constant α.

$$F_{obj_1} = \sum_{k=1:n_s} \left| P_s^k(0) - P_s^k(L) \right| \tag{4}$$

$$F_{obj_2} = Max\left|10 \times \log\left(\frac{P_s^k(L)}{P_s^k(0)}\right)\right| + \alpha \sum_{k=1:n_s} \left|10 \times \log\left(\frac{P_s^k(L)}{P_s^k(0)}\right)\right| \tag{5}$$

where $P_s^k(0)$ and $P_s^k(L)$ are the kth input and output signal powers, respectively, and α is an arbitrary constant that, here has a value of 0.05. We utilize each of these fitness functions according to our needs.

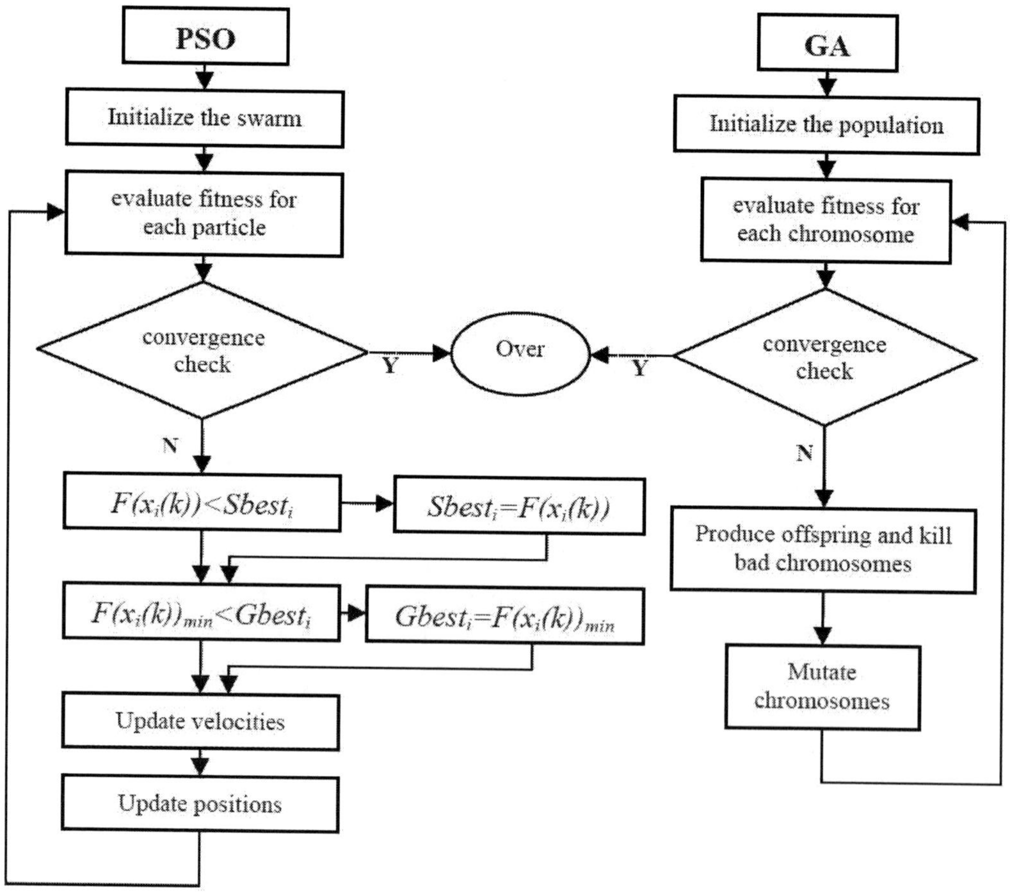

Figure 1. Comparison of basic flow charts of PSO and GA [30].

Our goal is to minimize F_{obj1} and F_{obj2}. If we use F_{obj1}, the overall power deviations from the input powers will be minimized. On the other hand, if we use F_{obj2}, we minimize both the gain variation and output power deviations. Note that we want the output powers to be the same as the input powers and that we prefer the gains to be zero. Therefore minimum ripples of output signal are obtained. The number of particles that form the stochastic initial swarm is chosen based on the dimension and expansion of the search space, where its typical value in different problems is nearly in the range from 20 to 50. During the iterations, after evaluating the fitness functions, a convergence test is done based on the error limits. Put another way, the number of iterations can be chosen in order to have a reasonable convergence of particles.

3. EDFA Equations and Simulations

Since optimization is a time consuming process, we need a fast and efficient model for EDFA to simulate its performance. Throughout this work, we adopt the Giles model [1] which is used in most cases of design works. Eliminating the effects of higher levels, a two-level, homogeneously broadened model for Er^{3+} ions are assumed, where the ground level is $^4I_{15/2}$ and the excited level is $^4I_{13/2}$. This two level model is enough to describe the propagation of optical power in the EDF, because, nearly all of the radiative transitions take place between these two levels. To work with the continuous spectrum of optical light, we divide the light into optical beams of frequency bandwidth Δv_k with optical power of P_k centered at frequency v_k and wavelength λ_k.So, we can consider the propagation equation as N ordinary differential equations, where N is the total number of pump and signal channels. In addition, the rate equation of upper-level $^4I_{13/2}$ population of Er^{3+} will be a summation of the emission and absorption contributions of all the light frequency components [19]. Emission and absorption cross sections of Al/P-silica fiber are utilized, which are adopted from Miniscalco investigation [50]. In addition, we choose fiber parameters, somehow, to have V number less than 2.405, so that we will have single mode fiber.

Also, we adopt the steady state condition $\partial n_2(r,z,t)/\partial t = 0$ and weakly guiding approximation $\Delta - (n_1 - n_2)/n_2 << 1$ to simplify the equations. Therefore, the equation that describes the optical propagation along z-axis is given as follows [19]:

$$\frac{dP_k(z)}{dz} = u_k \sigma_{ek} \int_0^a i_k(r) n_2(r,z,t)[P_k(z) + mhv_k \Delta v_k] 2\pi \, rdr - u_k \sigma_{ak} \int_0^a i_k(r) n_1(r,z,t) P_k(z) \, 2\pi \, rdr - u_k l_k P_k(z) \tag{6}$$

where $P_k(z)$is the optical power propagating in the core of EDF with frequency v_k, $u_k=+1$ and $u_k=-1$ demonstrate the forward and backward direction of pump schemes, σ_{ek} and σ_{ak} are the emission and absorption cross sections, a is the fiber radius, i_k is the normalized transverse mode intensity, n_1 and n_2 are the population of Er^{3+} ions in the ground level $^4I_{15/2}$ and metastable level $^4I_{13/2}$ states, respectively, $mhvk\Delta v_k$ specifies the amount of spontaneous emission that takes place from upper level population n_2 while Δv_k is the effective noise bandwidth, v_k is the noise power frequency, h is the Planck's constant, and here m is 2, because of the bidirectional ASE, while for pump and signals m is 0. Finally, l_k is the excess fiber loss. This equation is derived for a circularly symmetric fiber.

By solving the rate equations in the steady-state, metastable level population of Er^{3+} ions can be written as:

$$n_2(r,z) = n_t \frac{\Sigma_k (\tau\sigma_{ak} / h\nu_k) P_k(z) i_k(r)}{1+\Sigma_k (\tau(\sigma_{ak}+\sigma_{ek}) / h\nu_k) P_k(z) i_k(r)} \tag{7}$$

where τ is the metastable level lifetime. Note that saturation is involved in this formula, n_t is the erbium-doped concentration $n_t = n_1 + n_2$ and it has a uniform distribution over the core radius a. In addition, the formula of normalized transverse mode intensity considering the fundamental mode inside the core $r<a$ is given as [19]:

$$i_k(r) = \frac{1}{\pi a^2}\left[\frac{\upsilon_k}{V_k}\frac{J_0(u_k r/a)}{J_1(u_k)}\right]^2 \tag{8}$$

where $V_k = 2\pi a NA/\lambda_k$ is the V number at wavelength λ_k, $NA = \sqrt{n_1^2 - n_2^2}$ is the numerical aperture, and the approximate values for υ_k and u_k for $1 \le V_k \le 3$ are $\upsilon_k = 1.1428 V_k - 0.9960$ and $u_k = \sqrt{V_k^2 - \upsilon_k^2}$. $J_{0,1}$ are zeroth- and first-order Bessel functions. Also, the following normalization condition must be satisfied:

$$\int_0^a i_k(r) 2\pi r dr = 1 \tag{9}$$

Satisfying this normalization condition ensures us about propagation of total input power in the core radius. Inasmuch as the values of input pump and signal powers are known and we aim to find their output values, here, we have an initial value problem. In the cases of bidirectional pumping and considering back- and forward ASE, we have boundary value problem. Throughout this work, using some approximation, we solve pump, signal, and noise propagation equations as initial value problems, utilizing one-step predictor-corrector method (PCM) [36]. The EDF is divided into small segments and power at each segment is calculated using its own value at previous step and a prediction made at its current step. Similarly, the power is calculated step by step at each frequency. In order to solve Eq. (6) we rewrite it in the form below:

$$\frac{dP_k(z)}{dz} = P_k(z) F_k(z) + C_k(z) \tag{10}$$

where:

Table 1. Parameters required in numerical calculations [29]

Parameter	Value
Upper level life time (ms)	10
Input pump power (mW)	50
Step distance (m)	*L*/60
Cladding index	1.445
Fiber core radius (µm)	4.1
Excess fiber loss (dB/m)	0.03
Start frequency (THz)	186.80
End frequency (THz)	196.55
Number of channels	40
Frequency steps (THz)	0.25

$$F_k(z) = u_k[\sigma_{ek} + \sigma_{ak}]\int_0^a i_k(r) n_2(r,z,t) 2\pi\, r dr - u_k[\sigma_{ak} n_t + l_k] \tag{11}$$

$$C_k(z) = u_k \sigma_{ek} m h \nu_k \Delta\nu_k \int_0^a i_k(r) n_2(r,z,t) 2\pi\, r dr \tag{12}$$

Utilizing this notation, we can easily solve Eq. (6). Because the light amplification or attenuation has the exponential variations, we can accept exponential evolutions to simulate power propagations in fiber amplifiers [51]. Accordingly, Eq. (6) can be solved as:

$$P_k(z_{j+1}) = P_k(z_j)\exp\left((F_k(z_j) + C_k(z_j)/P_k(z_j))\Delta z\right) \tag{13}$$

where for pump and signal $C_k(z) = 0$, since in these cases $m = 0$.

Table 2. Ranges of parameters, chosen randomly, in the optimization algorithm [29]

Parameter	Range
Pumping wavelength, λ_p (nm)	1450-1500
Numerical aperture, NA	0.056287-0.16886
Erbium-doped radius, b (µm)	1.5-4.1
Input signal power, P_s (µW)	1-10
Er^{3+} concentration, n_t (m^{-3})	$(1\text{-}30)\times10^{24}$
Fiber length, L (m)	2-50

The pumping wavelength is about 1480 nm, which is adopted for third communication window and is speculated in two-level system. Pumping wavelength of 980nm is used when the amplifier is assumed to have a tree-level model.

Step index fiber and uniform distribution of Er^{3+} ions are used in our model to avoid complexity of using graded index fiber and considering radial effects [18]. In the numerical calculations, the data in Table 1 are used. As shown in Table 1, the system has 40 signal channels at frequencies 186.80 to 196.55 THz spaced 250 GHz/channel. Also, ASE noises are calculated at these channels. One forward pump is chosen to pump the amplifier while more than one pump in both directions can be included. Core radius a = 4.1 μm, to conform with current fiber data of Lucent technology products [19]. The range of the Er-doped radius b is smaller than that of fiber core.

Solving Eqs. (10)–(13), EDFA characteristics are known. Fitness function is defined by using the amplifier characteristics as following equation:

$$F = \Delta + \alpha G_c - \beta F_f \tag{14}$$

where Δ is 3 dB bandwidth while the middle hollow of the gain spectrum is neglected. G_c is the central signal gain at wavelength 1550 nm that must be maximized. G_c can be defined at other wavelengths, when we want to work in other bands like S- and L-band. F_f is the difference between the maximum and the average values of the signal gains in the bandwidth, which is included to have flatter gain spectrum. α and β (nm/dB) are the weighting factors. We define fitness function in some way that includes bandwidth, gain, and flatness. Based on the situation, weighting factors are evaluated. Here we assume α=β=1. Now we consider the fitness function F as sub-program for the main PSO program. Main program is written according to the algorithm described in the previous section. First, the main program spreads the points or the particles over the parameter ranges. Afterwards a sub-program provides fitness values by use of Eq. (14). Then the main program amends particle positions in each cycle to reach better fitness values. Therefore, throughout this evolutionary process, the global maximum will be found. Continuous PSO is used in this investigation and values of parameters, involved in optimization process, are chosen randomly over the ranges in Table 2. These reasonable ranges are chosen in some way to include likely optimum values according to the existing results in this context.

4. Raman Equations and Simulations

Propagation of the average optical power of pumps, signals, ASEs, and Rayleigh backscattering components are described by the following coupled ordinary differential equations in the steady state [34–36]:

$$\pm\frac{dP_i^{\pm}}{dz} = -\alpha_i P_i^{\pm} + \gamma_i P_i^{\mp} + P_i^{\pm}\sum_{\nu_j>\nu_i}\frac{g_{ji}}{\Gamma A_{eff}}\left[P_j^{+}+P_j^{-}\right]$$
$$-P_i^{\pm}\sum_{\nu_j<\nu_i}\frac{\nu_i}{\nu_j}\frac{g_{ij}}{\Gamma A_{eff}}\left[P_j^{+}+P_j^{-}\right] - 2h\nu_i P_i^{\pm}\sum_{\nu_j<\nu_i}\frac{\nu_i}{\nu_j}\frac{g_{ij}}{\Gamma A_{eff}}\left[1+\left(e^{\frac{h(\nu_i-\nu_j)}{KT}}-1\right)^{-1}\right]\Delta\nu$$
$$+2h\nu_i\sum_{\nu_j>\nu_i}\frac{g_{ji}}{\Gamma A_{eff}}\left[P_j^{+}+P_j^{-}\right]\left[1+\left(e^{\frac{h(\nu_j-\nu_i)}{KT}}-1\right)^{-1}\right]\Delta\nu \tag{15}$$

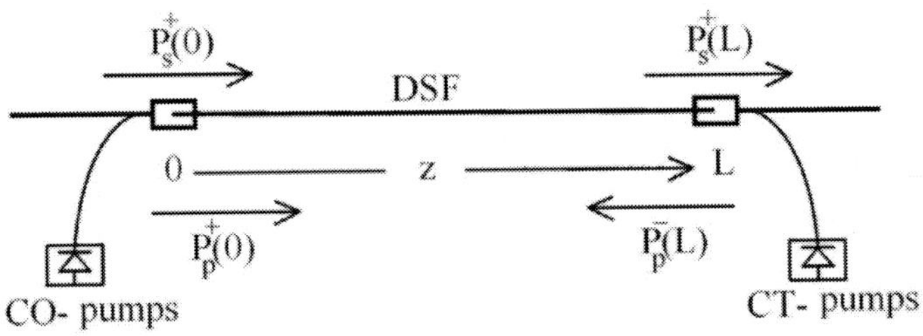

Figure 2. Propagating diagram of signal powers, co-propagating (CO) and counter-propagating (CT) pump powers [34].

where $P_{v_i}^{+}$ and $P_{v_i}^{-}$ are the average optical powers at frequency v_i, propagating in the +z and −z directions, respectively, in a frequency slot Δv around v_i. Also, α_i, γ_i, A_{eff}, h, K, and T are the attenuation coefficient, Rayleigh backscattering coefficient, effective area of optical fiber, Planck's constant, Boltzmann constant and temperature, respectively; Γ is a factor that shows polarization randomization effects with a value in the range from 1 to 2, and g_{ji} is the Raman gain parameter at frequency v_i due to the optical wave at frequency v_j. Summations should be calculated over the frequency ranges of $v_j > v_i$ or $v_j < v_i$ so that all the optical power components are taken into account, that is from v_0 to v_n with intervals of Δv, while Δv is the noise bandwidth. Note that the second and last terms of Eq. (15) are related to the Rayleigh backscattering and the ASE components, respectively.

Considering co-propagating and counter-propagating pumps, a two-point boundary value problem must be solved. If the amplifier has only forward pumps, the problem is an initial value problem, which is simpler to solve. In figure 2 the propagation of signal and pump powers in the forward and backward directions is depicted. In contrast to pump powers, signal powers propagate only in the forward direction. The values of signal and forward pump powers at $z=0$ and backward pump powers at $z=L$ are our first design assumptions.

In the computer simulation we try to find the values of optical powers at any other point in the middle of the fiber amplifier. In this paper, we have used the predictor–corrector method (PCM) based on the Adams–Bashforth–Moulton formula [36] to solve the equations, and the shooting algorithm is used to estimate the signal and forward pump powers at $z=L$ [37], which is the starting point of the power calculation. Using the initial counter-propagating pump powers and the estimated values of the signal and co-propagating pump powers at $z=L$, we start to calculate the optical powers. The first cycle of the calculation is completed when we arrive at $z=0$. Then the calculated values of the signal and co-propagating pump powers are compared with their initial values at $z=0$. If the desired accuracy is not achieved, the second calculation cycle will be started again at $z=L$. The shooting algorithm modifies the estimated values at each cycle until the convergence test is done. The convergence criterion in this work for calculated powers is 1% deviation the initial powers. That means for initial powers of 100 μW, the calculated power should be in the range from 99 to 101 μW.

To perform the four-step PCM, first the fiber amplifier length is divided into pieces, which are called steps. Then the four-step predictor method based on the Adams– Bashforth formula is used and the optical powers of the frequency beams are calculated at each step according to the data from the four previous steps. Also, different one-, two-, and three-step formulas are used to calculate the powers at the first, second, and third steps, respectively.

After that, a corrector method based on the Adams– Moulton formula is used to modify the predicted powers [36]. This corrector method is used to achieve a higher accuracy. The Raman gain efficiency and attenuation spectrum of the dispersion-shifted fiber (DSF) used as the amplification media of the DFRA in the computer model of this paper are shown in figures 4(b) and 2(a), respectively [45].

Table 3. Parameters of distributed fiber Raman amplifier used for optimization by PSO [30]

Parameters	Values
Rayleigh backscattering, γ (m^{-1})	7×10^{-8}
Temperature, T (°K)	300
Number of signal channels	57
Signal channels wavelength range (nm)	1524.5-1616.6
Signal channels frequency range (THz)	185.45-196.65
Signal channels spacing (THz)	0.2
Gain bandwidth, $\Delta\lambda$ (nm)	92.1
Input signal power per channel (μW)	100
Number of pumps	6 or 10
Pumps wavelength range (nm)	1417.5-1510.3
Pumps frequency range (THz)	198.5-211.5
Fiber length, L (km)	60
Number of steps	23

The Raman gain efficiency is formulated as $g_{ij}/(\Gamma A_{eff})$ in Eq. (15). Although the g_{ij} values are almost the same for different pump wavelengths, the A_{eff} of the DSF will increase at higher wavelengths [35]. The peak values of the Raman gain efficiency used in this paper are 0.7612 and 0.6206 l/W/km for pump wavelengths of 1420 and 1510 nm, respectively [35]. An almost linear reduction of the Raman gain efficiency peak value is assumed when the pump wavelength increases. Also, the fiber attenuation spectrum in figure 4(a) is used. We use MatLab programming to model the DFRA with six and ten co-propagating and counter-propagating pumps. The required times to run these models depend on the number of pumps, the pump powers, and the direction of the pumps. Moreover, the number of iterations needed to satisfy the convergence test is different for each case. Typical iteration numbers are from 2 to 20, and the required times for running programs with these numbers of iterations are 0.464 to 2.820 s, respectively. An AMD Athlon 2.71 GHz CPU with 1.00 Gbytes of RAM has been used to run these programs. Other parameters of our DFRA model are listed in Table 3.

5. Hybrid EDFA/FRA Simulations

Two configurations are proposed in figure 3 for hybrid EDFA/FRAs. The first module in this figure is a 60km standard SMF as the transmission media, the second module is a 10.5km DCF to compensate the dispersion of the light wave signal caused by the SMF, the third module is a 0.2 dB attenuation isolator (ISO) to avoid the propagation of the backward FRA pump and amplified spontaneous emission (ASE) powers in the DCF, the fourth module is a

60km DSF as a multi-wavelength backward-pumped FRA, the next module is a WDM coupler to couple the Raman pumps into the FRA, and the last segment is a C-band EDFA. Characteristics of the fiber modules are given in Table 4. The multi-wavelength FRA in figure 3 has six or 10 backward pumps.

The last segment is the EDFA because its noise is higher than the other parts, especially higher than the FRA. Therefore, at this position the overall noise figure is reduced. The overall span length of our proposed hybrid EDFA/FRA is 130.5km. The length of the DCF is chosen to compensate the dispersion caused by the SMF. Hence, total dispersion of the span is zero, because the FRA is assumed to be a DSF whose dispersion is zero. Moreover, the EDFA is short in comparison with the other parts and its dispersion is negligible.

To perform optimization, a computer programming model for a hybrid EDFA/FRA is needed, which should be written based on equations of EDFA [1, 29] and FRA [34]. The Giles model [1] is used in this work to model the EDFA and the two-level homogeneously broadened model is assumed for Er^{3+} ions. The continuous spectrum of optical light is considered as N discrete optical beams and the ordinary differential equations of EDFA will be written as N separate equations when N is the total number of pumps and signals. The model is studied under steady-state conditions and only the fundamental mode is propagated in all types of fiber.

The problem in the case of solving the differential equations of the EDFA is an initial value problem and the two-step predictor corrector method [29] is used to solve these equations and to simulate the EDFA. Again, in the case of the FRA, the discrete average optical powers of the pumps and signals are studied. Raman amplification equations are coupled ordinary differential equations. Considering the propagation of signals in the co-propagating direction and backward pumps in the counter-propagating direction, a two-point boundary value problem should be solved. In this work, a four-step predictor corrector method is used to evaluate the optical powers along the fiber [36] and a shooting algorithm is used to come up with two-point boundary value problem [37]. MatLab programming is used to simulate the hybrid EDFA/FRA, and the required time to accomplish one simulation is 2 to 3 s, based on the number of FRA pumps and the value of FRA pump powers. An AMD Athlon 2.71 GHz CPU with 1.00 GB of RAM has been used in this simulation.

Table 4. Characteristics of hybrid EDFA/FRA fiber modules [31]

Module	Fiber type	Attenuation (dB/km)	Dispersion (ps/(nm.km))	Length (km)
Transmission	SMF	Figure 4(a) [33]	16.5 [45]	60
Compensating	DCF	Figure 4(a) [33]	-95 [33]	10.5
FRA	DSF	Figure 4(a) [45]	0	60
EDFA	SMF	Figure 4(a) [33]	16.5	Variable (1 to 15 m)

The attenuation spectrum of the SMF [33], the DCF [33], and the DSF [45] used in the simulation is depicted in figure 4(a). Also, Raman gain efficiency [45] and the erbium absorption and emission cross section spectra of Al/P-silica fiber [50] are shown in figures 4(b) and 4(c), respectively. Other parameters used in modeling the EDFA and the FRA are listed in Table 5.

Table 5. EDFA and RFA parameters used in modeling [31]

FRA parameters	Value	EDFA parameters	Value
Rayleigh backscattering, γ (m^{-1})	7×10^{-8}	Upper level life time (ms)	10
Temperature, T (°K)	300	Cladding index	1.445
Number of pumps	6 or 10	Fiber core radius (μm)	4.1
Pumps direction	Backward	Excess fiber loss (dB/m)	0.03

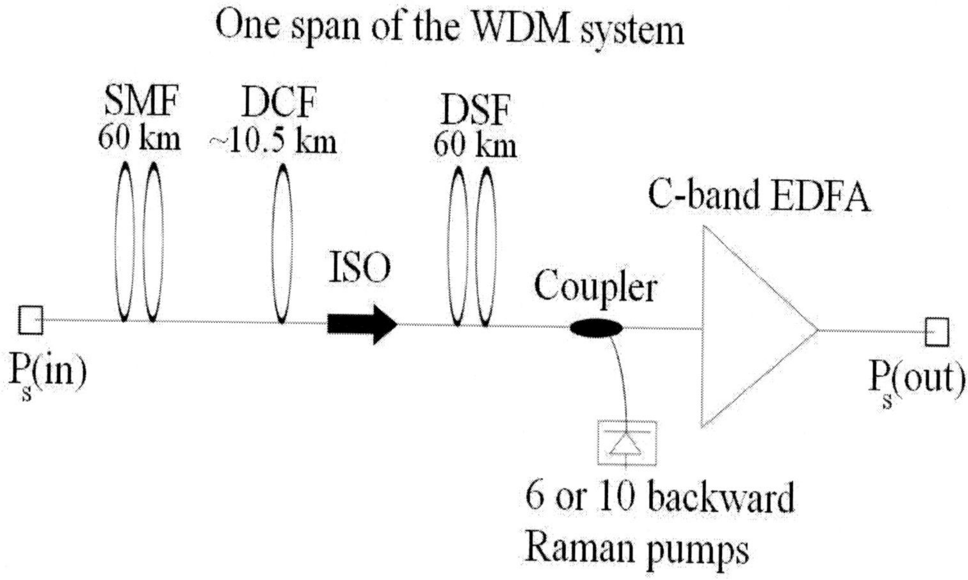

Figure 3. Configurations of hybrid EDFA/FRA with a six- or 10-backward-pumped FRA [31].

Table 6. Characteristics of the WDM transmission system [31]

WDM characteristics	**Value**
Number of signal channels	60
Signal wavelength range (nm)	1529.2-1627.1
Signal channels start frequency (THz)	184.25
Signal channels end frequency (THz)	196.05
Signal channels frequency space (THz)	0.2
Raman pumps wavelength range (nm)	1425-1515

In this paper, EDFA works in the C-band that is 1530–1570nm and its gain spectrum decreases at wavelengths of over 1570nm. The FRA works effectively in L and higher bands, that is, 1570–1630 nm. Then, adding these amplification bands in the hybrid EDFA/FRA there is a wide band of amplification in which the WDM signal channels are contained; that is 1530–1630nm.

The pumps of the FRA are spread in the wavelength range of 1425–1515nm at the equal frequency spaces, depending on the number of pumps. Characteristics of the WDM transmission system that we have used in the optimization process are listed in Table 6.

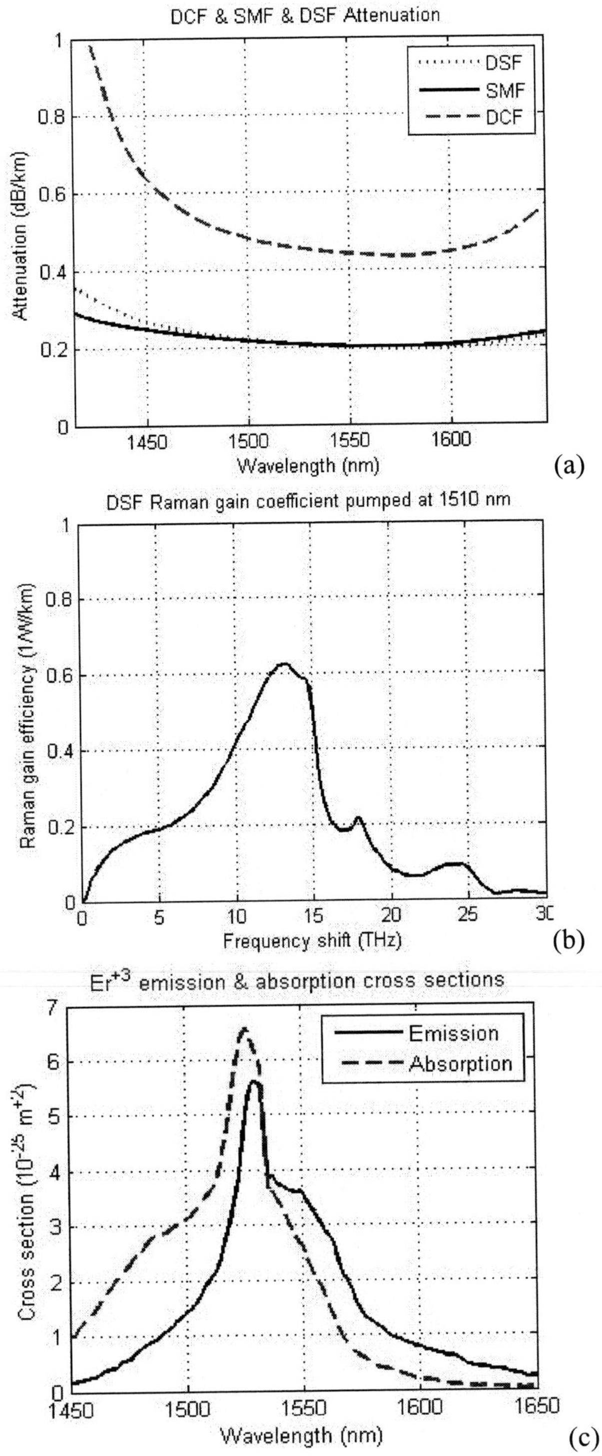

Figure 4. Characteristics of the fibers. (a) DCF, SMF, and DSF attenuation spectra, (b) Raman gain efficiency of DSF, and (c) emission and absorption cross sections of Er^{3+} doped in Al/P-silica SMF [31].

6. RESULTS OF EDFA OPTIMIZATION

Table 7 contains the results of the particle swarm optimization of EDFA for different pumping powers. In the modeling computer program, pumping power is a constant value. When the pumping power increases, the central signal gain will increase consequently and there is almost a linear relation between pumping power and the central signal gain. Optimal bandwidth will be nearly the same for different pumping powers with a little drop when pumping power increases. Optimized signal gain spectrum and its associated NF spectrum are shown in figures. 5(a) for 50mW pumping power.

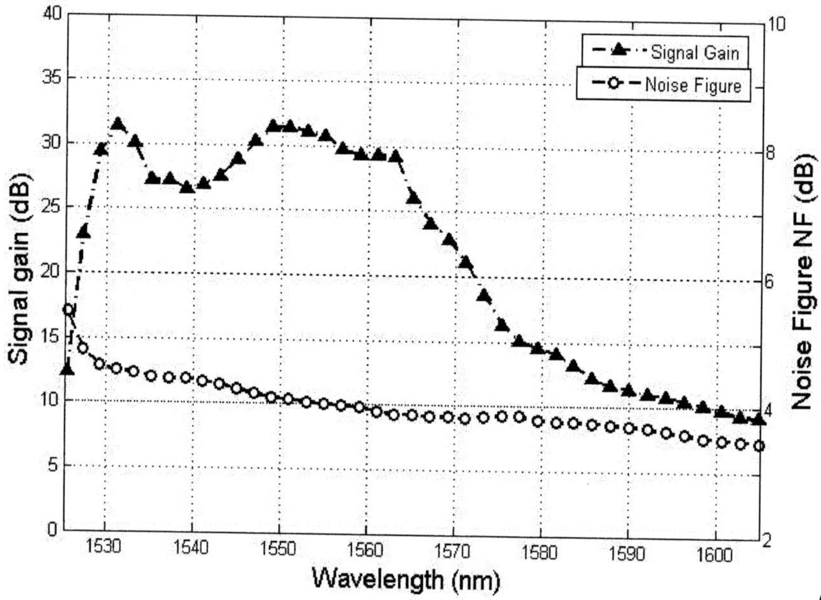

(a)

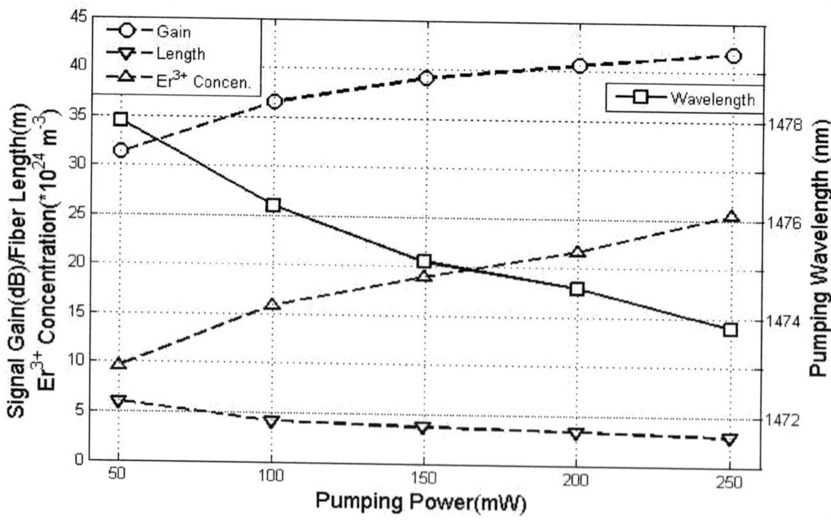

(b)

Figure 5. (a) Optimized signal gain spectrum and noise figure for 50mW pumping power and (b) Values of central signal gains, fiber lengths, Er^{3+} concentrations, and pumping wavelengths vs. pumping powers [29].

Also, optimal central signal gains, fiber lengths, Er^{3+} concentrations, and pumping wavelengths are shown in figure 5(b) vs. five tested pumping powers. As it is obvious, central signal gains and Er^{3+} concentrations must increase with pumping power while fiber lengths and pumping wavelengths must decrease. If we took Er^{3+} concentration as a constant value, fiber length would increase when the pumping power increases [20, 21].

Optimal input signal power converges to minimum value over its range that results in the highest possible gain spectrum whereas optimal Er-doped radius approaches its maximum value over its range, that is the fiber core radius, to use all of the pumping power. In the case of b = a, where a is one of the variables of optimization, optimized a will be too small, less than 1 μm, which the fiber can easily be damaged by high power pumps and also when a is small, NA will attain a high value that causes higher dispersion distortions. Since the value of a is constant NA does not have a great influence on the results. It is natural to have nuances between results obtained by different global searching methods when the number of parameters is high, but these slight differences are negligible.

The information we need to run the optimization program are given in Table 8. While in genetic algorithm, we need a large number of initial chromosomes to start the algorithm, it is not necessary in PSO to have a large number of particles. The state of being dynamic in PSO is higher than that of GA and each of the particles in PSO are able to find their ways based on their memories of themselves and their neighbors. Both GA and PSO are evolutionary population-based search methods that use a combination of rules to move from a set of points to other ones which are most likely the better positions. In contrast to GA, PSO has memory and is able to utilize the past events. This history of the best solutions is an important key in modifying the positions of particles.

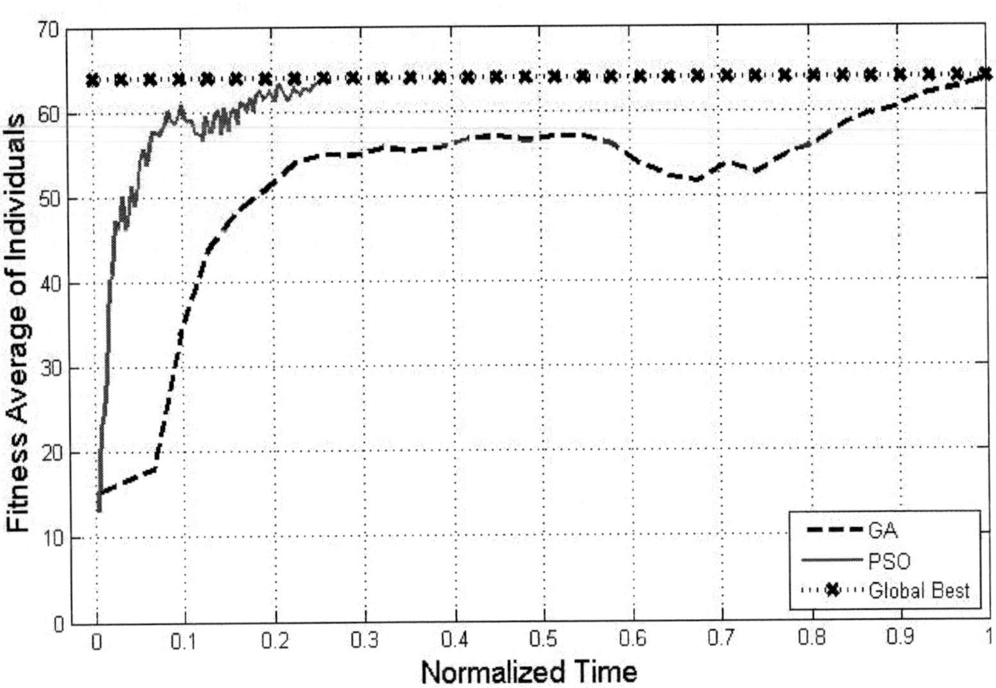

Figure 6. Convergence time comparison of PSO and GA [29].

Table 7. Results of the particle swarm optimization of the EDFA for different pumping powers [29]

Parameters	Values				
Pumping powers, P_p (mW)	50	100	150	200	250
Optimal pumping wavelengths, λ_p (nm)	1477.9	1476.2	1475.1	1474.6	1473.8
Optimal numerical apertures, NA	0.1630	0.1138	0.1165	0.1225	0.1596
Optimal erbium-doped radii, b (μm)	4.1	4.1	4.1	4.1	4.1
Optimal input signal powers, P_s (μW)	1.0	1.0	1.0	1.0	1.0
Optimal fiber lengths, L (m)	5.995	4.187	3.740	3.400	3.005
Optimal erbium Con. n_t ($\times 10^{24}$ m^{-3})	9.569	15.874	18.978	21.688	25.495
Central signal gains, G_s (dB)	31.34	36.55	39.01	40.61	41.87
Bandwidths, -3dB $\Delta\lambda$ (nm)	34.69	34.30	34.14	34.05	33.95

Table 8. Required information to run PSO program [29]

Data	Value
Number of particles	40
Number of iterations	70
Acceleration constant	2
Inertia weight, φ	0.4
Initial velocity	0
Initial personal and global best	0

Convergence times of PSO and GA algorithms are compared in figure 6. In this figure, the averages of fitness values pertaining to particles of PSO and chromosomes of GA are plotted versus normalized time and it can be seen that required time of convergence in PSO is much less than that of GA. Continuous GA with 80 chromosomes is used whereas its mutation and crossover rates are 0.5and 0.7, respectively. Choosing inertia weight and acceleration constant in PSO are of great importance. They must be picked up somehow that let PSO be both converged and able to find the global best. If we assume φ higher than 0.4, the algorithm may diverge and if smaller than 0.4 it may converge too fast that the optimization process falls into local maximum. Also, since the used random function has a uniform distribution, the best value for acceleration constant is 2.

There is evidence that when λ_p increases to the values over 1500 nm, amplifier 3 dB bandwidth will increase dramatically while this is accompanied with rises in the values of n_t and L. That is because when we increase the pumping wavelength, the gain spectrum also will shift to the higher wavelengths and at the expense of loosing gain, the bandwidth in the L-band will increase.

7. Results of FRA Optimization

The pumps may be placed equidistantly in the pump frequency range, but if we implement the optimization on the pump frequencies, they will no longer be at the same frequency spaces. In this paper we optimize the DFRA, considering optimization both with and without pump frequencies. Moreover, since the peak of the Raman gain spectrum occurs at about 13.2 THz lower than the pump frequency, all the pump frequencies should be higher than the signal frequencies, i.e., within a 13.2 THz frequency range above the highest signal frequency.

In the case of *Fobj1*, our goal is to find the optimized pump powers for both BW and BiD configurations, and we do not take pump frequencies into account as optimization variables. The only type of BiD structure that is included in this work has two FW pumps, and the other pumps are BW. The optimized output results are shown in figure 7 for 60 km (a) BiD and (b) BW amplifiers. In addition, the optimized pump powers and their wavelengths are given in Tables 9 and 10 for six- and ten-pumped amplifiers, respectively. All the pumps are BW except the pumps that are appended by FW in the tables. Values of *Fobj1* are also given in these tables in microwatts.

As shown in figure 7, BiD pumps result in relatively flatter output powers. Although the gain spectrum fluctuation has a faster variation versus the wavelength for ten pumped amplifiers, the peak values of these changes are lower in comparison with six-pumped amplifiers. Gain variation in the six-BiD-pumped amplifier is 1.21 dB, which is better than that of the six-BW-pumped amplifier, which is 1.25 dB. Also, ten-pumped amplifier gain variations are lower than six-pumped gain variations. In the case of 10-BiD and ten-BW-pumped amplifiers, obtained fitness values by *Fobj1* are 166 and 170 μW, respectively, and obtained gain variations are 0.79 and 0.65 dB, respectively.

Although the fitness value is lower for a BiD-pumped amplifier, as we like it to be, the gain variation is higher for a BiD-pumped amplifier. That is because *Fobj1* does not take gain variation into account. So we need to define another fitness function. A smaller *Fobj1* means the output powers are nearer to the initial signal powers, which is what we desire in designing DFRAs, but the gain variations are still high. Therefore we define the fitness function *Fobj2*, which results in lower gain variations.

In the case of *Fobj2*, we consider the maximum gain variation and the overall gain deviations from zero in decibels. Frequencies of pumps are also included in addition to pump powers, and signal gains of channels are shown in Figure 8 for the case of *Fobj2*. Tables 11 and 12 contain the optimization results when *Fobj2* is the considered fitness function. Considering pump frequencies in optimization, the gain variation is 0.80 dB in a six-BW-pumped amplifier, which is better than 1.09 dB when we ignore frequency to optimize this amplifier. Also in a ten-BW-pumped amplifier the gain variations improve from 0.62 to 0.49 dB when we consider pump frequencies as optimization variables. The results of the *Fobj2* case, when we ignore pump frequencies from optimization, are not shown in Tables 11 and 12, but the results are shown in figure 8.

The results in Tables 9–12 are derived by PSO, which helped us to find the global minimum very fast. To write the PSO program, we have assumed 40 particles, each containing 6 or 10 elements, for six- or ten-pumped amplifiers, respectively. The number of elements will be twice that when we consider pump frequencies in addition to pump powers that are 12 and 20 elements for sixand ten-pumped amplifiers, respectively. In the program, instead of inserting a convergence check, we have adjusted the number of iterations to achieve the appropriate precision.

Three to five hundred iterations are enough to reach the optimum answer, given that particles with more elements need higher iteration numbers than particles with fewer elements. For each particles element, the pump power is chosen randomly in the range from 5 to 300 mW or the pump frequency is chosen randomly in the range from 197.8 to 212.0 THz. In the case of *Fobj2*, the gain variations are better than in the case of *Fobj1*.

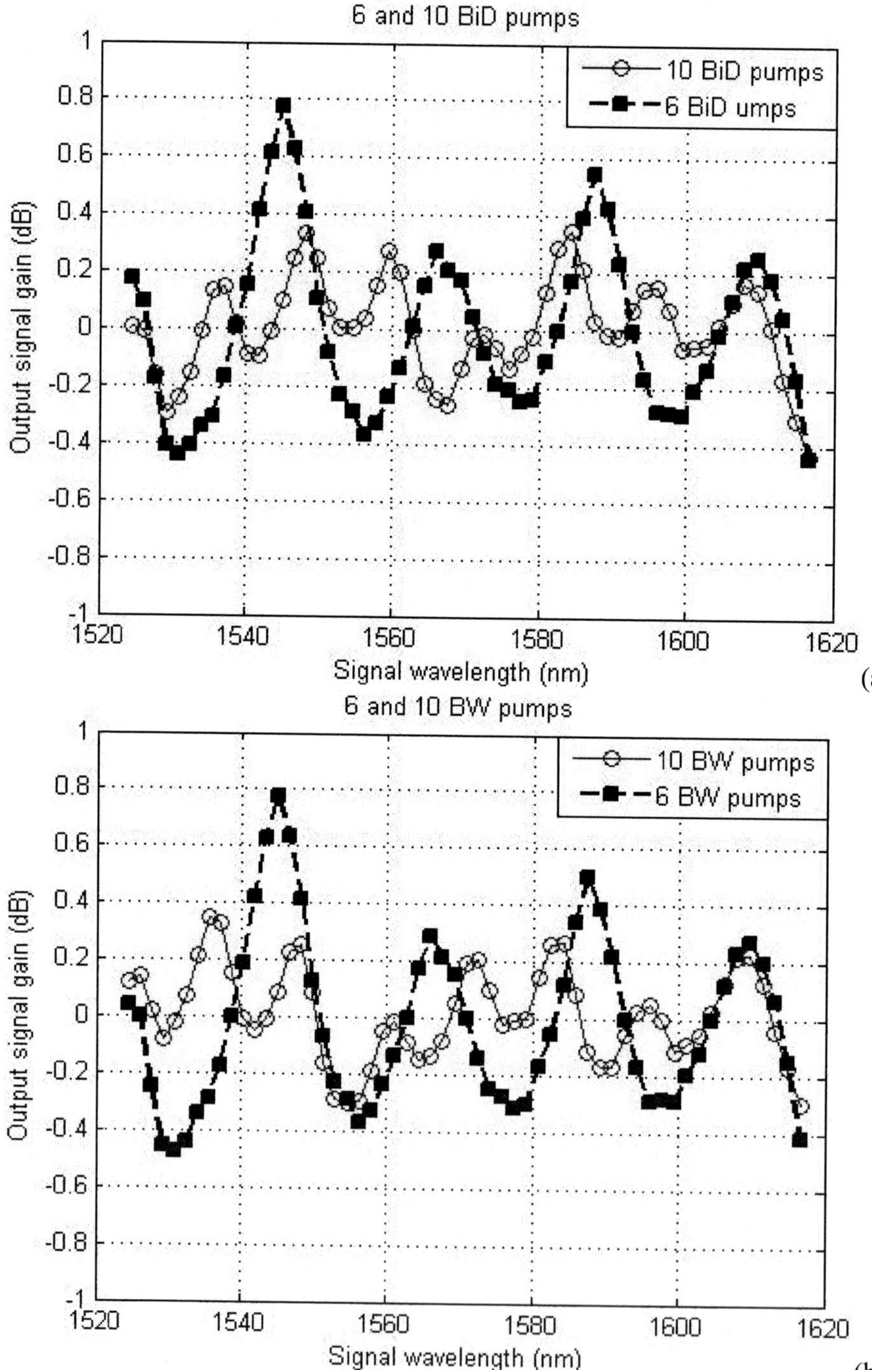

Figure 7. Output signal powers of DFRA for (a) six- and ten-BiD-pumped and (b) six- and ten-BW-pumped 60 km fiber amplifiers [30].

Table 9. Optimized pump powers for 60 km six-BW-pumped and six-BiD-pumped DFRAs considering *Fobj1* [30]

Parameters	Values						Fitness
Pump number	1	2	3	4	5	6	(μW)
Wavelengths (nm)	1417.5	1435.1	1453.2	1471.8	1490.8	1510.3	-
Optimized BW powers (mW)	185.3	126.4	60.7	53.4	15.5	33.2	324.24
Optimized BiD powers (mW)	179.8	117.2	66.1 (FW)	57.7	32.7 (FW)	39.3	320.25

Table 10. Optimized pump powers for 60 km ten-BW-pumped and ten-BiD-Pumped DFRAs considering *Fobj1* [30]

Parameters	Values										Fitness
Pump number	1	2	3	4	5	6	7	8	9	10	(μW)
Wavelengths (nm)	1417.5	1427.2	1437.1	1447.1	1457.3	1467.6	1478.1	1488.7	1499.4	1510.3	-
Optimized BW Powers (mW)	129.7	99.2	78.7	28.5	25.2	42.1	22.3	5.1	11.4	29.6	170.47
Optimized BiD Powers (mW)	129.0	80.7	51.0	49.0	26.7 (FW)	37.5	38.6 (FW)	5.3	40.2 (FW)	32.2	166.43

Table 11. Optimized pump powers and frequencies for 60 km six-BW-pumped DFRAs considering *Fobj2* [30]

Parameters	Values						Fitness
Pump number	1	2	3	4	5	6	(dB)
Optimized wavelengths (nm)	1420.6	1435.1	1450.6	1469.6	1496.7	1514.9	0.8783
Optimized powers (mW)	186.6	101.4	58.8	61.0	41.8	15.5	

ASE due to spontaneous Raman scattering is the most important source of noise in DFRAs. Another source of noise is double Rayleigh backscattering (DRBS). The initial FW and BW ASE powers are so small at the input of the amplifier, and we assume them to be −100 dBm per channel. The propagations of FW and BW noises for a ten- BiD-pumped amplifier, containing ASE and DRBS, are shown in figure 9. The amount of ASE noises at the output of the amplifier is not very dependent on the tiny initial values that we choose until the initial values are very small.

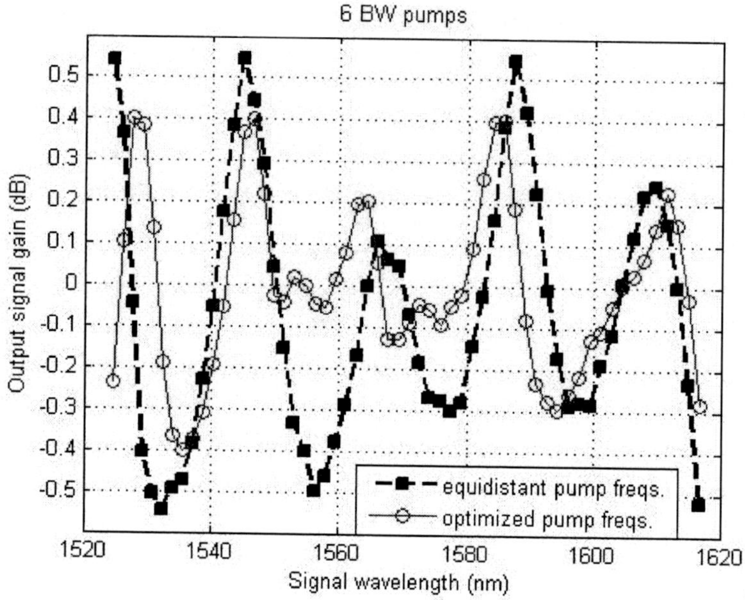

(a)

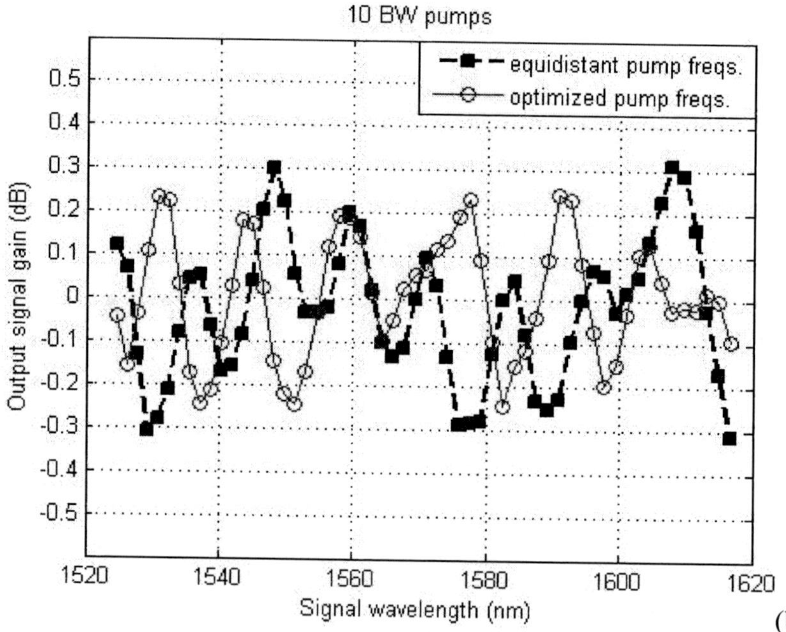

(b)

Figure 8. Signal gains of DFRA for (a) six BW pumps and (b) ten BW pumps, considering and ignoring frequencies in optimization [30].

The results in Tables 9–12 are derived by PSO, which helped us to find the global minimum very fast. To write the PSO program, we have assumed 40 particles, each containing 6 or 10 elements, for six- or ten-pumped amplifiers, respectively. The number of elements will be twice that when we consider pump frequencies in addition to pump powers that are 12 and 20 elements for sixand ten-pumped amplifiers, respectively. In the program, instead of inserting a convergence check, we have adjusted the number of iterations to achieve the appropriate precision. Three to five hundred iterations are enough to reach the optimum answer, given that particles with more elements need higher iteration numbers than particles with fewer elements. For each particles element, the pump power is chosen randomly in the range from 5 to 300 mW or the pump frequency is chosen randomly in the range from 197.8 to 212.0 THz. In the case of *Fobj2*, the gain variations are better than in the case of *Fobj1*.

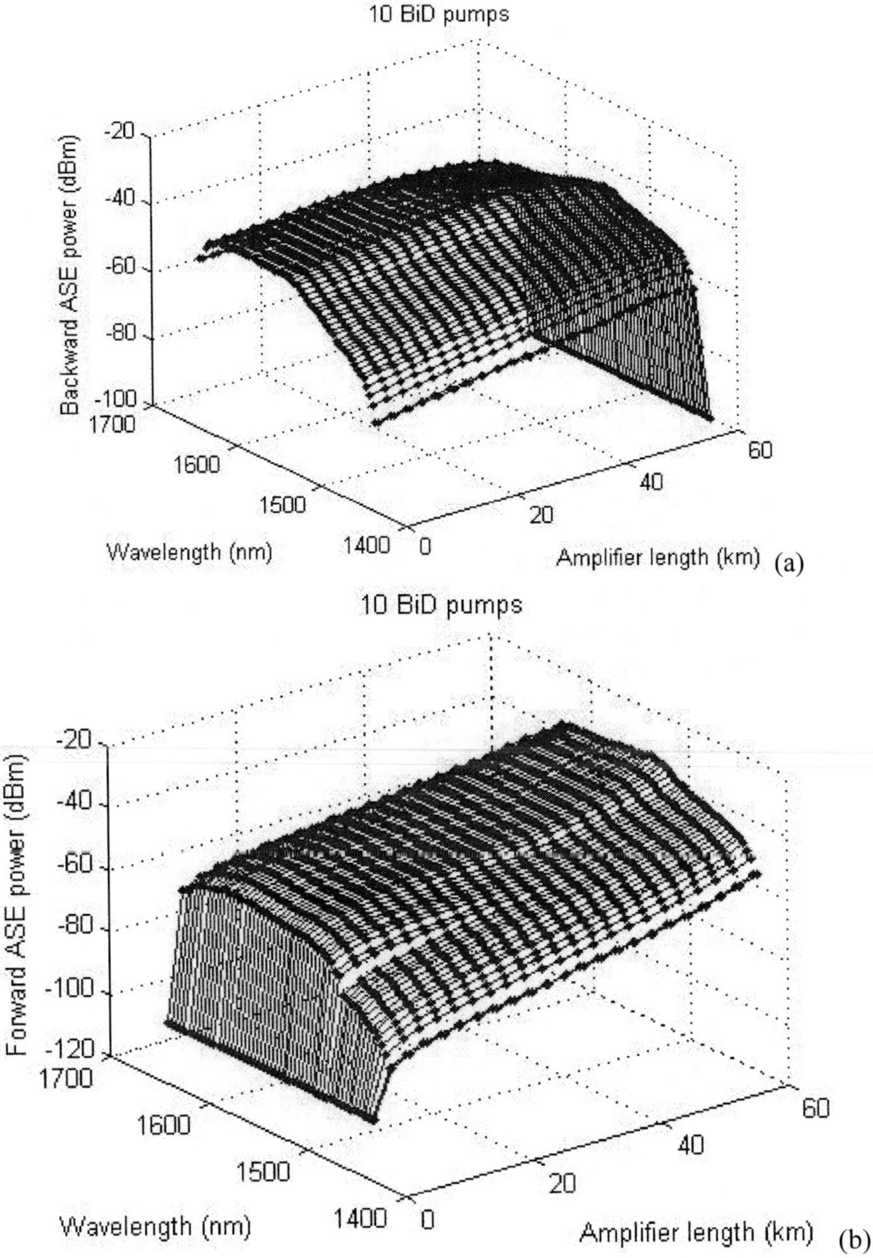

Figure 9. ASE and DRBS noises for a 60 km BiD amplifier: (a) backward noise powers and (b) forward noise powers propagating through the amplifier [30].

ASE due to spontaneous Raman scattering is the most important source of noise in DFRAs. Another source of noise is double Rayleigh backscattering (DRBS). The initial FW and BW ASE powers are so small at the input of the amplifier, and we assume them to be −100 dBm per channel.

The propagations of FW and BW noises for a ten- BiD-pumped amplifier, containing ASE and DRBS, are shown in figure 9. The amount of ASE noises at the output of the amplifier is not very dependent on the tiny initial values that we choose until the initial values are very small.

Table 12. Optimized pump powers and frequencies for 60 km ten-BW-pumped DFRAs considering *Fobj2* [30]

Parameters	Values										Fitness
Pump number	1	2	3	4	5	6	7	8	9	10	(dB)
Optimized wavelength (nm)	1414.3	1423.1	1433.3	1447.6	1457.8	1462.9	1474.8	1484.5	1494.6	1512.0	0.5804
Optimized power (mW)	136.9	119.1	78.9	39.5	6.9	39.1	32.9	7.9	6.0	27.1	

Since ASE and DRBS do not have a considerable effect on the propagation of signal powers, we can neglect both of them in the optimization process in order to promote the speed of optimization. A comparison between PSO and GA speeds of convergence, toward the minimum value, is shown in figure 10.

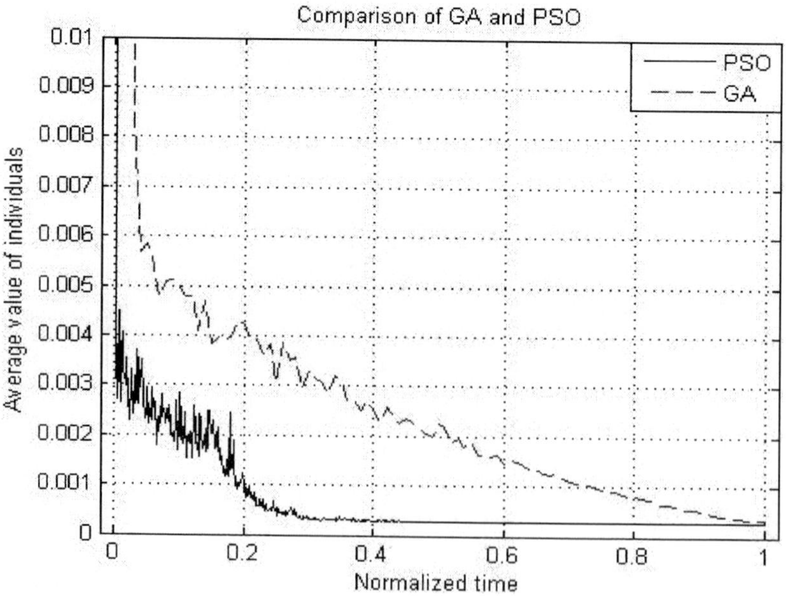

Figure 10. Comparison of GA and PSO speeds of convergence to the global solutions by plotting the fitness mean values versus the normalized time [30].

Table 13. Variables of the optimization and their ranges [31]

Variables to be optimized	Range
EDFA pumping wavelength, λp (nm)	1450-1500
EDFA Er^{3+} concentration, n_t (m^{-3})	$(1\text{-}30)\times10^{24}$
EDFA fiber length, L (m)	1-50
EDFA pump power (mW)	100-500
FRA pump powers (mW)	5-350

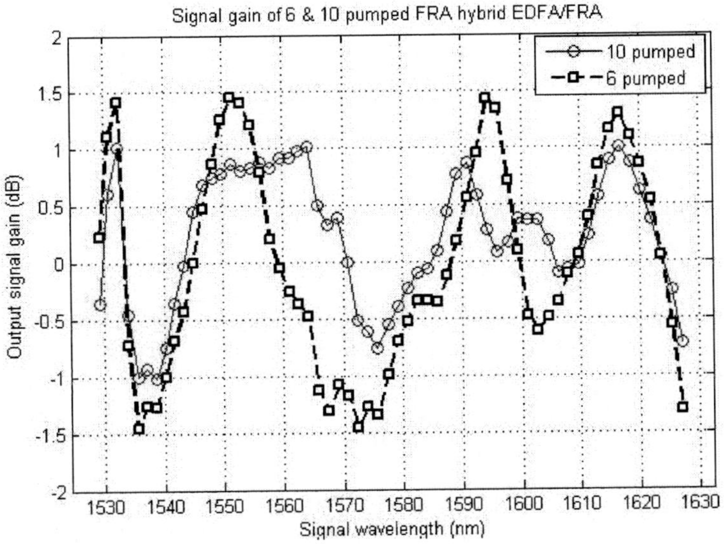

Figure 11. Optimum signal gain spectra of hybrid EDFA/FRAs with six- and 10-pumped FRAs [31].

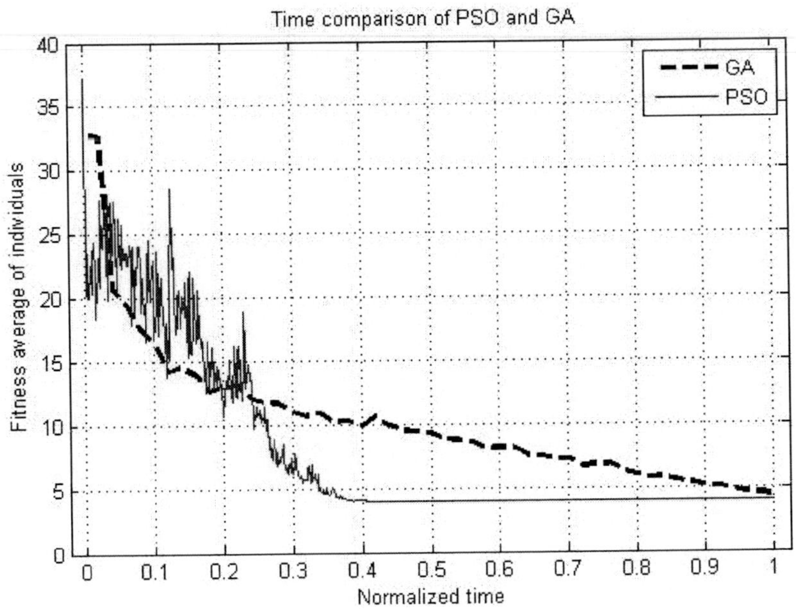

Figure 12. Comparison of the time consumption of PSO and the GA [31].

In this diagram the fitness mean values of the particles and chromosomes are shown versus nor- malized time for PSO and GA, respectively, and the evolutionary process of optimization is shown. It is shown that PSO can be up to three times, faster, in comparison with GA. The time required to run a PSO program depends on several factors such as the amplifier pumping scheme and the numbers of particles and iterations in the optimization program. For example, the times to optimize six-BW-pumped and ten-BiD-pumped amplifiers are 85 and 425 min, respectively, by an AMD Athlon 2.71 GHz CPU with 1.00 Gbytes of RAM.

8. Results of Hybrid EDFA/FRA Optimization

8.1. Optimization Ignoring FRA Pump Wavelengths

Optimized variables are obtained after running the PSO program. As was mentioned in Section 3, from the numerous parameters affecting the hybrid EDFA/FRA characteristics, just some of them are involved in the optimization process; they are listed in Table 13. In addition to the variables in Table 13, some other variables can be included in the optimization process that have been ignored. The optimum signal gain spectra of hybrid EDFA/FRAs with sixand 10-pumped FRAs are shown in figure 11 versus wavelength. In the case of a 10-Raman-pumped hybrid EDFE/FRA, the gain spectrum variation is lower than the case of the six-pumped hybrid EDFA/FRA. A large number of FRA pumps allow the overall gain spectrum of the hybrid amplifier to be more accurately tailored. This hybrid amplifier is a distributed amplifier and we prefer the network gain to be 0 dB; that means the same amount of initial signal power will be transferred to the end of the amplifier.

Table 14. Values of the parameters and characteristics of hybrid EDFA/FRA with 6- or 10-backward-pumped FRA optimized by PSO [31]

Variables	Values for 10 pumps FRA HA	Values for 6 pumps FRA HA
EDFA pump wavelength, λ_p (nm)	1479.8	1456.4
EDFA Er^{3+} concentration, n_t (m^{-3})	9.991×10^{24}	9.988×10^{24}
EDFA fiber length, L (m)	1.127	1.185
EDFA pump power (mW)	482.1	468.2
FRA pump power, P_1 (mW)	147.0 (at 1425.0 nm)	69.6 (at 1425.0 nm)
FRA pump power, P_2 (mW)	185.4 (at 1434.5 nm)	254.0 (at 1442.1 nm)
FRA pump power, P_3 (mW)	69.8 (at 1444.1 nm)	33.2 (at 1459.7 nm)
FRA pump power, P_4 (mW)	86.1 (at 1453.8 nm)	213.5 (at 1477.7 nm)
FRA pump power, P_5 (mW)	19.4 (at 1463.6 nm)	48.7 (at 1496.1 nm)
FRA pump power, P_6 (mW)	124.1 (at 1473.6 nm)	122.7 (at 1515.0 nm)
FRA pump power, P_7 (mW)	82.3 (at 1483.8 nm)	-----
FRA pump power, P_8 (mW)	6.7 (at 1494.0 nm)	-----
FRA pump power, P_9 (mW)	6.3 (at 1504.4 nm)	-----
FRA pump power, P_{10} (mW)	71.9 (at 1515.0 nm)	-----
Fitness (dB)	1.0168	1.4542

There are 60 signal channels distributed in the bandwidth of 97.9nm in the wavelength range from 1529.2 to 1627.1 nm. Gain spectrum variation of a hybrid EDFA/FRA with six- and 10-pumped FRAs are 2.9074 and 2.0336 dB, respectively. When we add up the number of Raman pumps, the Raman gain spectrum will be more capable of compensating the EDFA gain spectrum variation and it will decrease. The type of hybrid EDFA/FRA should be chosen based on the demands of gain spectrum flatness and affordable costs. The values of parameters and characteristics of a hybrid EDFA/FRA with six and 10-backward-pumped FRAs optimized by PSO are presented in Table 14. Fitness values are also listed in this table.

In this optimization stage, the frequencies of the FRA pumps are predefined with the same frequency spaces, and the frequencies of the FRA pumps can be obtained by the PSO method. Here we have ignored pump frequencies as variables of the optimization to avoid a large number of variables. When the number of variables gets very large, the optimization may not find the global optimization.

Considering the fact that the number of variables is quite large, PSO acted very well in comparison with the other evolutionary algorithms, such as GA, which is also popular. Figure 12 shows a comparison of the time consumptions of PSO and GA. The PSO can be converged to the global optimized values almost three times faster than GA. The reason is that less computational work is needed for PSO in comparison with GA. The time required to run the optimization programs for 6- and 10-pump FRA hybrid EDFA/FRAs are 6:20 and 6:45 h, respectively, using an AMD Athlon 2.71 GHz CPU with 1.00 GB of RAM and MatLab programming.

8.2. Optimization Including FRA Pump Wavelengths

In this paper we have simulated 4 different configurations for hybrid EDFA/FRA. Two configurations for 108.5 km and two for 174 km link with 6- and 10-pumped-FRA. Also, we survey the results for both input signal powers of 1 or 10 mW. A group of optimization variables or the group of simulation inputs which are introduced in Table 15 are known as a particle Pi in the PSO. We have n=40 particles in our optimization process and it means that we have 40 groups of optimization variables which are randomly spread in our optimization space. Each one of the optimization variables makes a dimension of the optimization space. Particle positions xi are arrays which are composed of the optimization variable values. We also have n=40 particle positions xi in our optimization process that need to be optimized.

At each iteration, we insert the inputs which are the optimization variable values or the particle positions to the main hybrid EDFA/FRA simulation computer program or fitness function. Then the simulation will result in the output or the fitness value F(xi) which is the maximum signal network gain deviation from zero. We consider the deviation from zero to take into account both the negative and positive gain values. According to the PSO steps which are introduced in the previous section, new values for optimization variables are selected at the end of each iteration. We have 24 and 16 optimization variables for 10- and 6-pumped FRA hybrid amplifiers. So, the particle positions are composed of 24 or 16 elements.

To reach the flattest gain spectrum for the hybrid EDFA/FRA, it is better to set the EDFA pump wavelength to have a C-band EDFA. Because of the shape of the EDFA gain spectrum, it is preferred to have amplification at lower wavelengths by EDFA and at higher wavelengths by FRA. In the optimization process, the EDFA pump power is inclined to attain high values.

So, it is better to choose 1480 nm pumping wavelength for the EDFA, although it has higher noise compared to 980 nm ones.

Also, in high power schemes it is more efficient to use 1480 nm pumping wavelength. The radius of the Er^{3+} doped area at EDFA is equal to the radius of the fiber used as EDFA that is 6.5 μm. An EDFA with a small core radius will result in better gain spectrum characteristics, but its fabrication is hard and it is more susceptible to damage in facing high EDFA pump power. Also a small core radius EDFA is rare in the market. Er^{3+} ion concentration has to be chosen in a confined area, since low Er^{3+} concentration would elongate the EDFA and consequently create higher variation of the gain spectrum. Also, high Er^{3+} concentration will increase unfavorable intra- ion interactions [52]. The main source of noise of EDFAs is ASE. The level of noise in FRAs is lower than that in EDFAs. In this paper, we have assumed an acceptably low level for noise, and to speed up the optimization process, we have not simulated noises. Lower input signal power in EDFAs will result in higher and flatter gain spectra, but input signal power should not be so low that it could not be retrieved from the noise.

The designed amplifier is a distributed amplifier, and it is better to define net gain for it. We prefer the signal powers at the end of the link to be the same as their values at the beginning of the link. So, the ideal net gain of our hybrid EDFA/FRA is zero. We have assumed the EDFA to be polarization independent as it almost is, but we consider the polarization dependency of FRA by including a factor in our simulations. Amplification in the FRA is due to the nonlinear phenomenon of stimulated Raman scattering (SRS). By using a high number of pumps in FRAs with different wavelengths and powers, the gain spectrum is tailored and the EDFA gain variation is compensated. The amplification band can be chosen simply by adjusting the pump wavelengths at appropriate distances.

The noise transition from pumps to signals occurs less in the counter-propagating pump scheme, hence it is more advantageous to place the pumps at the end of the FRA and increase the length of link span. We have to appoint the input signal powers to somehow avoid harmful interference with noise due to the low level of the input signal power, and unfavorable nonlinear effects due to the high level of the input signal powers.

All the optimization results, obtained by PSO, for different schemes and configurations are given in Tables 16-19. It is obvious that gain spectrum variations will be lower if we choose 10-pumped configurations instead of 6-pumped ones, or if we choose 108.5 km links instead of 174 km ones, and also if we choose 1mW of input signal power instead of 10 mW. As listed in Table 16, gain spectrum variations of 1.392 and 1.043 dB are attained for 108.5 km, 6- and 10-pumped-FRA hybrid EDFA/FRA, respectively with 1mW input signal powers.

Table 15. Variables of the optimization of the EDFA/FRAs and their ranges [32]

Variables to be optimized	Range
EDFA pumping wavelength, λ_p (nm)	1450-1500
EDFA Er^{3+} concentration n_t (m^{-3})	$(1\text{-}30)\times10^{24}$
EDFA fiber length L (m)	1-10
EDFA pump power (mW)	100-800
FRA pump wavelengths (nm)	1420-1520
FRA pump powers (mW)	5-600

Also, as listed in Table 17, gain spectrum variations of 1.864 and 1.794 dB are attained for 174 km, 6- and 10-pumped-FRA hybrid EDFA/FRA, respectively with 1mW input signal powers.

Using the results listed in the Tables 16 and 17, optimized signal gain spectra of 108.5 and 174 km 10-pumped-FRA hybrid EDFA/FRA are shown in Figure 13(a), 108.5 and 174 km 6-pumped-FRA hybrid EDFA/FRA are depicted in Figure 13(b), 6- and 10-pumped-FRA 108.5 km hybrid EDFA/FRA is pictured in Figure 14(a), and 6- and 10-pumped-FRA 174 km hybrid EDFA/FRA is pictured in figure 14(b), all for 1 mW input signal powers. These figures present good visual comparison of the optimized gain spectra of different schemes.

It is common to select the FRA pump wavelengths in an equidistant form, but here we search the FRA pump wavelengths through an optimization process which will result in better outcomes for gain spectrum variations

Table 16. Values of the optimized parameters of 6- and 10-pumped-FRA, 108.5 km hybrid EDFA/FRA with 1mW of input signal power, obtained by PSO [32]

Variables	Values for 10 pumps FRA HA	Values for 6 pumps FRA HA
EDFA pump wavelength, λ_p (nm)	1478.9	1460.4
EDFA Er^{3+} concentration, n_t (m^{-3})	7.2×10^{24}	8.4×10^{24}
EDFA fiber length, L (m)	1.139	1.247
EDFA pump power (mW)	338.5	283.6
FRA pump power, P_1 and wavelength, λ_1	114.2 mW and 1420.9 nm	102.4 mW and 1430.3 nm
FRA pump power, P_2 and wavelength, λ_2	261.2 mW and 1431.5 nm	350.5 mW and 1435.8 nm
FRA pump power, P_3 and wavelength, λ_3	177.3 mW and 1440.5 nm	18.6 mW and 1467.8 nm
FRA pump power, P_4 and wavelength, λ_4	15.9 mW and 1458.2 nm	117.9 mW and 1476.6 nm
FRA pump power, P_5 and wavelength, λ_5	64.3 mW and 1468.1 nm	94.0 mW and 1487.5 nm
FRA pump power, P_6 and wavelength, λ_6	23.9 mW and 1476.8 nm	74.2 mW and 1520.9 nm
FRA pump power, P_7 and wavelength, λ_7	71.4 mW and 1481.1 nm	-----
FRA pump power, P_8 and wavelength, λ_8	24.8 mW and 1489.5 nm	-----
FRA pump power, P_9 and wavelength, λ_9	13.4 mW and 1500.9 nm	-----
FRA pump power, P_{10} and wavelength, λ_{10}	43.6 mW and 1519.3 nm	-----
Fitness (dB)	0.521	0.696
Gain spectrum variations (dB)	1.043	1.392

Table 17. Values of the optimized parameters of 6- and 10-pumped-FRA, 174 km hybrid EDFA/FRA with 1mW of input signal power, obtained by PSO [32]

Variables	Value for 10 pumps FRA HA	Value for 6 pumps FRA HA
EDFA pump wavelength, λ_p (nm)	1460.4	1460.7
EDFA Er^{3+} concentration, n_t (m^{-3})	10.1×10^{24}	14.7×10^{24}
EDFA fiber length, L (m)	1.017	1.012
EDFA pump power (mW)	799.9	265.1
FRA pump power, P_1 and wavelength λ_1	435.6 mW and 1428.7 nm	342.2 mW and 1431.4 nm
FRA pump power, P_2 and wavelength λ_2	307.2 mW and 1438.9 nm	434.9 mW and 1435.6 nm
FRA pump power, P_3 and wavelength λ_3	97.5 mW and 1442.5 nm	115.5 mW and 1465.5 nm
FRA pump power, P_4 and wavelength λ_4	10.7 mW and 1450.0 nm	159.5 mW and 1478.6 nm
FRA pump power, P_5 and wavelength λ_5	7.9 mW and 1466.1 nm	50.2 mW and 1487.7 nm
FRA pump power, P_6 and wavelength λ_6	146.9 mW and 1469.0 nm	30.6 mW and 1518.1 nm
FRA pump power, P_7 and wavelength λ_7	123.0 mW and 1482.6 nm	-----
FRA pump power, P_8 and wavelength λ_8	16.4 mW and 1490.8 nm	-----
FRA pump power, P_9 and wavelength λ_9	5.8 mW and 1501.9 nm	-----
FRA pump power, P_{10} and wavelength λ_{10}	24.1 mW and 1519.3 nm	-----
Fitness (dB)	0.897	0.932
Gain spectrum variations (dB)	1.794	1.864

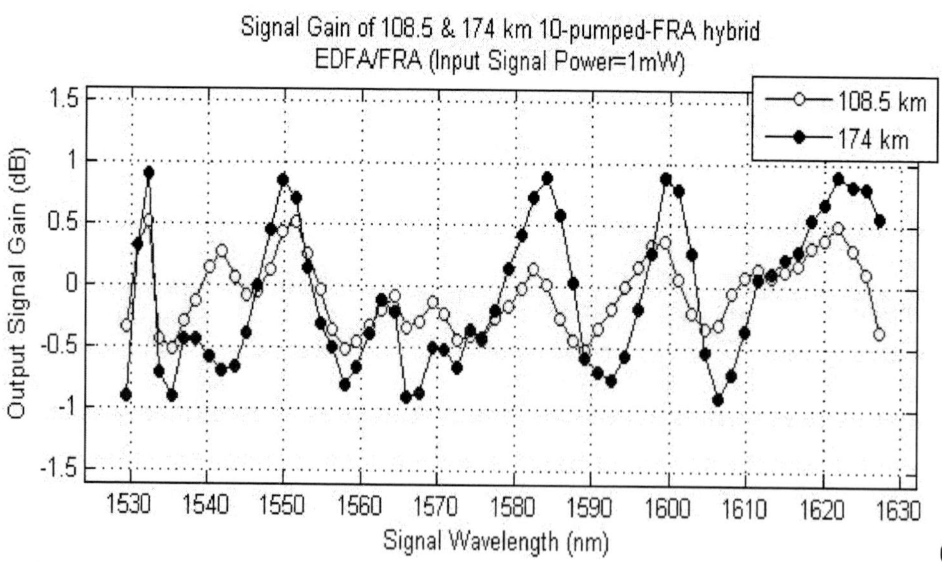

(a)

Figure 13. (Continued).

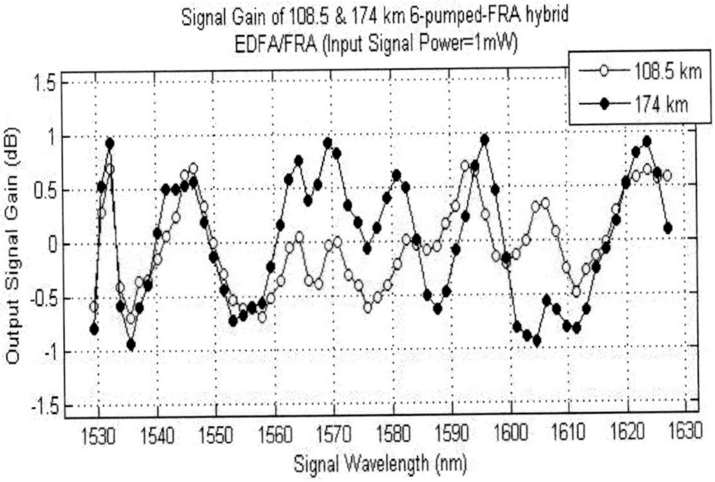

(b)

Figure 13. Optimum signal gain spectra of (a) 108.5 and 174 km 10-pumped-FRA hybrid EDFA/FRA and (b) 108.5 and 174 km 6-pumped-FRA hybrid EDFA/FRA (both for input signal power of 1mW) [32].

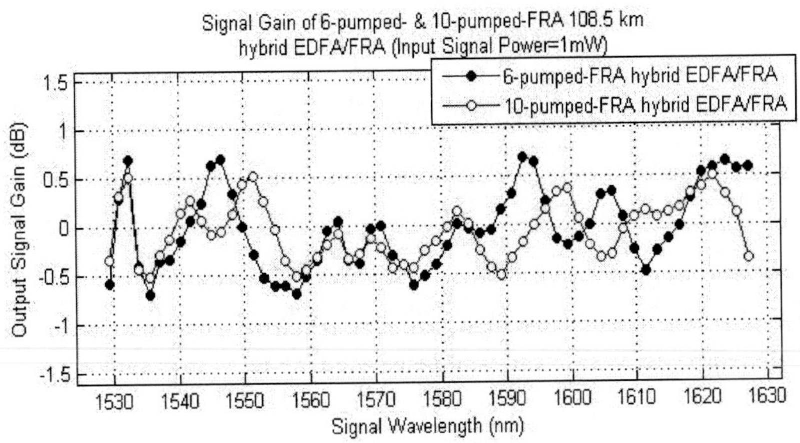

(a)

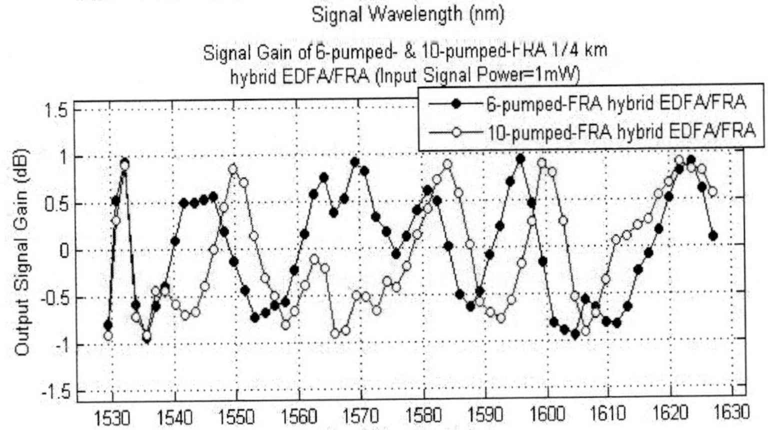

(b)

Figure 14. Optimum signal gain spectra of (a) 6- and10-pumped- FRA 108.5 km hybrid EDFA/FRA and (b) 6- and10-pumped-FRA 174 km hybrid EDFA/FRA (both for input signal power of 1mW) [32].

Table 18. Values of the optimized parameters of 6- and 10-pumped-FRA, 108.5 km hybrid EDFA/FRA with 10mW of input signal power, obtained by PSO [32]

Variables	Values for 10 pumps FRA HA	Values for 6 pumps FRA HA
EDFA pump wavelength, λ_p (nm)	1475.8	1478.9
EDFA Er^{3+} concentration, n_t (m^{-3})	14.1×10^{24}	18.8×10^{24}
EDFA fiber length, L (m)	1.021	1.007
EDFA pump power (mW)	799.7	796.7
FRA pump power, P_1 and wavelength, λ_1	141.4 mW and 1422.9 nm	329.9 mW and 1422.6 nm
FRA pump power, P_2 and wavelength, λ_2	390.0 mW and 1430.0 nm	342.5 mW and 1434.8 nm
FRA pump power, P_3 and wavelength, λ_3	205.3 mW and 1439.5 nm	6.1 mW and 1451.0 nm
FRA pump power, P_4 and wavelength, λ_4	10.1 mW and 1449.3 nm	98.4 mW and 1474.9 nm
FRA pump power, P_5 and wavelength, λ_5	9.3 mW and 1459.5 nm	51.7 mW and 1487.8 nm
FRA pump power, P_6 and wavelength, λ_6	12.4 mW and 1469.5 nm	43.3 mW and 1515.6 nm
FRA pump power, P_7 and wavelength, λ_7	104.9 mW and 1478.9 nm	-----
FRA pump power, P_8 and wavelength, λ_8	33.1 mW and 1489.1 nm	-----
FRA pump power, P_9 and wavelength, λ_9	11.5 mW and 1509.1 nm	-----
FRA pump power, P_{10} and wavelength, λ_{10}	21.2 mW and 1519.2 nm	-----
Fitness (dB)	0.691	0.863
Gain spectrum variations (dB)	1.382	1.717

For example, gain spectrum variations of 2.91 dB for a 130.5 km, 6-pumped-FRA hybrid EDFA/FRA, equidistant FRA pump wavelengths [31] has become 1.864 dB for 174 km, 6-pumped-FRA hybrid EDFA/FRA, by optimization of FRA pump wavelengths and also gain spectrum variations of 2.03 dB for a 130.5 km, 10-pumped-FRA hybrid EDFA/ FRA, by equidistant FRA pump wavelengths [31] has become 1.794 dB for 174 km, 10-pumped-FRA hybrid EDFA/FRA, by optimization of FRA pump wavelengths, which are notable improvements.

For most cases, the wavelength spectral distance between the FRA pump wavelengths, presented in Tables 16-19, are longer than 8 nm. At the closest point it will reach to 2.3 nm.

By using narrow spectral width laser diodes at the places where the pump wavelengths are near the adjacent pump wavelengths, the pump powers will not interfere with each other and sound operation is guaranteed. Typical spectral width of a laser diode is about 0.1-10 nm. Figure 15 shows a comparison between the convergence speeds of PSO and GA optimization processes which are carried on the optimization of one of the configurations of our WDM systems. It is apparent that PSO acts faster and more efficiently in searching for the optimum values for a system under test.

Table 19. Values of the optimized parameters of 6- and 10-pumped-FRA, 174 km hybrid EDFA/FRA with 10mW of input signal power, obtained by PSO [32]

Variables	Values for 10 pumps FRA HA	Values for 6 pumps FRA HA
EDFA pump wavelength, λ_p (nm)	1476.6	1476.2
EDFA Er^{3+} concentration, n_t (m^{-3})	17.8×10^{24}	8.3×10^{24}
EDFA fiber length, L (m)	1.217	2.788
EDFA pump power (mW)	799.9	798.9
FRA pump power, P_1 and wavelength, λ_1	560.7 mW and 1428.1 nm	559.0 mW and 1428.1 nm
FRA pump power, P_2 and wavelength, λ_2	303.3 mW and 1432.1 nm	468.5 mW and 1435.5 nm
FRA pump power, P_3 and wavelength, λ_3	173.8 mW and 1439.5 nm	26.5 mW and 1467.7 nm
FRA pump power, P_4 and wavelength, λ_4	7.4 mW and 1449.2 nm	118.1 mW and 1475.6 nm
FRA pump power, P_5 and wavelength, λ_5	65.4 mW and 1468.3 nm	54.4 mW and 1487.3 nm
FRA pump power, P_6 and wavelength, λ_6	104.5 mW and 1478.5 nm	14.9 mW and 1516.8 nm
FRA pump power, P_7 and wavelength, λ_7	7.2 mW and 1487.8 nm	-----
FRA pump power, P_8 and wavelength, λ_8	15.0 mW and 1489.1 nm	-----
FRA pump power, P_9 and wavelength, λ_9	8.5 mW and 1508.2 nm	-----
FRA pump power, P_{10} and wavelength, λ_{10}	10.4 mW and 1516.7 nm	-----
Fitness (dB)	1.011	1.064
Gain spectrum variations (dB)	2.021	2.127

CONCLUSION

In this chapter, we have introduced EDFAs, FRAs, and hybrid EDFA/FRAs which are three kinds of the most important optical fiber amplifiers used in long-haul optical communication systems. The equation systems of these amplifiers have been introduced and used for modeling. Different simulation methods have been introduced and the simulations of the different configurations have been carried on.

The optimization process has been performed on the amplifier models and optimum parameters have been obtained. PSO method is used as the optimization device and its efficiency is compared with the genetic algorithm. It has been shown that PSO will converge faster toward the global best answers. Particle swarm optimization (PSO) allows us to include a large number of parameters in the optimization process, so that we can simultaneously optimize as much as 24 parameters. Using the efficient PSO algorithm, pump wavelengths of FRA are optimized in addition to the FRA pump powers, EDFA pump power and wavelength, Er^{3+} concentration, EDFA length, input signal power, fiber numerical aperture, and erbium-doped area of fiber cross section. This results in obtaining the flattened-gain amplifiers like hybrid amplifiers with gain spectrum variations of as low as 1.043 dB for a 108.5 km and 1.794 dB for a 174 km, 10-pumped-FRA hybrid EDFA/FRA, respectively.

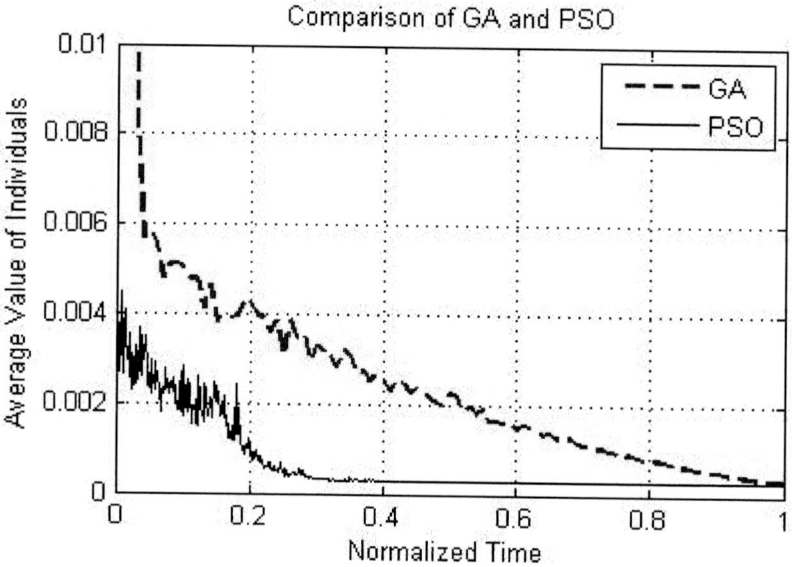

Figure 15. Comparison of the convergence speeds of the PSO and GA toward the optimum values versus the normalized executing time [32].

REFERENCES

[1] A. M. Saleh, R. M. Jopson, J. D. Evankow, and J. Aspell, "Modeling of gain in erbium-doped fiber amplifiers," *IEEE Photon. Technol. Lett.* , Vol. 2, No. 10, pp. 714-717, 1990.

[2] Carena, V. Curri, and P. Poggiolini, "On the optimization of hybrid Raman/erbium-doped fiber amplifiers," *IEEE Photon. Technol. Lett.* Vol. 13, No. 11, pp. 1170-1172, 2001.

[3] Mowla and N. Granpayeh, "A novel design approach for erbium-doped fiber amplifiers by particle swarm optimization," *Progress in Electromagnetic Research, PIER M*, Vol. 3, pp. 103–118, 2008.

[4] Mowla and N. Granpayeh, "Design of a flat gain multi-pumped distributed fiber Raman amplifier by particle swarm optimization," *J. Opt. Soc. Am. A*, Vol. 25, No. 5, pp. 3059-3066, 2008.

[5] Mowla and N. Granpayeh, “Optimum design of a hybrid erbium-doped fiber amplifier/fiber Raman amplifier using particle swarm optimization,” *Appl. Opt.* Vol. 48, No. 5, pp. 979-984, 2009.

[6] P. Engelbrecht, *Fundamental of Computational Swarm Intelligence*, (John Wiley and Sons, New York, USA, 2005), Chap. 16.

[7] P. Zhang, X. W. Chen, J. H. Yan, Z. G. Guan, S. He, and H. Y. Tam, "Optimization and fabrication of stitched long-period gratings for gain flattening of ultrawide-band EDFAs," *IEEE Photon. Technol. Lett.* Vol. 17, No. 12, pp. 2559-2561, 2005.

[8] H. Choi, H. H. Park, and M. J. Chu, "New pumped wavelength of 1540-nm band for long-wavelength-band erbium-doped fiber amplifier (L-band EDFA)," *J. Quantum Electron.* Vol. 39, No. 10, pp. 1272-1280, 2003.

[9] Cheng and M. Xiao, "Optimization of a dual pumped L-band erbium-doped fiber amplifier by genetic algorithm," *IEEE J. Lightwave Technol.* Vol. 24, No. 10, pp. 3824-3829, 2006.

[10] Cheng and M. Xiao, "Optimization of an erbium-doped fiber amplifier with radial effects," *Opt. Commun.* Vol. 254, No. 4-6, pp. 215-222, 2005.

[11] Cheng, "A global design of an erbium-doped fiber and an erbium-doped fiber amplifier," *Opt. Laser Technol.* Vol. 36, No. 8, pp. 607-612, 2004.

[12] Cheng, Z. Xu, and C. Sui, "A novel design method: a genetic algorithm applied to an erbium-doped fiber amplifier," *Opt. Commun.* Vol. 227, No. 4-6, pp. 371-382, 2003.

[13] H. Yeh, C. C. Lee, and S. Chi, "S- plus C-band erbium-doped fiber amplifier in parallel structure," *Opt. Commun.* , Vol. 241, No. 4-6, pp. 443-447, 2004.

[14] Headley and G. P. Agrawal, *Raman Amplification in Fiber Optical Communication Systems*, (Elsevier, Burlington, USA, 2005), pp. 33-163.

[15] L. Chang, L. Wang, and Y. J. Chiang, "A dual pumped double-pass L-band EDFA with high gain and low noise," *Opt. Commun.* Vol. 267, No. 1, pp. 108-112, 2006.

[16] M. Hung, N. K.Chen, Y. Lai, and S. Chi, "Double-pass high-gain low-noise EDFA over S- and C+L-bands by tunable fundamental-mode leakage loss," *Opt. Express*, Vol. 15, No. 4, pp. 1454-1460, 2007.

[17] R. Giles, and E. Desurvire, "Modeling erbium-doped fiber amplifiers," *IEEE J. Lightwave Technol.*, Vol. 9, No. 2, pp. 271-283, 1991.

[18] Yau and S. Crozier, "A genetic algorithm/method of moments approach to the optimization of an RF coil for MRI applications-theoretical considerations," *Progress In Electromagnetic Research*, PIER 39, pp. 177-192, 2003.

[19] H. Afkhami, A. Mowla, N. Granpayeh, and A. R. Hormozi, “Wideband Gain Flattened Hybrid Erbium-doped Fiber Amplifier/Fiber Raman Amplifier,” *J. Opt. Soc. Korea,* Vol. 14, No. 4, pp. 342-350, 2010.

[20] H. Masuda and S. Kawai, “Wide-band and gain-flattened hybrid fiber amplifier consisting of an EDFA and a multi-wavelength pumped Raman amplifier,” *IEEE Photon. Technol. Lett.* Vol. 11, No. 6, pp. 647-649, (1999).

[21] H. Wei, Z. Tong, and S. Jian, "Use of a genetic algorithm to optimize multistage erbium-doped fiber amplifier system with complex structures," *Opt. Express*, Vol. 12, No. 4, pp. 531-544, 2004.

[22] J. Bromage, “Raman Amplification for Fiber Communications Systems,” *IEEE J. Lightwave Technol.* Vol. 22, No. 1, pp. 79-93, 2004.

[23] J. C. Martin, "Erbium transversal distribution influence on the effectiveness of a doped fiber: optimization of its performance," *Opt. Commun.* Vol. 194, No. 4-6, pp. 331-339, 2001.

[24] J. D. A. Castanon, I. O. Nasieva, S. K. Turitsyn, N. Brochier, and E. Pincemin, "Optimal span length in high-speed transmission systems with hybrid Raman-erbium-doped fiber amplification," *Opt. Lett.* Vol. 30, No. 1, pp. 23-25, 2005.

[25] J. Kennedy and R. C. Eberhart, "Particle swarm optimization," Proc. IEEE Int. Conf. on Neural Networks, 1942-1948 (1995).

[26] J. Kennedy and R. C. Eberhart, *Swarm Intelligence,* Morgan Kaufmann Publishers, Academic Press, 2001, Chap. 7.

[27] J. T. Ahn, and K. H. Kim, "All-optical gain-clamped erbium-doped fiber amplifier with improved noise figure and freedom from relaxation oscillation," *IEEE Photon. Technol. Lett.* , Vol. 16, No. 1, pp. 84-86, 2004.

[28] L. L. Yi, L. Zhan, C. S. Taung, S. Y. Luo, W. S. Hu, Y. K Su, and Y. X. Xia, "Low noise figure all-optical gain-clamped parallel C+L band erbium-doped fiber amplifier using an interleaver," *Opt. Express*, Vol. 13, No. 12, pp. 4519-4524, 2005.

[29] L. L. Yi, L. Zhan, J.H. Ji, Q. H. Ye, and Y. X. Xia, "Improvement of gain and noise figure in double-pass L-band EDFA by incorporating a fiber Bragg grating," *IEEE Photon. Technol. Lett.* , Vol. 16, No. 4, pp. 1005-1007, 2004.

[30] L. L. Yi, L. Zhan, W. S. Hu, Q. Tang, and Y. X. Xia, "Tunable gain-clamped double-pass erbium-doped fiber amplifier," *Opt. Express*, Vol. 14, No. 2, pp. 570-574, 2006.

[31] L. Lucci, R. Nesti, G. Pelosi, and S. Selleri, "Phase centre optimization in profiled corrugated circulator horns with parallel genetic algorithms," *Progress In Electromagnetic Research*, PIER 46, pp. 127-142, 2004.

[32] M. E. Marhic and D. E. Nikonov, "Low third-order glass-host nonlinearities in erbium-doped waveguide amplifiers," *Proc. of SPIE*, Vol. 4645, 193, 2002.

[33] M. H. Ahmed, M. Shalaby, and F. M. Misk, "Combined erbium and Raman amplification at 1.55 μm in submarine links using backward pumping at 1.48 μm," *J. Opt. A: Pure Appl. Opt.* Vol. 7, No. 4, pp. 659-666, 1998.

[34] M. N. Islam, *Raman Amplifiers for Telecommunications I* and *II*, (Springer, New York, USA, 2004).

[35] M. Pal, M. C. Paul, A. Dhar, A. Pal, R. Sen, K. Dasgupta, and S. K. Bhadra, "Investigation of the optical gain and noise figure for multi-channel amplification in EDFA under optimized pump condition," *Opt. Commun.* Vol. 273, No. 2, pp. 407-412, 2007.

[36] M. Yamada, A. Mori, K. Kobayashi, H. Ono, T. Kanamori, K. Oikawa, Y. Nishida, and Y. Ohishi, "Gain-flattened tellurite-based EDFA with a flat amplification bandwidth of 76 nm," *IEEE Photon. Technol. Lett.* , Vol. 10, No. 9, pp. 1244-1246, 1998.

[37] P. C. Becker, N. A. Olsson, and J. R. Simpson, *Erbium-Doped Fiber Amplifiers: fundamentals and technology*, (Academic Press, San Diego, USA, 1999), Chapter 1.

[38] R. C. Eberhart and J. Kennedy, "A new optimizer using particle swarm theory," *Proc. 6th Int. Symposium on Micro Machine and Human Science*, Nagoya, Japan, 39-43 (1995).

[39] R. L. Haupt and S. E. Haupt, *Practical Genetic Algorithms*, John Wiley and Sons, Chaps. 2 and 3, 2004.

[40] R. Singh, Sunanda, and E. K. Sharma, "Gain flattening by long period gratings in erbium doped fibers," *Opt. Commun.* , Vol. 240, No. 1-3, pp. 123-132, 2004.

[41] S. Namiki and Y. Emori, "Ultrabroad-Band Raman Amplifiers Pumped and Gain-Equalized by Wavelength-Division-Multiplexed High Power Laser Diodes," *IEEE J. Select. Topics Quantum Electron.* Vol. 7, No. 1, pp. 3-16, 2001.

[42] T. C. Liang, and S. Hsu, "The L-band EDFA of high clamped gain and low noise figure implemented using fiber Bragg grating and double-pass method," *Opt. Commun.* Vol. 281, No. 5, pp. 1134-1139, 2007.

[43] W. J. Miniscalco, "Erbium-doped glasses for fiber amplifiers at 1500 nm," *IEEE J. Lightwave Technol.* Vol. 9, No. 2, pp. 234-250, 1991.

[44] X. Liu and B. Lee, "Optimal design for ultra-broad-band amplifier," *IEEE J. Lightwave Technol.* Vol. 21, No. 12, pp. 3446-3455, 2003.

[45] X. Liu and B. Lee, "A fast and stable method for Raman amplifier propagation equations," *Opt. Express*, Vol. 11, No. 18, pp. 2163-2176, 2003.

[46] X. Liu and B. Lee, "Effective shooting algorithm and its application to fiber amplifiers," *Opt. Express*, Vol. 11, No. 12, pp. 1452-1461, 2003.

[47] X. Liu and Y. Li, "Optimizing the bandwidth and noise performance of distributed multi-pump Raman amplifiers," *Opt. Commun.* Vol. 230, No. 4-6, pp. 425-431, 2004.

[48] X. Liu, "Optimization for various schemes of distributed fibre Raman amplifiers," *J. Opt. A: Pure Appl. Opt.* Vol. 6, No. 11, pp. 1017-1026 2004.

[49] Y. B. Lu, and P. L. Chu, "Gain flattening by using dual-core fiber in erbium-doped fiber amplifier," *IEEE Photon. Technol. Lett.* , Vol. 12, No. 12, pp. 1616-1617, 2000.

[50] Y. B. Lu, P. L. Chu, A. Alphones, and P. Shum, "A 105-nm ultrawide-band gain-flattened amplifier combining C- and L-band dual-core EDFAs in a parellel configuration," *IEEE Photon. Technol. Lett.* , Vol. 16, No. 7, pp. 1640-1642, 2004.

[51] Z. Li, C. L. Zhao, Y. J. Wen, C. Lu, Y. Wang, and J. Chen, "Optimization of Raman/EDFA hybrid amplifier based on dual-order stimulated Raman scattering using a single-pump," *Opt. Commun.* Vol. 265, No. 2, pp. 655-658, 2006.

[52] Z. Q. Meng, "Autonomous genetic algorithm for functional optimization," *Progress In Electromagnetic Research*, PIER 72, pp. 253-268, 2007.

In: Optical Amplifiers
Editor: Galina Nemova

ISBN 978-1-61209-835-7

Chapter 7

APPLICATION OF OPTICAL AMPLIFIERS IN LASER GAIN MEASUREMENTS: EXPERIMENTAL AND THEORETICAL INVESTIGATIONS

Akbar Hariri*
Laser and Optics Research Department,
Nuclear Science and Technology Research Institute,
North Kargar Avenue, Tehran, Iran

ABSTRACT

Pulsed optical amplification, although has shown its profound application in generation of single mode, high power lasers through master oscillator-power amplifiers in experiments related to laser fusion facilities, it has also shown its ability to be used in other scientific and industrial activities. In this review details of experimental and theoretical investigations for gain measurements using oscillator-amplifier (OSC-AMP) N_2-laser systems will be presented. The OSC-AMP laser systems proved their abilities for gain measurements and also for further applications on understanding the gain behavior when the interest is oriented toward the study of effects of laser geometrical configurations, current density, input voltage, etc. on the gain-parameter, where these areas of research have not been explored thoroughly yet. The work is related almost completely to the author's publications in the recent and past years, although many related topics including Nd: YAG and CVL lasers will be given for clarifying the subject. Simplicity of N_2-laser designs, ease of operation, flexibility for geometrical modifications, and wide ranges of collection of the published theoretical and experimental related works make the N_2-laser as a good candidate for studying the laser behavior. Here, after reviewing the previous research works on the gain measurement, a model based on introducing an equivalent OSC system to apply for a real OSC-AMP system is presented, where without ignoring some terms in the relevant rate equations; it was possible to predict gain-values for N_2-lasers. Although the introduced model is

* E-mail: akbar_hariri@yahoo.com

simple, but it shows that it can be used to explain the experimental measurements for N_2-lasers. The model was successfully applied for CVL lasers and the approach may find its suitability for other types of laser systems.

Keywords: Oscillator-amplifier, N_2-laser, Nd:YAG and CVL laser

1. INTRODUCTION

Amplification of laser optical pulses using a laser arrangement consist of combination of an oscillator (OSC) and an Amplifier (AMP) to operate as an OSC – AMP laser system can be used to measure laser parameters, known as gain coefficient or small signal gain, g_0, and saturation energy density, E_s, of the amplifying medium. So far, the experimental trends concerning the measurement of these parameters have been directed toward introducing the measured values of the laser parameters for conventional laser designs, and for a long time from the engineering point of view, introducing a laser device having a higher gain – value and output power compared to previous devices, was the principle goal of the researchers dealing with laser engineering. In this track, various designs and configurations along with the operational optimization of the devices were introduced. In fact, from the subject understanding point of view, less serious attention was paid for better realization of the laser behavior when these parameters were subject to some changes. The reason is that performing the experiments and measurements consume time and also the information appeared in the literature are not sometimes quite suitable for conveying the information to those who are interested in to challenge with the subject from the theoretical point of view. In addition, except some analytical relations which are useful to be applied experimentally for data reduction, and sometimes they come across with the measurements, the amplifier theories are usually not solvable analytically, so their justifications with the experiment are hard to be realized. Also, it has to point out that for the process of data reduction in an experiment, presence of suitable analytical relations will greatly help the researchers for better realization of the processes involved and this area of theoretical works has attracted less attention throughout the years.

The experimental approach which is presented here is applicable to different laser media, but suitability of gas lasers, in particular, when one type of gas is used has proved its capability. With the gas media, we can vary the inversion density, where it cannot be easily provided, for example, in solid state lasers. Among various types of gas lasers, nitrogen lasers for the $C^3\Pi_u \rightarrow B^3\Pi_g$ transition at 337.1 *nm* seems to be a good candidate for experimental and theoretical studies and that is due to their simplicity, ease of operation and gas handling. Also, historically, for a long time these lasers have been under the extensive studies both experimentally and theoretically. Thus, the available information in the fields of laser construction, observations, measurements and theoretical works can be helpful for the researchers dealing with the subject. Furthermore, as N_2-lasers are categorized as self – terminating gas lasers, pursuing the research with these lasers can be useful to be applied to other analogous and more sophisticated devices in this category such as eximer, CVL, and X – ray lasers.

In this chapter after reviewing the past history of gain measurements corresponding to a number of research works, a particular attention has been paid to the author's works on N_2-lasers, including the experimental and theoretical considerations. Upon collection of the past experimental measurements on the N_2-laser gain – value, a model based on applying an equivalent OSC laser system instead of using a complicated real OSC – AMP laser arrangement is introduced. The model is based on imposing some theoretical and experimental conditions on the rate equations and experimental measurements. It was found that the model is able to predict correctly the influence of the laser geometrical configuration, including electrodes length, l_{AMP}, and gap separation, d_{AMP}, on the gain – value. The system in this respect is operating under the moderate current density condition and at a constant input operating voltage of 14 kV. For the gain prediction upon using the deduced analytical expression, it was found that is not necessary to consider the laser operational gas pressure, where the effect of the operational gas pressure is already included in the electron temperature formalism through the E/p –value, and that appears in the simulation study used for the calculating the gain - parameter. In this respect, based on our observations that the gain – value is almost linearly proportional to the input voltage, the correction due to applying different input voltages can be made for finalizing the prediction. For the geometrical consideration of the laser, at least as far as the gain – parameter is concerned, it was realized that transition from one – dimensional analysis of the proposed model to a two dimensional consideration is unavailable. Thus in this respect, the laser has been viewed as a two dimensional device with two independent gain values corresponding to the optical axes, g_{0z}, and the direction parallel to electrodes gap separation, g_{0y}, respectively. With this realization a complete analytical expression for the geometrical gain – parameter has been obtained. The model is able to predict the gain – value of other TE/TEA N_2-laser designs operating at different input voltages, while still keeping the lasers in moderate current density condition. The predictions for the gain – values corresponding to other researchers' works are rather excellent.

The following subjects will be covered in this review article. In section 2, the theory of pulse amplification as appeared in the literature and is suitable to be used in experimental measurements for the data reduction will be presented. The goal in this section is to familiarize the reader with the formalisms to be used in the experimental measurements. The general treatment of the gain-measurements which is commonly applied in any laboratory, with the least requirement for the experimental measuring devices is given in section 3. In this section amplification and gain measurements in solid state lasers, and N_2-lasers of different types will be introduced. These two types of lasers, i.e., solid –state lasers and gas lasers, have different pumping mechanisms, so the procedure for amplifications, although are similar, but they are different for the experimental arrangement to obtain proper desynchronization time requirements between two sections of OSC and AMP laser systems. The last part in this review is devoted to introducing an equivalent OSC laser to be applied instead of a real OSC-AMP system. This model facilitates the difficulties that one faces with handling an OSC-AMP laser system mathematically.

Upon extending the understanding of the model, it was possible to introduce an analytical expression for the gain-value, where in this respect the geometrical configuration of the N_2-laser system was of our concern. With performing the experimental measurements upon using limited ranges of the AMP electrodes lengths and gap separations, the model is able to predict gain-values corresponding to different laser devices appeared in the literature. The model was also applied for CVLs and the agreement with the measurement was excellent.

2. Theoretical Background on Optical Amplifier

Development of optical amplifiers theories and laser theories occurred almost at the same time. The earliest work for the pulse amplification is a paper by Frantz and Nodvik [1], while a complete picture of optical maser amplifiers for media with homogeneous and inhomogeneous line broadening in the continuous wave (cw) operation was presented by Rigrod [2].

The analytical expressions appeared in these two papers, up to present time, are commonly applied for lasers of different active media. Different aspects of the idea have also been spread out and appeared in varieties of text books and reviews [3 – 7].

The formulations that are given in [1, 2], although are given in a simple form of a two – level system, but so far, it seems that no other analytical expression has been found to show its suitability to be applied for data reduction corresponding to gain measurements leading to evaluate laser parameters.

That is due to complication of rate equations in a real optical amplifier. So, in order to have a closed – form solution to the photon transport equation, many terms such as pumping rate and spontaneous emission from the upper state level to the lower – level have to be ignored. Although derivations of the Frantz – Nodvik equation, for solving some practical applications of the equation such as zig – zag optical path laser amplifiers have been reported [8], but in some complicated problems such as multi-pass amplifiers applying numerical analyses shown to be unavoidable [9].

For presenting a rather complete picture of the optical amplifier to be used as a technique for gain measurement, or arranging setups for the pulse amplification, it is necessary to derive the basic formulations in this respect. For this purpose the theoretical procedure appeared in [3] will be presented here briefly.

2.1. Solution to the Rate Equations

The following two equations in a real laboratory frame of reference are giving the photon – density, $n_{ph}(z,t)$, and inversion population density $N(z,t) = N_2(z,t) - N_1(z,t)$, in an amplifying medium, where $N_{2,1}$ are referring to upper and lower population densities, respectively. In deriving these equations, the effects of the upper – state lifetime and pumping rate are ignored. Thus, in these equations only the stimulated emission rate with the corresponding cross – section, σ_{st}, are used. The equations are as follows

$$\frac{\partial n_{ph}(z,t)}{\partial t} + c\frac{\partial n_{ph}(z,t)}{\partial z} = c\sigma_{st} N(z,t) n_{ph}(z,t) \tag{2.1}$$

$$\frac{\partial N(z,t)}{\partial t} = -c\sigma_{st} N(z,t) n_{ph}(z,t) \tag{2.2}$$

where, the z – axis corresponds to the propagation direction of the pulse entering the amplifying medium, $n_{ph}(z,t)$ is the photon – density and is assumed that it is z – dependent. c is the light velocity in the amplifying medium. For solving these equations a coordinate system that moves with the forward traveling pulse has been proposed. The prime on the characters indicates that equations are given in the moving frame of reference, where they are related as

$$z' = z \text{ and } t' = t - \frac{z}{c} \tag{2.3}$$

With $n_{ph}(z',t') \equiv n_{ph}(z,t)$, and $N(z',t') \equiv N(z,t)$ [3], and $c = dz/dt$ we can write Eqs. (2.1) and (2.2) in the moving frame of reference

$$\frac{\partial n_{ph}(z',t')}{\partial z'} = \sigma_{st} N(z',t') n_{ph}(z',t') \tag{2.4}$$

$$\frac{\partial N(z',t')}{\partial t'} = -c\sigma_{st} N(z',t') n_{ph}(z',t') \tag{2.5}$$

Eq. (2.4) can be integrated to give

$$\int_{n_{ph}^{out}(t')}^{n_{ph}^{in}(t')} \frac{dn_{ph}}{n_{ph}} = \int_{z'=0}^{z'=l} \sigma_{st} N(z',t') dz' \tag{2.6}$$

Or, if we proceed further, Eq. (2.6) leads to

$$n_{ph}^{out}(t') = n_{ph}^{in}(t')\, e^{\sigma_{st} N_{tot}(t')} = G(t') n_{ph}^{in}(t') \tag{2.7}$$

where $N_{tot}(t') = \int_0^l N(z',t')dz'$; $G(t') = \exp[\sigma_{st} N_{tot}(t')]$ is the power gain at any instant within the pulse; $n_{ph}^{in}(t')$ and $n_{ph}^{out}(t')$ are photon – densities at the input plane and output plane of the amplifier, measured in the moving frame of reference. l is the length of amplifying medium.

Eq. (2.5) can be integrated to yield

$$\frac{\partial}{\partial t'} \int_0^l N(z',t')dz' = -c \int_0^l \frac{\partial n_{ph}(z',t')}{\partial z'} dz'$$

or

$$\frac{d}{dt'}N_{tot}(t') = -c[n_{ph}^{out}(t') - n_{ph}^{in}(t')] = -cn_{ph}^{in}(t')\{\exp[\sigma_{st}N_{tot}(t')] - 1\}$$

$$= -cn_{ph}^{out}(t')\{1 - \exp[-\sigma_{st}N_{tot}(t')]\} \tag{2.8}$$

where Eq. (2.4) and definition of $N_{tot}(t')$ have been used.

For the input energy density, E_{in}, and output energy density, E_{out}, we can write

$$E_{in}(t') = \int_{t_0'}^{t'} c\hbar\omega n_{ph}^{in}(t')dt' \tag{2.9}$$

and

$$E_{out}(t') = \int_{t_0'}^{t'} c\hbar\omega n_{ph}^{out}(t')dt' \tag{2.10}$$

Substituting $n_{ph}^{in}(t')$ and $n_{ph}^{out}(t')$ from two last expressions given by Eq. (2.8) into Eqs. (2.9) and (2.10), respectively, we obtain

$$E_{in}(t') = \int_{N_0}^{N_{tot}(t')} \hbar\omega \frac{\frac{d}{dt'}N_{tot}(t')}{1 - \exp[\sigma_{st}N_{tot}(t')]} dt' \tag{2.11}$$

and

$$E_{out}(t') = \int_{N_0}^{N_{tot}(t')} \hbar\omega \frac{\frac{d}{dt'}N_{tot}(t')}{\exp[\sigma_{st}N_{tot}(t')] - 1} dt' \tag{2.12}$$

where

$$N_0 = \int_{z=0}^{z=l} N(z, t_0)dz \tag{2.13}$$

is the initial inversion at a time t_0 prior to the arrival of the input pulse into the amplifier. By defining $E_s = \hbar\omega/\sigma_{st}$, known as saturation energy density, and integrating Eq. (2.11) gives

$$\frac{E_{in}}{E_s} = \ln\left\{\frac{\left(n_{ph}^{out}/n_{ph}^{in}\right)(1-G_0)}{G_0\left(1-n_{ph}^{out}/n_{ph}^{in}\right)}\right\} \tag{2.14}$$

where $G_0 = \exp[\sigma_{st} N_0]$ is the initial single - pass power gain of the amplifier. Rearranging Eq. (2.14) leads to

$$G(t) = \frac{n_{ph}^{out}}{n_{ph}^{in}} = \frac{G_0 e^{E_{in}/E_s}}{(1-G_0)+G_0 e^{E_{in}/E_s}} \tag{2.15}$$

If we evaluate $E_{out}(t')$ using Eq. (2.10) and applying the last expression given in Eq. (2.8), we obtain, analogously

$$G(t) = \frac{n_{ph}^{out}}{n_{ph}^{in}} = \left\{1+(G_0-1)e^{-E_{in}/E_s}\right\} \tag{2.16}$$

By equating $G(t)$ appeared in Eqs. (2.15), and (2.16), we will obtain the following equation relating G_0 to E_{out}, E_{in}, and E_s.

$$G_0 = \frac{e^{E_{out}/E_s}-1}{e^{E_{in}/E_s}-1} \tag{2.17}$$

or equivalently, it leads to the following equation

$$E_{out} = E_s \ln\left\{1+G_0\left[e^{E_{in}/E_s}-1\right]\right\} \tag{2.18}$$

If we assume $N(z,t_0)$ is not varying with the z – direction, then from Eq. (2.13) we have $N_0 = N(t_0)l$, and $G_0 = \exp[\sigma_{st} N(t_0)l]$. With the definition of small-signal gain, given by $g_0 = \sigma_{st} N(t_0)$, then $G_0 = \exp[g_0 l]$ can be used in Eq. (2.18).

If we introduce the passive loss, by presenting the loss coefficient per unit length, α_L, the Frantz – Nodvik model will be modified as

$$\int_{E_{in}}^{E_{out}} \frac{dE}{\left\{E_s\left[1-e^{E/E_s}\right]-\alpha_L E/g_0\right\}} = \ln G_0 \tag{2.19}$$

which reduces to the Frantz – Nodvik equation when $\alpha_L = 0$, but for $\alpha_L > 0$, the equation must be solved by numerical integration. Eq. (2.19) is known as Avizonis and Grotbeck equation [5]. The equation is a nonlinear relation between the variables E_{in}, E_{out}, and G_0.

Because of this nonlinearity, the external parameters, i.e., passive transmission, input and output energy density, must be carefully measured.

For operation in the steady state condition for homogeneously broadened media, where the saturated small signal gain, g, is given by $g = g_0/[1+I/I_s]$, where the I_s is the saturation intensity, g_0 is the unsaturated gain coefficient, $I_s = E_s/\tau_f$, and τ_f is the upper – state lifetime, it not difficult to calculate the following formula

$$\ln(E_{out}/E_s) = g_0 l - (E_{out} - E_{in})/E_s \qquad (2.20)$$

Eqs. (2.18) and (2.20) are the relations that are commonly applied for data reduction, and a few experimental examples for the validity of these expressions are given, as appeared in the literature.

2.2. Examples of Experimental Observations

As a first example, we are referring to a report given by Banic et al. for the gain and saturation energy measurements in a KrF OSC – AMP laser system [10]. The OSC section of their system was operating with two different modes of pulse durations of 2 *nsec* and 10 *nsec*, applied in different experiments. Their data were fitted to two models: the short – pulse amplification given by Frantz and Nodvik equation, i.e., Eq. (2.18), and the steady – state model, given by Eq. (2.20). From the slope of the linear portions of their experimental measurements, i.e., when in Eq. (2.17) we let $\exp[E_{in}/E_s] J\ 1 + E_{in}/E_s$, and then we get $E_{out}/E_{in} = G_0 = \exp(g_0 l)$, the power gain of 650 and 1400 for 2 and 10 *ns* pulses for $l = 45\,cm$ amplifying medium were obtained. This equation gives $g_0(\text{for } 2\,n\text{sec}) = 0.145\,cm^{-1}$, and $g_0(\text{for } 10\,n\text{sec}) = 0.160\,cm^{-1}$. They believe that when the duration of the laser pulse is much shorter than the upper state radiative lifetime, the exchange of the energy in the laser medium do not have time to affect the amplification during the pulse and molecule interaction. Thus, with $g_0(2\,ns)$ and by using Eq. (2.18), they obtained $E_s = 3.2\ mJ/cm^2$, and $I_s = E_s/\tau = 1.6\ MW/cm^2$, where τ is the pulse duration. For $g_0(10\,ns)$, Eq. (2.18) gives $E_s = 26\ mJ/cm^2$ and $I_s = 1.7\ MW/cm^2$, which was quite different from their previous analysis. Also, upon applying the cw model, given by Eq. (2.20), the laser parameters of $E_s = 39\ mJ/cm^2$ and $I_s = 2.6\ MW/cm^2$, ~ 50% larger than short pulse approximation were obtained. They concluded that for the 10 *ns* input pulse, neither of two models are valid for describing the amplification, since the laser pulse width is comparable to the laser upper – state lifetime. This example shows clearly that the absence of an accurate theory of pulse propagation in a specific medium leads to different results that they might be different at least by a factor of 2.

The second example is given for the CVL laser, categorized as a self – terminating metal vapor laser. In an experiment reported by Hargrove et al. they used three optical extraction models: a pulsed propagation formalism, simple steady – state inhomogeneously broadened and homogeneously broadened formulas, and they found that the saturation of both 510.6 and 578.2 *nm* lines can be best fitted by the steady – state formula appropriate to homogeneously broadened transition[11]. The same approach for the pressure dependence of the gain coefficient in a CVL laser was applied by Behrouzinia et al [12]. As CVL is operating at high repetition rates of a few kHz, but it seams a suitable model for evaluating the gain coefficient is the steady – state formula, given by Eq. (2.20).

As a third example we are referring to [13 and 14], where in their work, the nitrogen laser amplifiers were operating at low pressure (TE) and atmospheric pressure (TEA), respectively. As the upper state lifetime of N_2 – laser ~ 40 *ns*, then by injecting an optical short pulse of a few *ns* duration into the AMP section and by appropriately adjusting the desynchronization condition between OSC – AMP sections, the laser parameters, g_0 and E_s, using Eq. (2.18), can be obtained. Their measurements are: $g_0 = 0.277\ cm^{-1}$, $E_s = 104.6\ \mu J/cm^2$ at 56 *torr* [13], and $g_0 = 0.22\ cm^{-1}$, $E_s = 54.55\ \mu J/cm^2$ at 760 *torr*, i.e., at atmospheric pressure [14]. They noticed that at the low input energy density of $E_{in} < 10^{-3}\ \mu J/cm^2$ their measurements give higher gain – values compared to those which are predictable by Eq. (2.18), and the measurements for the logarithmic scale of E_{in} do not present a plateau as would be expected from the theoretical curves. On the other hand, at high input energy densities, the data corresponding to g_0 – measurements are slightly higher than those predicted by the theory. They concluded that: (a) in the intermediate and low – energy areas of E_{out} vs. E_{in} curves, the upper state depopulation by the ASE tends to decrease the output energy density, and (b): the pumping process tends to increase the energy density of the amplified pulse compared to that which is given by Eq. (2.18). Their observations confirm this fact that near saturation, the approximations used for deriving Eq. (2.18) are not completely valid and also the assumption that population of the upper and lower laser level is not affected by the pumping is no longer justified.

These examples are indicating that as long as a complete theoretical picture in this subject is not available to be used for data reduction, the experimental workers are facing with difficulties for their data analyses. Thus in this respect, introducing some sophisticated analytical expressions would be quite helpful to give more accurate results in deducing laser parameters.

3. Measurements: General Treatment

To measure laser parameters, g_0 and E_s, we need to have a laser system consist of an OSC and an AMP to work as an OSC – AMP laser system. If two sections of the OSC and AMP are not properly time-adjusted, or if they work independently, we are not expecting to reach at a complete amplification, or even approaching the chance of occurrence any amplification. An accurate occurrence of the amplification corresponds to the realization of an instant when a proper desynchronization time condition between these two sections takes place. Thus, an important portion of the process of amplification is devoted for obtaining the proper desynchronization time condition. In fact, in a 4 – level laser system, for example, injection of the laser pulse from the OSC output section into the AMP section must occur when the amplifier reaches at its maximum population inversion, otherwise, at some instant prior to this time when the AMP is in its ground state, the injected pulse will travel through the AMP section without bearing any amplification. Or, in the case when the injected pulse comes into the AMP section with some time – delay so late that the amplifier has lost its population by decaying to the lower energy level, then upon the interaction of the input pulse with a system having its population in the lower energy level, absorption of the input laser pulse, instead of amplification, will occur. The cited process can be observed using fast photo – diodes with an oscilloscope. The power or energy measurements of the amplified output pulses will be also helpful to keep the system in its complete desynchronization time condition. Two examples in this regard for Nd:YAG and nitrogen lasers will be useful to illustrate the process of amplification.

3.1. Amplification in Solid State Lasers

(a) Single-Pass Amplification

Theory of laser amplification for solid – state amplifiers, in particular from the point of view of experimental works and laser engineering has been described completely by Koechner [5]. In these lasers, for example, in Nd: YAG, and Nd: glass lasers, if for pumping of the active media flash – lamps are used, the desynchronization time condition is accomplished electronically, because the time duration for the pumping event is usually long of a few tens of μs . Referring to our work for Nd: YAG laser, Figure 3.1 shows the OSC – AMP laser system used for g_0 and E_s measurements in a single – pass amplification [15 – 16]. The triggering units are used for triggering the flash – lamps to pump Nd: YAG laser rods corresponding to the OSC and AMP sections. They are fired when the electric pulses which are injected to the triggering units have already been controlled electronically through a time- delay generator unit. The OSC is a Q–switched flash–lamp pumped Nd: YAG laser using a combination of a rotating prism and a 50% reflecting mirror to work as an active Q-switching element as well as the cavity front mirror, as shown in the figure. The OSC is delivering a giant pulse of ~ 30 *ns* FWHM time duration. An interactivity of 50% transmission mirror has also been used to avoid multimode operation through making a three-mirror resonator. The AMP laser rod dimensions are 54 *mm* in length and 6.3 *mm* in diameter. The arc length of the AMP flash – lamp is 75 *mm*. The flash – lamp and laser rod are mounted in a total reflecting cavity with an elliptical cross – section. For operation of the OSC – AMP system, the AMP is fired slightly before the OSC. The output energy from the OSC section is ~ 17 *mJ.* The capacitance corresponding to the AMP electric – charging unit is 80 μF , holding an energy of 62 *J.* The flash – lamp of the AMP section has a FWHM of 56 μs time duration.

Figure 3.2 shows the output energy extracted from the AMP with respect to time – delay between the electric pulses injected into the OSC and AMP flash-lamps triggering units. For the gain and saturation energy measurements, the output energy density, E_{out}; and input energy density (i.e. energy density of the pulse that enters into the AMP), E_{in} , have been measured. For reducing the input energy entering into the AMP, neutral density filters calibrated at 1064 nm are used. The beam cross – sections of the input pulse entering into the AMP and extracting from AMP are equal (~ 0.31 cm^2). The AMP is operating with flash – lamp charging energies of 160 and 90 *J* in two separate measurements.

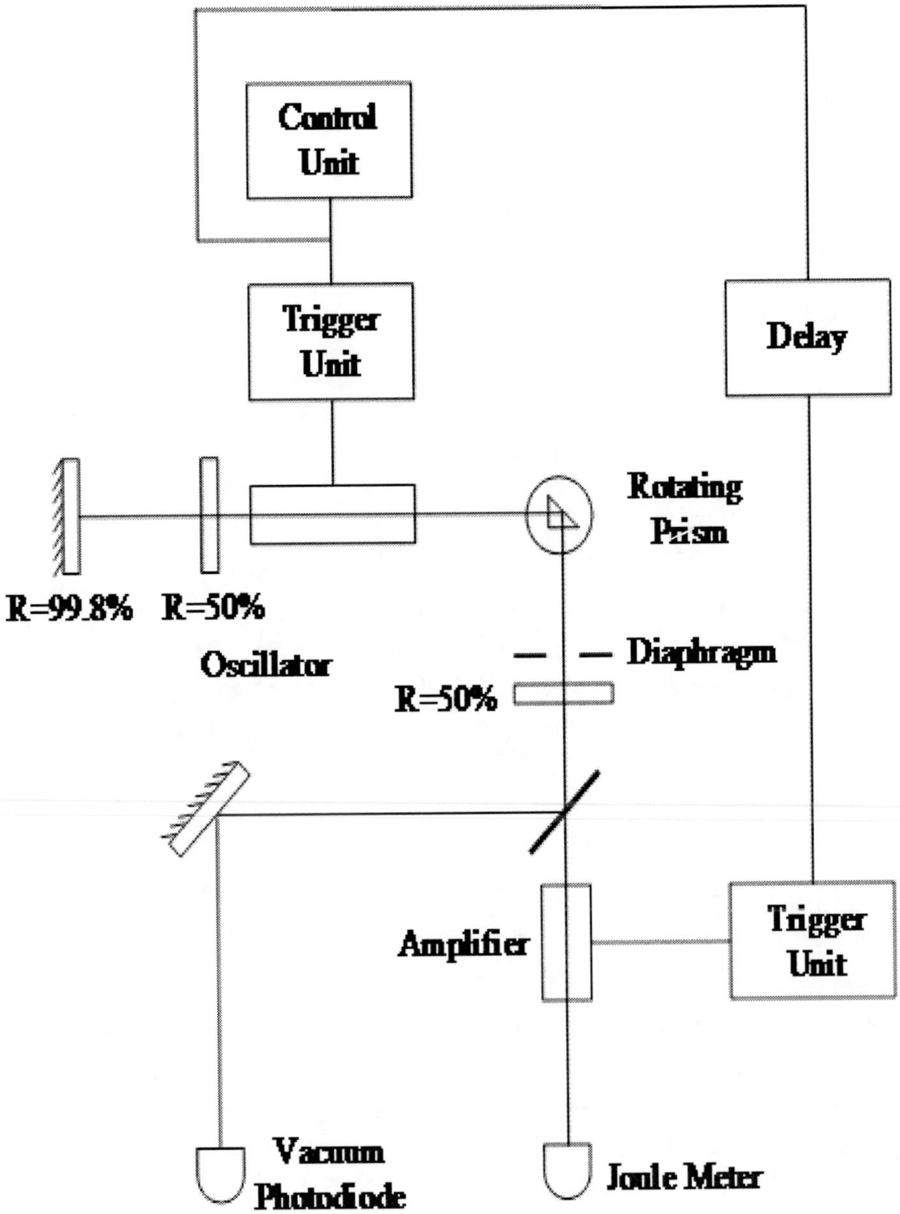

Figure 3.1 The schematic configuration of a Nd:YAG, OSC-AMP laser system, used for gain and saturation energy measurements. An electronic control unit is used for generating a variable delay-time between the triggering units of the OSC and AMP sections. The OSC is a Q-switched Nd:YAG laser with a ~ 30 *ns* FWHM time duration.

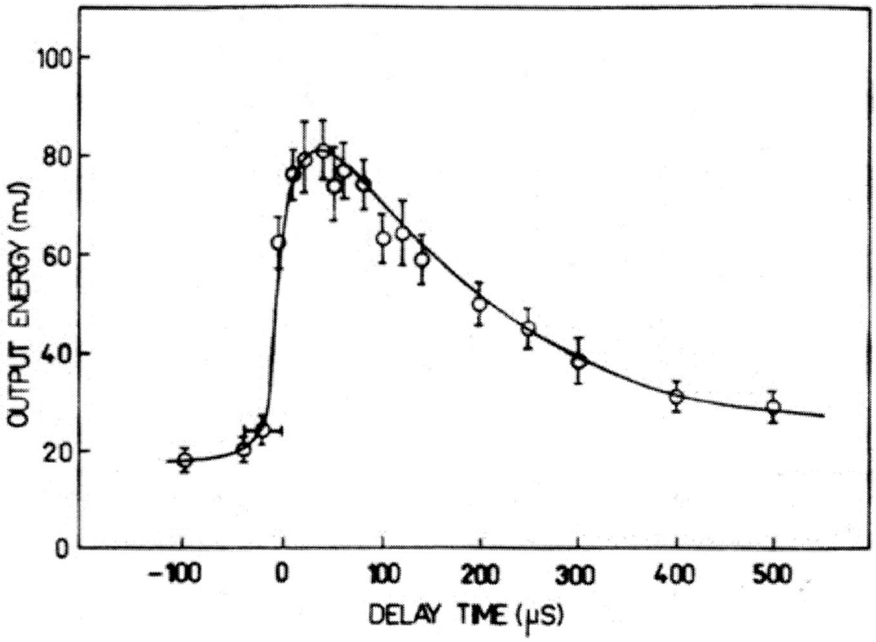

Figure 3.2. The output energy extracted from the AMP, versus the electric delay-time between OSC and AMP. The average input pulse energy to the AMP is 17.5 *mJ*.

A simple way for the data analysis, as given in Figure 3.3, is to evaluate g_0 from the data corresponding to the low input energy density, where as shown by Eq. (2.17) or (2.18), E_{out}/E_{in} J $\exp(g_0 l)$ is valid for E_{in}/E_s ? 1. On the other hand, for E_{in}/E_s @1, which corresponds to the saturation parts of the curve, we can use Eq. (2.18) to have

$$E_{out} = E_{in} + g_0 E_s l \tag{3.1}$$

By drawing a straight line of $E_{out} = E_{in}$, i.e., C in the figure, the corresponding vertical distance from the saturated portion of the curve to the line given by $E_{out} = E_{in}$ gives rise to evaluating $g_0 E_s l$, and consequently for a given *l*, and the calculated g_0,the corresponding value of E_s will be obtained. Also, as pointed out in Ref. [13], since the left – hand side of Eq. (2.17) is independent of E_{in} and E_{out}, the best value of E_s corresponds to the value that minimizes the standard deviation of the right – hand side of Eq. (2.17), when the experimental data are used for the calculation. For example for curve (A) which corresponds to a stored energy $\varepsilon_{store} = 160\,J$, by introducing

$$f(E_s, E_{out}, E_{in}) = \frac{\exp(E_{out}/E_s) - 1}{\exp(E_{in}/E_s) - 1} = \exp(g_0 l) \tag{3.2}$$

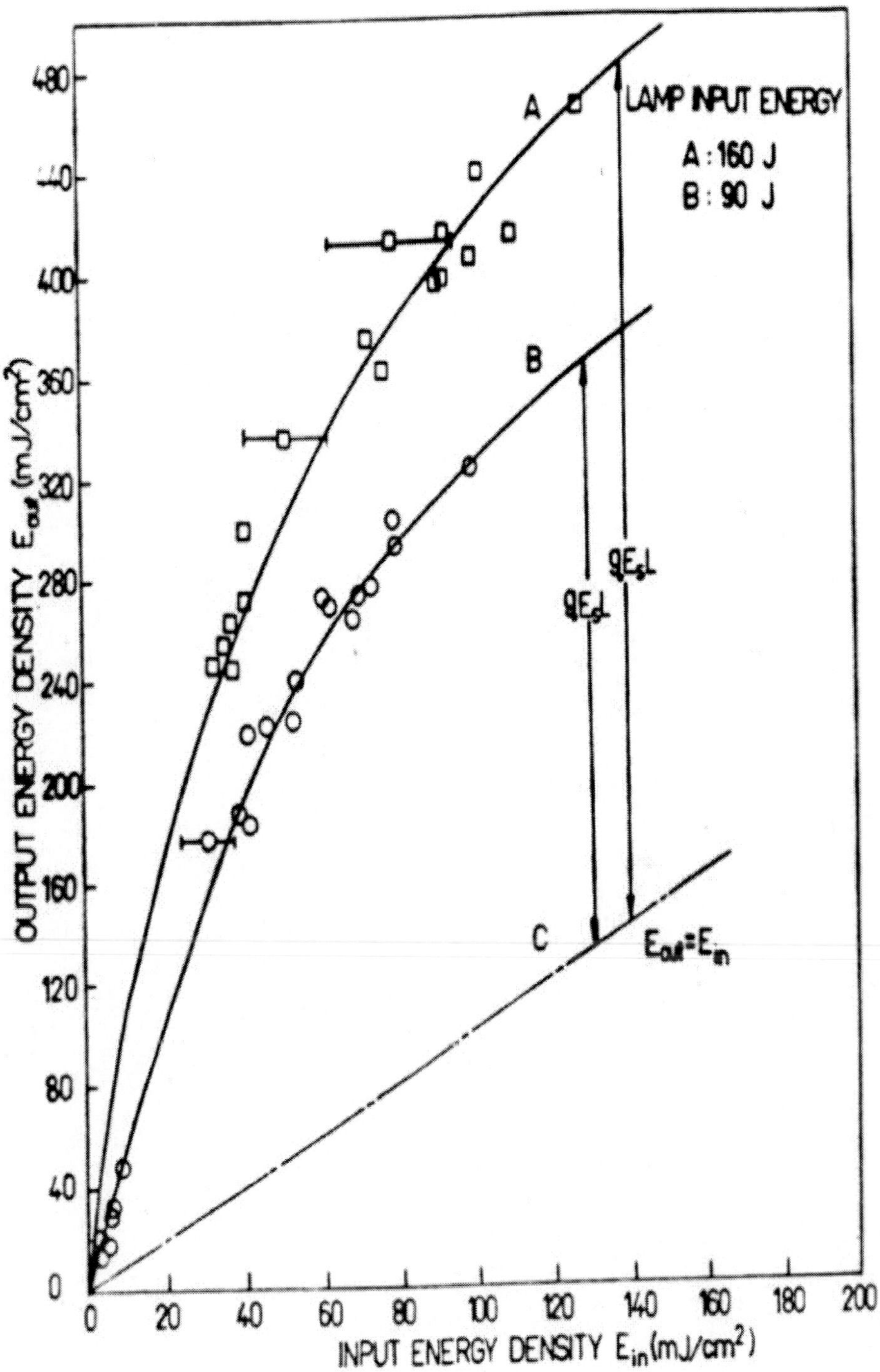

Figure 3.3. Output energy density, E_{out}, versus input energy density, E_{in}, measured under two AMP operational conditions: (A) The plot corresponding to 160 *J* lamp input energy; and (B) corresponds to 90 *J* lamp input energy.

The calculated values for the selected points are:

$$f(E_s, 407, 93.3) = f(E_s, 363, 77) \Rightarrow E_s = 244\, mJ/cm^2 \quad , \quad g_0 = 0.41\, cm^{-1}$$
$$f(E_s, 410, 79) = f(E_s, 253, 35.2) \Rightarrow E_s = 277\, mJ/cm^2 \quad , \quad g_0 = 0.46\, cm^{-1}$$
$$f(E_s, 253, 35.2) = f(E_s, 377, 72.6) \Rightarrow E_s = 183\, mJ/cm^2 \quad , \quad g_0 = 0.49\, cm^{-1}$$

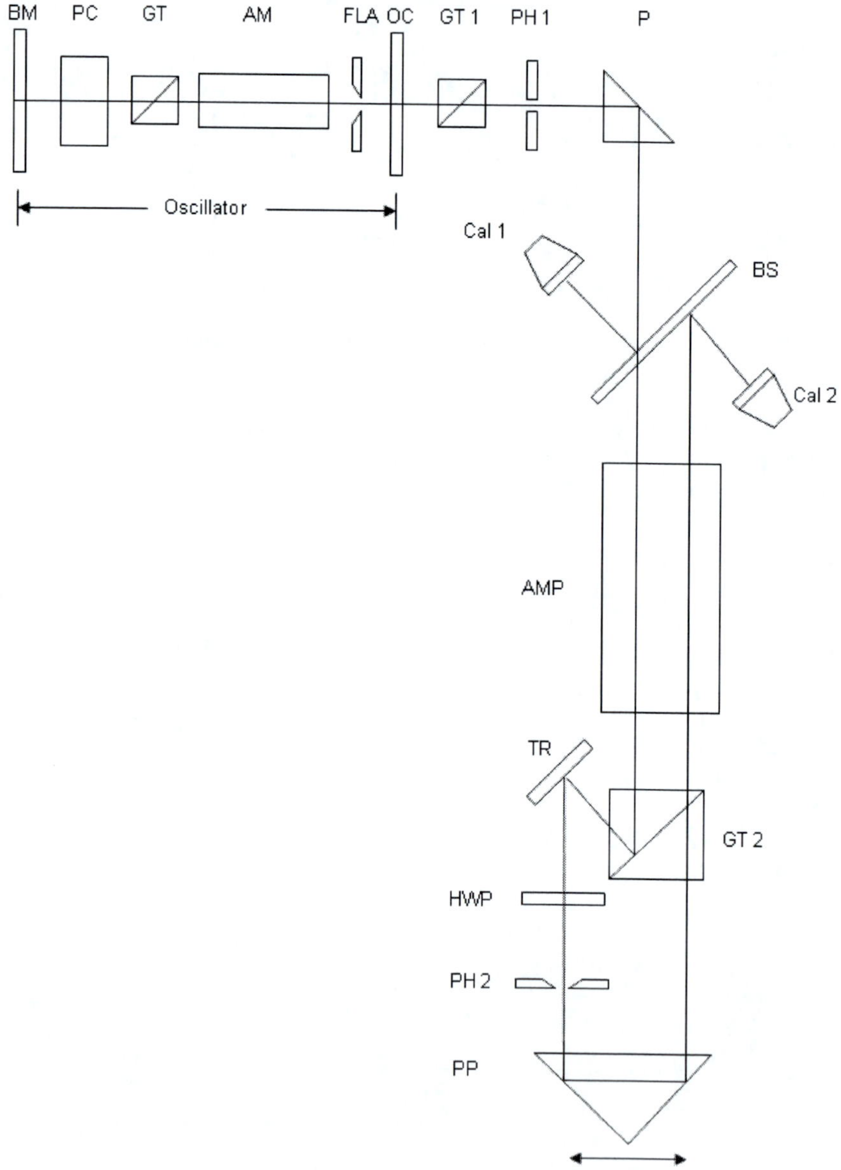

Figure 3.4. An experimental setup arranged for a double-pass amplification study in a Nd:YAG amplifier medium: GT1,2 are Glan Thomson prisms, PC is referring to Pockels' cell used for Q-switching, Cal1,2 are Joule meters, BM and OC are back mirror, and output coupler, HWP is a half-wave plate, TR is a total reflecting mirror, and PP is a reflecting prism.

The average values and the corresponding standard deviations are:

$$\overline{E}_s = 218\, mJ/cm^2 \quad , \quad \sigma_s = 25.7\, mJ/cm^2$$
$$\overline{g}_0 = 0.45\, cm^{-1} \quad , \quad \sigma_g = 0.033\, cm^{-1}$$

The corresponding average values of E_s and g_0 for curve (B) for $\varepsilon_{store} = 90\,J$ along with the standard deviations are as follows

$$\overline{E}_s = 215.8\, mJ/cm^2 \quad , \quad \sigma_s = 29.5\, mJ/cm^2$$
$$\overline{g}_0 = 0.364\, cm^{-1} \quad , \quad \sigma_g = 0.016\, cm^{-1}$$

This observation is indicating that by applying a higher input energy to the flash-lamp, the gain – value increases. It is shown experimentally and by the calculation that g_0 increases almost linearly and then approaches to a saturation level when the input energy to the flash – lamp increases [5].

The tedious work of the past days for evaluating the laser parameters is evident in this example. Now a day, however, by applying least squares fitting of experimental data to the given analytical expressions upon using personal computers, the required parameters can be readily obtained. This will be shown in the related section in this review.

(b) Double-pass Amplification in Nd: YAG Amplifier

At this stage it deserves to describe multi-pass amplification, where it is used for amplification of low energy pulses and high energy pulses with large cross – sections up to ~ 50 cm in diameter. The method improves the amplifier efficiency and in particular it is important to be used in high energy applications, because when a system is operating in its highest efficiency, the number of expensive amplifier stages will be reduced. One way of increasing the efficiency is to choose an appropriate optical pass of the input pulse when it travels through the AMP section in its second – pass. In a single – pass amplification a fraction of the stored energy in the AMP is extracted out from the AMP and a less amplification will be experienced if the second – pass amplification occurs in the same direction as in the first – pass. Thus, this leads to an overall reduction of the extraction efficiency. Referring to [17a], an optical arrangement corresponding to a double – pass amplification is shown in Figure 3.4. The OSC is operating in a free running and also in the Q – switched modes of operation with a 20 *ns* pulse width. No observable difference was found for the results when each of these two laser beams was injected into the AMP, because the 20*ns* Q – switched pulse width is much longer than the time duration for the light that passes through the AMP. A Glan – Thomson Prism GT1 is used to have a birefringence compensated polarized beam from the OSC. The output beam from the OSC section has a beam diameter of 3 *mm*, and the AMP laser rod has a diameter of 9 *mm* and a length of 100 *mm*. A pinhole PH1, shown in the figure, with either 1.8 or 2.5 *mm* diameter is used for reducing the diameter of the input beam entering into the AMP, and also it improves the beam quality. The beam after reflection from prism P, and passing through an uncoated beam splitter BS is injected into the AMP. Two Joule meters, cal1 and cal2, are used to monitor the energies corresponding to the input beam enters into the AMP, and the output laser beam extracted out from the AMP. The Polarizer GT2 is oriented so that the transmitted beam is reflected and then directed toward a half– wave plate and a pinhole PH2 upon reflection by a total reflection mirror, TR. The diameter of the second beam is determined by the pinhole PH2, and for this experiment it is chosen to be 1 *mm*. the half – wave plate, HWP, is set with its axes at 45° with respect to the field polarization direction, where the field polarization experiences a 90° rotation. Thus, the input laser beam will have its second pass through the AMP. The prism PP, when is translated appropriately, carries two duties: firstly, it keeps the parallelism between two beams in the first – and second – pass propagation through the AMP; and secondly, it is used to preserve a given distance between two beams for future studies of the gain – values with respect to distance between the first – and second – pass propagation through the AMP. The output energy corresponding to a double – pass amplification with respect to distance between the first – and second – pass directions is shown in Figure 3.5 (a and b). In Figure 3.5 (a) the input energy enters into the AMP, $E_{in} = 10\, mJ$, and in Figure 3.5 (b) it is 19.5 *mJ.* The flash – lamps in each case are operating with 72 and 98 Joules input stored energies, respectively. In Figure 3.6 a plot corresponding to behavior of the measured gain – parameter with respect to distance between two beams is introduced. The plot shows that when two beams get close to each other we have a reduction in the gain value. Thus, in order to avoid the presence of this attenuation, some appropriate directional adjustment for the second – pass amplification is needed.

From the theoretical point of view if we apply Frantz and Nodvik analytical expression, as given by Eq. (2.18), for a single – pass propagation through the amplifier, and considering T_{12} as the power transmission corresponding to all the optical elements corresponding to OSC and AMP sections. Then, we can write for the first – pass amplification

$$E_{out1}(r) = E_s \ln\left[1 + e^{g_0(r)l}\left(e^{E_{in1}/E_s} - 1\right)\right] \tag{3.3}$$

and for the second - pass we have

$$E_{in2}(r) = T_{12} E_{out1}(r) \tag{3.4}$$

Also, for the second-pass beam amplification, we can write

$$E_{out2}(r) = E_s \ln\left[1 + e^{g_0(r)l}\left(e^{E_{in2}(r)/E_s} - 1\right)\right] \tag{3.5}$$

Eq. (3.5) can be re- arranged to give

$$e^{E_{in2}(r)/E_s} = 1 + e^{-g_0(r)l}\left(e^{E_{out2}(r)/E_s} - 1\right) \tag{3.6}$$

By substituting $E_{in2}(r)$ from Eq. (3.4) into Eq.(3.5), and by finding the term $\exp(E_{out2}(r)/E_s)$ from this equation, we get

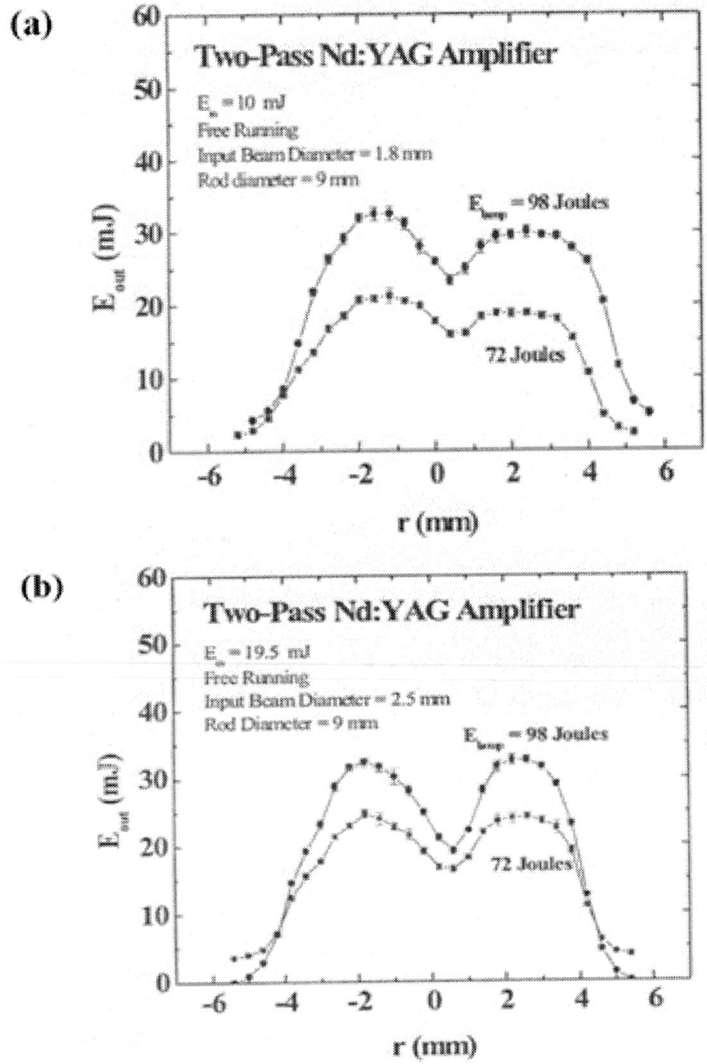

Figure 3.5. The output energy extracted from AMP with respect to distance between the first and the second-pass beams. Measurements have been carried out for: (a) $E_{in} = 10\ mJ$, and (b) 19.5 *mJ* input energy. The input laser beam diameter in (a) is 1.88 *mm* and in (b) is 2.5 *mm*.

$$e^{E_{out2}(r)/E_s} = 1 + e^{g_0(r)l}\left(-1 + e^{T_{12}E_{out1}(r)/E_s}\right)$$

With the use of Eq. (3.3), we will get an expression relating the energy density out from the AMP in the second- pass, i.e. $E_{out2}(r)$, in terms of the energy density enters into the AMP in the first – pass, i.e. E_{in1},

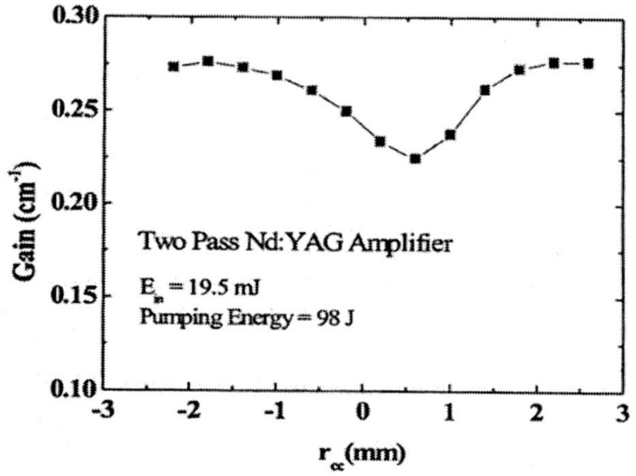

Figure 3.6. The gain behavior with respect to distance between two beams, for 98 *J* pumping energy. The input beam diameter is 2.5 *mm*, and the input energy to the AMP is 19.5 *mJ*.

$$E_{out2}(r) = E_s \ln\left[1 + e^{g_0(r)l}\left\{\left[1 + e^{g_0(r)l}\left(e^{E_{in1}/E_s} - 1\right)\right]^{T_{12}} - 1\right\}\right] \qquad (3.7)$$

For $T_{12} = 1$, Eq. (3.7) simplifies to

$$E_{out2}(r) = E_s \ln\left[1 + e^{2g_0(r)l}\left(e^{E_{in1}/E_s} - 1\right)\right] \qquad (3.8)$$

indicating that in this case the double – pass amplification is equal to $2l$ optical – pass with an amplifier gain coefficient of $g_0(r)$.

We can further define the extraction efficiency in single – and double – pass amplifications as

$\eta_1 = \dfrac{E_{out1}(r) - E_{in1}}{g_0(r)E_s l}$ and $\eta_2 = \dfrac{E_{out2}(r) - E_{in2}(r)}{g_0(r)E_s l}$ where the overall efficiency will be

$$\eta = \eta_1(r) \cdot \eta_2(r) \qquad (3.9)$$

The corresponding values of η_1, η_2, and η with respect to distance between two beams are shown in Figure 3.7. The results are indicating that in a complete overlapping of two beams, the amplifier has its minimum efficiency, and consequently the extraction energy is minimized. It is expected that when the input pulse has a pulse width comparable to the time duration for light passing through the amplifier, or even larger, the same behavior should be observable in other solid – state media such as Nd: glass lasers.

Eq. (3.8) which is referring to the use Frantz and Nodvik equation for light propagation in one direction (say, +z direction) through the amplifier, can be further modified in a double-pass amplification having flux densities of $n_{ph}^{(1)}$ and $n_{ph}^{(2)}$ propagate along the z-axis in the positive and negative directions, respectively [17b]. For this purpose the rate equations given by Eqs. (2.1) and (2.2), are written as

$$\frac{\partial n_{ph}^{(1)}(z,t)}{\partial t}+c\frac{\partial n_{ph}^{(1)}(z,t)}{\partial z}=c\sigma_{st}N(z,t)n_{ph}^{(1)}(z,t) \tag{3.10 a}$$

$$\frac{\partial n_{ph}^{(2)}(z,t)}{\partial t}-c\frac{\partial n_{ph}^{(2)}(z,t)}{\partial z}=c\sigma_{st}N(z,t)n_{ph}^{(2)}(z,t) \tag{3.10 b}$$

$$\frac{\partial N(z,t)}{\partial t}=-c\sigma_{st}N(z,t)\left(n_{ph}^{(1)}(z,t)+n_{ph}^{(2)}(z,t)\right) \tag{3.10 c}$$

A change of variable $z'=z$ and $t'=t-z/c$, as given in section 2, reduces (3.10 a) – (3.10 c) to

$$\frac{\partial n_{ph}^{(1)}}{\partial z'}=\sigma_{st}Nn_{ph}^{(1)} \tag{3.11 a}$$

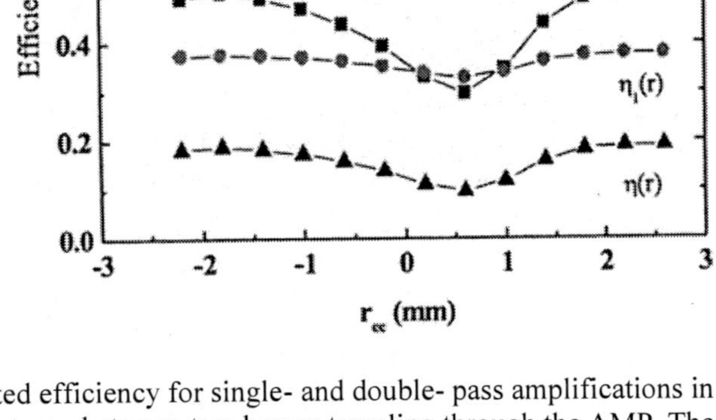

Figure 3.7. The extracted efficiency for single- and double- pass amplifications in a Nd:YAG laser amplifier versus the distance between two beams traveling through the AMP. The overall extraction efficiency is $\eta=\eta_1\eta_2$, and is also shown in the figure.

$$2\frac{\partial n_{ph}^{(2)}}{\partial t'} - c\frac{\partial n_{ph}^{(2)}}{\partial z'} = c\sigma_{st} N n_{ph}^{(2)} \quad (3.11\text{ b})$$

$$\frac{\partial N}{\partial t'} = -\sigma_{st} N \left(n_{ph}^{(1)} + n_{ph}^{(2)}\right) \quad (3.11\text{ c})$$

where the prime notations on the letters are referring to the moving frame of reference; and $N(z',t') = N(z,t) \equiv N$; $n_{ph}^{(1,2)}(z',t') = n_{ph}^{(1,2)}(z,t) \equiv n_{ph}^{(1,2)}$ have been used. Details of calculation and the approximation used for the final results appeared in [17 b]. The following result for the output energy density is given by

$$E_{out2} = \left(\frac{E_s}{2}\right) \ln\left[1 + \exp(2g_0 l)\left\{\exp\left(\frac{E_{in1}}{E_s/2}\right) - 1\right\}\right] \quad (3.13)$$

Comparison of Eq. (3.8) and Eq. (3.13) shows that for applying two wave propagation for two overlapping beams, E_s changes to $E_s/2$, but the gain coefficient, g_0, for a two-pass amplification of a total length 2*l*, remains the same.

3.2 Amplification in Transversely Excited (TE) Nitrogen Lasers

Long channel N2-laser Amplifier

In principal the same requirement mentioned in the previous section has to be considered for obtaining the laser parameters in N_2-lasers, except with the realization of this fact that in nitrogen lasers fast excitation circuitries, such as Blumlein circuit are used. Thus, we are facing with a time adjustment of a few *ns* desynchronization time between OSC and AMP sections. To obtain a proper desynchronization time between these two sections, generally 3 different approaches are used. One method is to use an OSC and an AMP laser system operating at appropriate pressures with respect to each other. In a reported experiment it was shown that when two N_2-lasers of equal designs are operating as an OSC – AMP system, difference in the lasing time between two OSC and AMP sections is proportional to the difference in the laser gas pressure [18]. The other method is to change the gap separation between electrodes in the OSC section. This adjustment needs a fine movement of electrodes during the laser operation, so the design must be considered in the laser fabrication. The last method is to position appropriately the reflecting mirrors used for directing the output beam from the OSC section into the AMP. In an experiment with the need of preserving the laser geometrical configuration, applying the gas pressures adjustment and mirrors displacement are to be considered for measurements. An experimental setup which was used for the gain measurement in a long channel amplifier is shown in Figure 3.8 [19]. A $107 \times 116\, cm^2$ commercial printed circuit board is double – sided copper with a thickness of 1.5 *mm* fiberglass insulation. The board has been cut to produce two excitation circuits of

$49 \times 34\ cm^2$ for the OSC; and $107 \times 62\ cm^2$ for the AMP sections. A 25-*mm* copper strip is removed from sides of the printed board to prevent arcing. Also 25- and 45- *mm* strips of copper are removed from central parts of the OSC, and AMP sections, respectively, to produce three capacitors of 1.8, 5.0 and 7.4 *nF* each. The OSC excitation section on the circuit board is in a shape of a double trapezoid with the shorter and longer sides of 34 and 38 *cm*, respectively. In order to minimize the overall circuit inductance, a single spark gap, SG, working at atmospheric N_2 pressure is directly solder to the circuit board, close to the OSC – section. The impedance of the spark gap in this arrangement is minimized, but due to direct contact of the sparks with the fiberglass insulator, the system operational lifetime is limited at the high repetition rates. The gas pressures for the input of each laser channel are adjusted by two input values, one for the OSC and the other for the AMP. The gases are exhausted from the exit channels of the OSC and AMP sections and are controlled by two independent pressure gauges. The N_2 gas of 99.99 % purity is flowed slowly through the laser channels. The gauges have been calibrated against a MKS Baratron of type 220 BA. For the OSC section, a design was adopt to utilize the corona type discharge. For the electrodes of the OSC flat plates of aluminum with a thickness of 10 *mm* and a length of 34 *cm* were used. One electrode is held into a U – shaped slot and tightened by screws, while the other is capable of being adjusted horizontally with two screws fixed at both ends of the electrode.

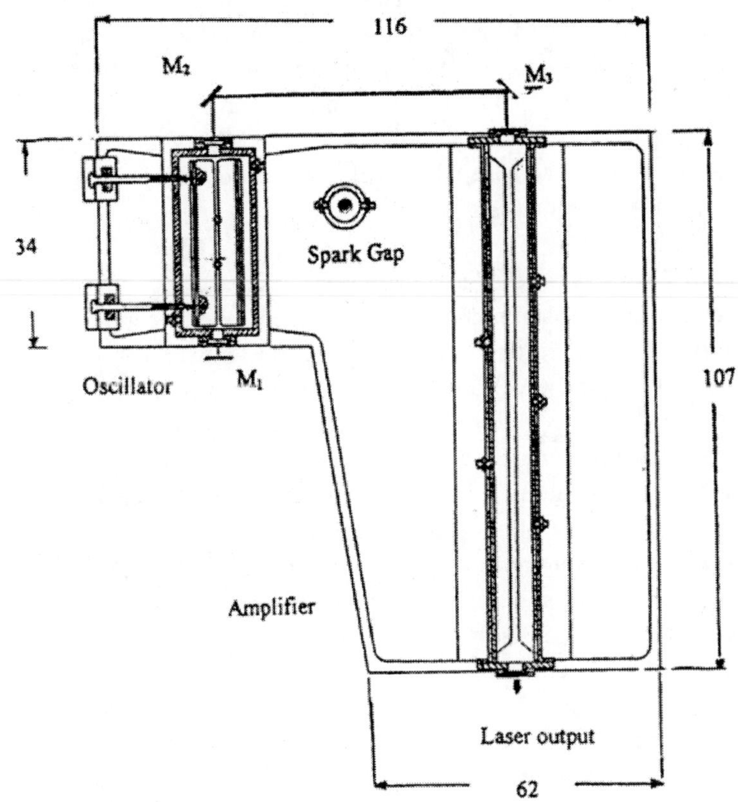

Figure 3.8. Top view of the TE-TE N_2-laser used for gain measurement in a long channel amplifier. All units are in *cm*.

The electrodes are mounted mechanically on aluminum holders and copper plates, where the later plates are soldered to the copper of the circuit board. The electrodes assembly, with electrodes fixed at a height of 35 *mm* above the printed board, are confined in a vacuum tight rectangular enclosure of Plexiglas with a thickness of 2 *cm* such that the lateral side of the circuit board is carried out and tightened to the lower section of the enclosure against an O – ring. Two quartz windows of 5 *mm* thickness and diameter of 4.5 *cm* are mounted at each end of the enclosure. The N_2 gas is flowed from the top cover and exhausted from both sides. When this system was used as an oscillator with a total capacitance of 4.7 *nF*, under an optimal condition of 350 *torr* gas pressure and the pulse repetition rates of 10 – 50 *Hz*, the output pulse power in excess of 70 *kW* with the pulse width of 2.5 *ns* was obtained [20].

The AMP part of system is a long channel laser of a length of 1*m* and electrodes gap separation of 1.5 *cm*. The electrodes lengths are 94 *cm* each and are made of cylindrical rods of stainless steel. The ends of each electrode are cut at some angle to avoid arcing. Two shaped aluminum plates of 1*m* long are holding the electrodes in parallel. For the lower and upper parts of the enclosure, two plates of 1*m* long and 2 *cm* thick Plexiglas are used. All 4 plates are tightened with O – ring. A number of pins, made of brass, are mounted on the upper part of the enclosure to work as preionizers. The ends of the enclosure are sealed with two quartz windows. The enclosure is vacuum tight for holding the gas pressure. When this system was operating as an OSC, the peak output power of 200 *kW* and pulse width of ~ 6 *ns* FWHM were obtained. These two laser channels were set on their positions on the circuit board to operate as an OSC – AMP laser system, as shown in Figure 3.8. In this arrangement, the smaller laser channel works as an OSC, while the longer laser channel is intended to work as an AMP. The optical cavity of the OSC is formed by an Al – coated mirror M_1 and a flat fused silica mirror, M_2. Two Al – coated mirrors M_3 and M_4 are used to direct the OSC beam into the AMP section. They can also be used for the necessary time – delay adjustment between the OSC and AMP sections.

General Aspect of Experimental Measurements

The procedure for the measurements that appears this review can be commonly applied to any type of lasers. Two important measurements are: power (or energy) measurements corresponding to the OSC and AMP output beams and the other are the time behavior of these two laser beams, in particular, when we are in the process of obtaining the condition for the required desynchronization time between OSC and AMP sections. For the time adjustment it is necessary to observe the output laser pulses while adjusting the operational gas pressures. As the procedure is generally applied in all measurements including the laser parameters appeared in this paper, therefore repeating the procedure in other relevant sections is avoided.

For the OSC – AMP laser systems appeared in this review, the laser pulses were recorded with a fast detection system consisting of a vacuum photodiode with 100 *ps* rise time (ITL, Model TF1850), and a fast oscilloscope (Model 7904 Tektronix with a 7A10 Plug – in unit of 500 *MHz* band width). The rise time corresponding to this combination is ~ 0.8 *ns*. With this combination a real time behavior of fast optical pulses was visualized and proved its fidelity compared with the digital monitoring equipment.

For the output energy measurements from the OSC as well as from the AMP a photodiode with a read – out micro - ammeter, calibrated at 337.1 *nm* wavelength against a disk calorimeter was used. For the photodiode calibration a laser beam from a TE laser operating at 337.1 *nm* was split into two separated beams. One beam was directed toward the disk calorimeter (Scientech, Model 360001) and displayed by a laser meter (Scientech, Model 1362), and the other was incident on the photodiode. By attenuating the TE N_2-laser beam by the use of various neutral density filters, two variable laser beams of equal intensities were produced. The average power measured by the calorimeter was recorded and the laser pulse energy was calculated by dividing by the pulse repetition rate. The measured output energy versus photodiode read – out was almost linearly correlated and the plot was used throughout the energy measurements when the laser output energy from the laser system was not high enough to be measured directly by the disk calorimeter.

The spatial intensity distribution of the laser output from both OSC and AMP were determined by scanning a pinhole (0.22 *mm* diameter) across the laser beam with the calibrated photodiode in conjunction with various neutral density filters. This measurement is important for determining the beam cross-sections corresponding to the output beams from OSC and AMP sections.

Gain and Saturation Energy Density Measurements

For the gain measurements, the output – input energy density (E_o and E_i) were determined. For this purpose E_o and E_i were calculated using $E_o = \varepsilon_o / A_o$ and $E_i = \varepsilon_i / A_i$, where ε_o and ε_i are the measured energy extracted from the amplifier and energy that enters into the amplifier, A_o and A_i are their corresponding beam cross – sections. For variation of the laser energy entering the AMP, several calibrated neutral density fillers were used. g_0 and E_s were obtained by the method of least – squares fit of the experimental points to Frantz and Nodvik equation given by Eq. (2.18), using 4.15 MB nonlinear fitting program (Origin, version 3 and 5). The results of measurement corresponding to [19], where an AMP of an effective length of 94 *cm* was used to measure the gain parameter is given in Figure 3.9. The laser pulses corresponding to different gas pressures are also given to clarify the effect of gas pressure for obtaining a complete amplification, as shown in figures 3.10 (a)-(c). At the input voltage of 17 *kV* when the system was operating at ~ 150 *Hz*, an average output power of > 30 *mW* from the OSC-AMP laser system with a long channel AMP was obtained. The power measurement corresponding to this mode of operation is given in figures 3.11. The results of g_0 and E_s measurements for $l_{AMP} = 94\ cm$ are tabulated in Table 1.

Table 1. Experimental results of g_0 and E_s parameters in an AMP with $l_{AMP} = 94 cm$ in a TE – TE OSC – AMP N_2-laser system

Specifications	$g_0(cm^{-1})$	$E_s(\mu J cm^{-2})$	curve
$p_{OSC} = 190 mbar, p_{AMP} = 84 mbar, V_0 = 14 kV$	0.064	65	(A)
$p_{OSC} = 190 mbar, p_{AMP} = 52 mbar, V_0 = 14 kV$	0.064	42	(B)
$p_{OSC} = 170 mbar, p_{AMP} = 72 mbar, V_0 = 12 kV$	0.068	25	(C)

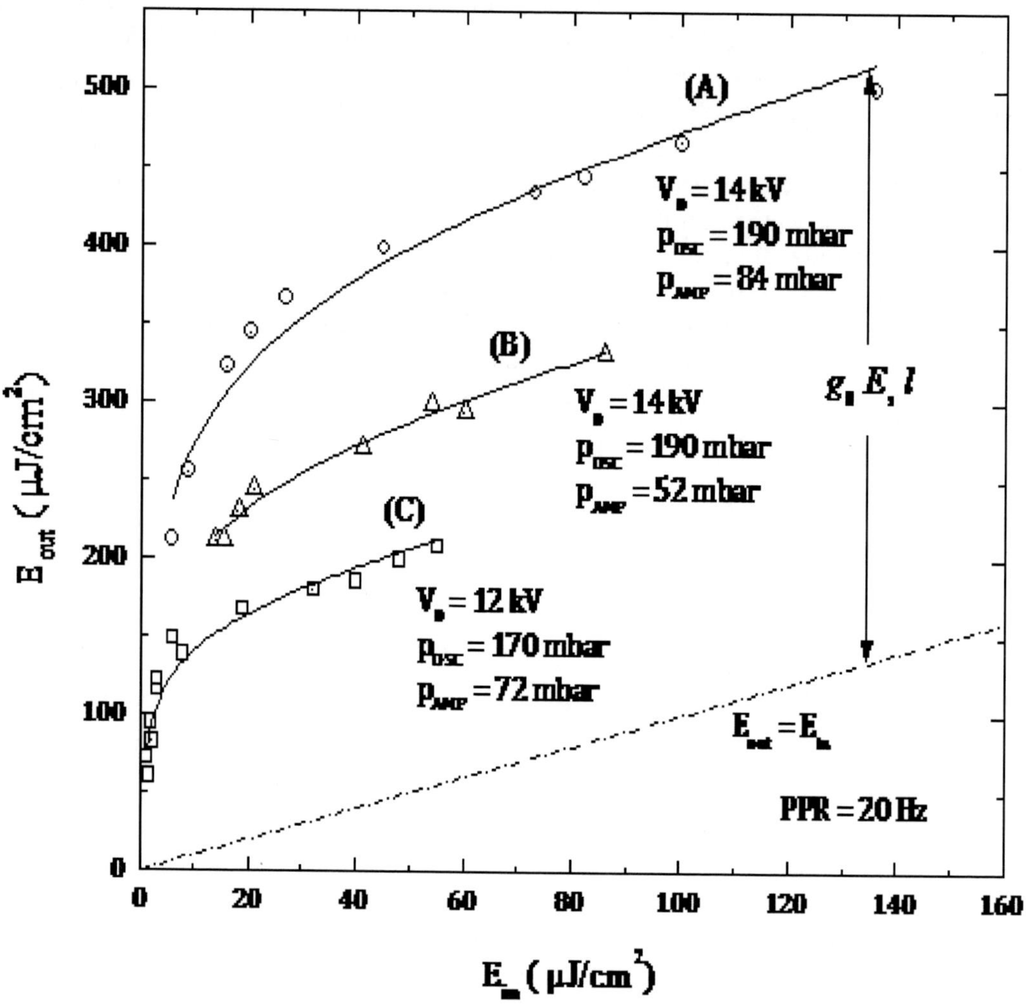

Figure 3.9. Plot of output energy density vs. input energy density for an AMP of 94 *cm* electrodes length. The operational gas pressures and input voltages are given in the inset of the figure.

As it is shown from the corresponding curves of E_{out} vs. E_{in}, as given in figures 3.9, 3.10(c) and Table 1, curve A with $p_{OSC} = 190\,mbar$, and $p_{AMP} = 84\,mbar$ gives the highest saturation energy density and that is due to a complete delay – time adjustment. By reducing p_{AMP} to $52\,mbar$, as it is indicated in curve B, a drastic reduction in E_s – parameter will appear. This experiment has shown that by selecting a long channel amplifier, the g_0 – parameter reduces to a small quantity and stays at its value more or less constant. The effect of external parameters such as gas pressure and input voltage shows itself on the values of the saturation energy density. With a further progress in the examples that will be given in this review, it proves that the geometrical configuration of N_2 – laser amplifiers plays an important role in the gain value.

(a)
(b)
(c)

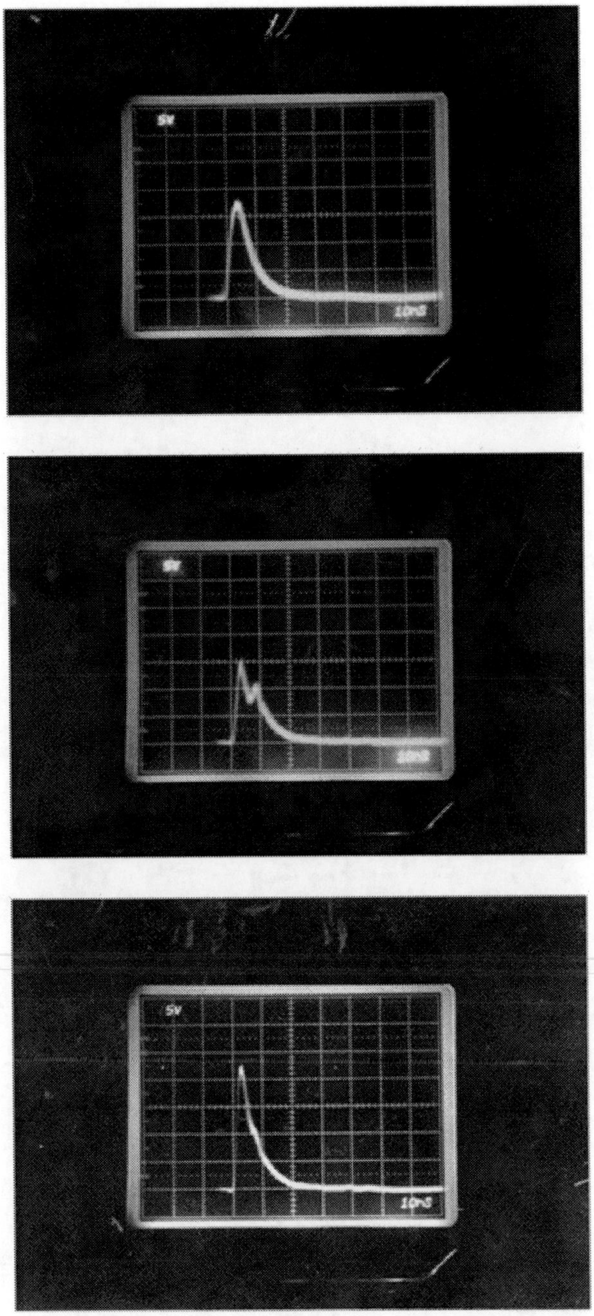

Figure 3.10. Typical laser output pulse shapes from the AMP: (a) when the output pulse from the OSC is not entering the AMP, i.e., the pulse is due to ASE from the amplifier. The AMP gas pressure is 90 *mbar* and $V_0 = 14\ kV$. The horizontal scale is 10 *ns*; (b) the AMP output pulse showing an incomplete amplification. $p_{OSC} = 110\ mbar$, and $p_{AMP} = 80\ mbar$; (c) the pulse shape shows a complete amplification at $p_{OSC} = 190\ mbar$ and $p_{AMP} = 80\ mbar$. The OSC and AMP electrodes gap separations are 5 and 15 *mm*, respectively.

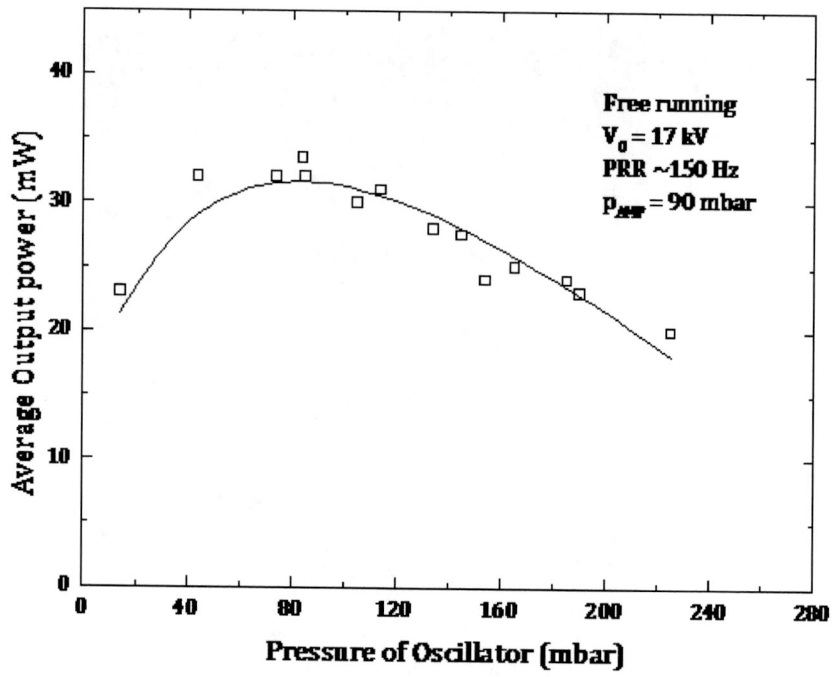

Figure 3.11. Plot of output average power vs. OSC gas pressure when the OSC-AMP is operating in the free running mode of ~ 150 *Hz* at 17 *kV* input voltage. $\overline{P} = 30\ mW$ for $p_{OSC} = 80\ mbar$.

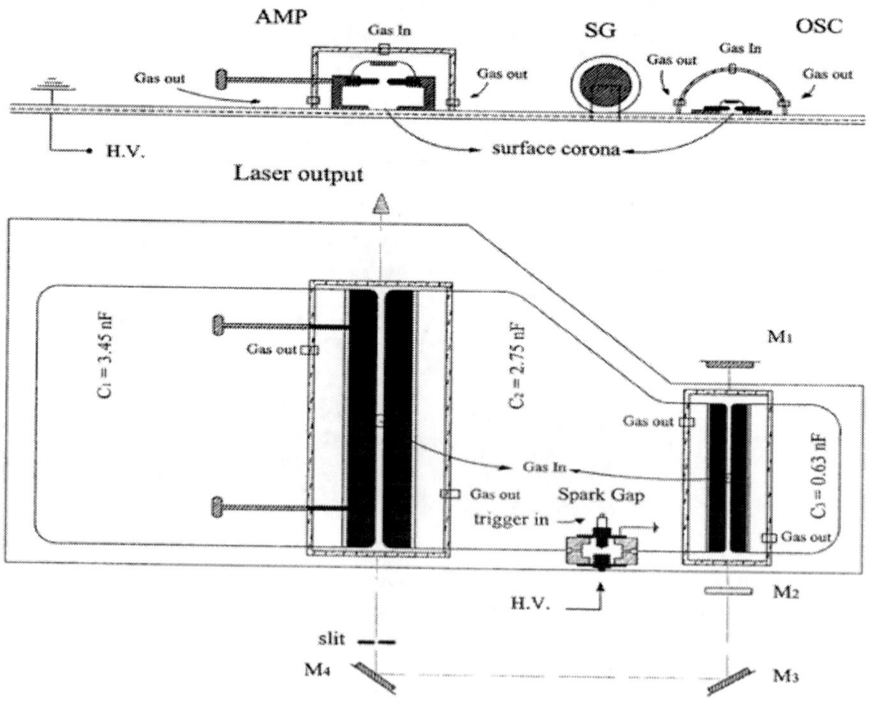

Figure 3.12. Cross-sectional and top views of the OSC-AMP N_2-laser system with $l_{AMP} = 31\,cm$.

(b) Short channel N_2-laser Amplifier

The schematic diagram of the laser system consists of a TE oscillator and a TE amplifier, both set to operate using a flat – plate Blumlein circuit with corona – type discharge preionizers is shown in Figure 3.12 [21]. The excitation circuit in this discharge is a $93 \times 44\, cm^2$ commercial printed circuit board. The board is divided to produce two excitation circuits of $50 \times 44\, cm^2$ for the AMP and $43 \times 24\, cm^2$ for the OSC sections. 30- and 23-*mm* strips of copper are removed from the central parts of the AMP and OSC circuit board to produce three capacitors of 3.45, 2.75, and 0.63 *nF* each. These areas serve as corona UV sources of radiation. The AMP section is the same as the OSC part that introduced in part (a) of this section. The electrodes are positioned at a height of 35 *mm* above the surface corona discharge, consequently it can be realized that for the gain and saturation energy measurements the corona preionization in the AMP section works as a rather mild (or weak) preionizing mechanism, and it is used mainly for maintaining a uniform gas discharge throughout the experiment. The corona – electrode distance for the OSC was chosen to be 10 *mm*.

The electrodes of both AMP and OSC sections are flat plates and are made of aluminum. For the AMP section, the thickness of the electrodes is 10 *mm* with the total and effective lengths of 34 and 31 *cm*, respectively. The electrodes gap separation can be varied from 3 to 11 *mm*, but in most of our measurement it has been fixed at 7 *mm*. A thickness of 6 *mm* and a length of 19 *cm* are chosen for the OSC electrodes. The gap separation of electrodes for the OSC is fixed at 5 *mm*. The shape of the electrodes has the profile of a semicircle of 5 *mm* and 3 *mm* radii for the AMP and OSC, respectively. A single spark-gap working at atmospheric N_2 gas pressure is directly soldered to the upper section of the circuit board at a distance of 24.5 and 13.2 *cm* from the AMP and OSC centerlines of the laser beam. The rest of the design corresponding to the present TE – TE laser system is similar to the design introduced in section 3.2 (a).

Gain and Saturation Energy Density Measurements

The Effect of Corona on g_0- and E_s- Values

Typical laser pulse shapes from the OSC and AMP outputs operating at pressures of 110 and 100 *torr*, respectively, are given in Figure 3.13. For obtaining the optimal OSC – AMP desynchronization time at a given OSC gas pressure of 110 *torr*, the AMP gas pressure was varied, while the output energy from the AMP was monitored. Typical measurements for this observation are shown in Figure 3.14 a – c, where the output energy density, E_o, is plotted versus the AMP gas pressure for three different operational conditions: when the corona preionization is effectively involved in the laser operation (Figure 3.14 a); it is masked by a UV – absorbing glass plate (Figure 3.14 b); and it is totally masked by a black insulation (Figure 3.14c). The corresponding E/p_{AMP} values (where E is the electric field and p_{AMP} is the amplifier gas pressure) at the peak points of curves b and c show reductions by factors of 1.14 and 1.33 with respect to the peak value of curve a (i.e., for the case when corona preionization is used). The shift in the AMP gas pressure toward the lower pressure upon eliminating the corona effect is also evident in this figure.

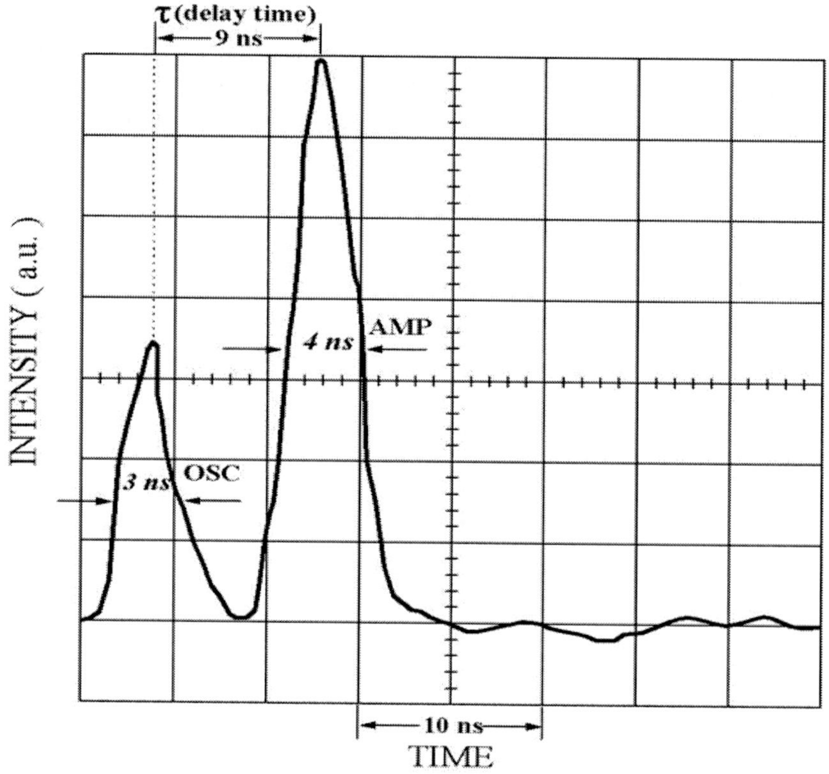

Figure 3.13. Typical laser pulse shapes obtained from OSC and AMP outputs in a TE-TE laser with $l_{AMP} = 31\,cm$. The OSC and AMP are operating at the gas pressures of 110 and 100 *torr*, respectively. The corresponding FWHM of laser pulses are 3 and 4 *ns*.

For the gain measurements, the output – input energy densities (E_o and E_i) for the OSC gas pressure of 110 *torr* and for different AMP gas pressure of 80 and 60 *torr* in a linear – linear plot of E_o versus E_i for the case of applying UV corona preionization in the gas discharge are shown in Figure 3.15.

The solid and dashed lines in this figure correspond to Frantz and Nodvik equation given by Eq. (2.18). g_0 and E_s were obtained by the method of least squares fit of the experimental points by the proposed theoretical model. Due to the presence of amplified spontaneous emission (ASE) in the AMP, the data points corresponding to the input energy densities lower than $10^{-1}\,\mu J/cm^2$ have not been taken into account. Also the data points corresponding to the high input energy density, due to the pumping mechanism are eliminated. The deduced values for the gain and saturation energy density show that by reducing the AMP gas pressure from 80 torr to 60 torr (where at the AMP gas pressure of 60 torr, we do not have a proper desynchronization time between the OSC and AMP sections), we have smaller values for the laser parameters. Thus, a proper delay time adjustment gives correct values for the laser parameters. The same measurements have also been carried out for the cases of blocking the corona by a UV absorbing glass plate and a black insulation. Results of these measurements along with the laser operational conditions are given in Table 2.

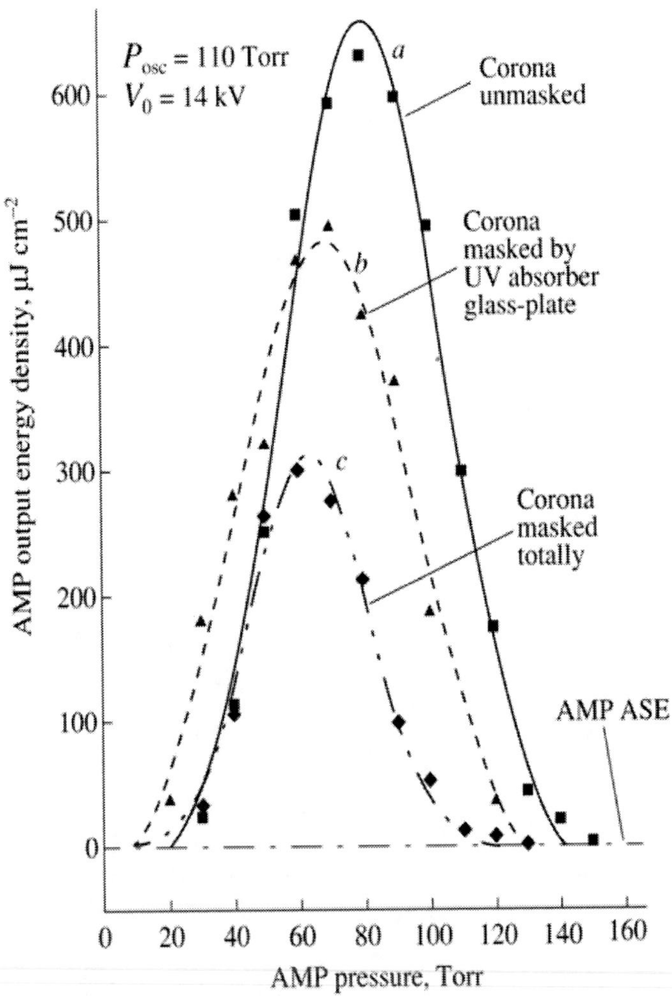

Figure 3.14. Plots of measured output energy densities versus AMP gas pressure for: (a) corona is involved in the laser operation; (b) corona is masked by a UV-absorbing glass plate; and (c) corona is masked totally by black insulation.

Table 2. Experimental results for g_0, E_s, and E_o measured for different states of corona preionization. $p_{OSC} = 110 torr$, and $V_0 = 14 kV$, $PRR = 10 Hz$ in a N_2-laser

Experimental condition	$p_{AMP}(torr)$	$g_0(cm^{-1})$	$E_s(\mu Jcm^{-2})$	$E_o(\mu Jcm^{-2})$	E/p_{AMP} $(Vtorr^{-1}cm^{-1})$
Unmasked corona	80	0.263	100.35	632.28	250.00
Corona is masked by a UV absorber glass plate	70	0.222	87.27	493.09	285.71
Corona is masked totally	60	0.201	60.52	298.43	333.33

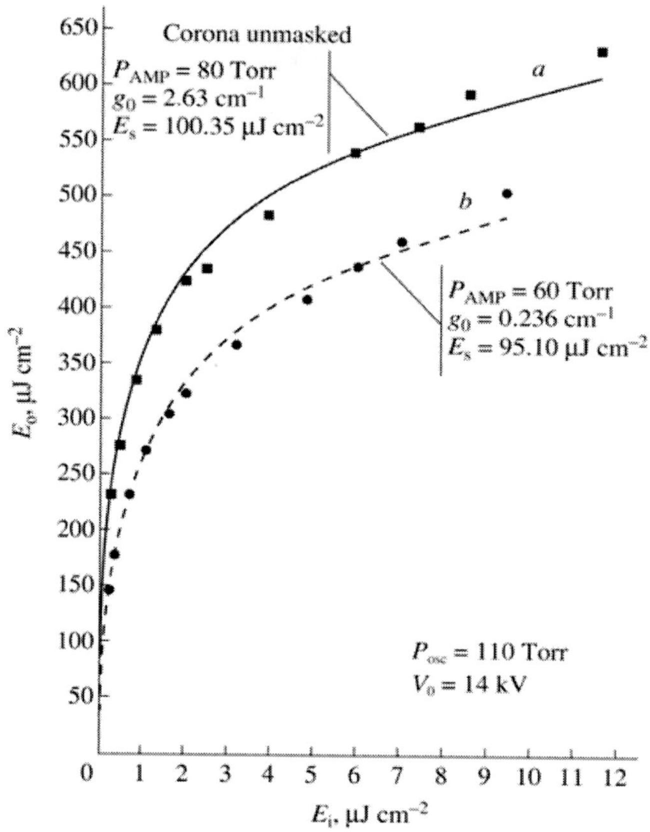

Figure 3.15. Typical linear-linear plots of output-input energy densities for: (a) 80 *torr* AMP gas pressure; and (b) 60 *torr* AMP gas pressure. $p_{OSC} = 110\ torr$ in both cases.

As it is shown in the Table, the measured values of E_o, g_0, and E_s are reduced upon blocking corona preionization with different absorber. Also, the AMP gas pressures decrease and consequently the E/p_{AMP} - values which is related to the electron temperature, T_e, increase.

The Effect of He on g_0- and E_s- Values

Gain measurements have also been carried out for different gas mixtures of He and N_2. For these measurements, the OSC – AMP laser system was allowed to operate at 110 and 100 *torr* of N_2 gas pressure, while the He buffer gas was mixed with the N_2 before entering the AMP. Typical results of the measurements for 64 and 300 *torr* of He are shown in Figure 3.16 curves (a and b). For a range of 0 – 150 *torr* of He gas pressure, it was observed that the measured values for the small – signal gain and saturation energy density remain almost constant at the corresponding values of $0.303 \pm 0.01\, cm^{-1}$ and $88.5 \pm 3\, \mu J cm^{-2}$, respectively, as shown in Figure 3.17. Above 150 torr of He buffer gas pressure, drastic reduction in g_0 and E_s values were obtained. Our observation verifies the fact that, in a laser system with a non-uniform discharge condition, addition of He, even in a small percentage, has an important effect on uniformizing the discharge and the shot – to – shot pulse reproducibility. In the present system having an excellent discharge uniformity, we found no substantial enhancement on the output energy and pulse – to – pulse reproducibility when a few *torr* of He was added to the gas discharge, which is consistent with Armandillo's observation [22].The maximum output energy enhancement was at most 10% when He with higher gas pressure (<150 *torr*) was added to the gas mixture. The behavior of the normalized output energy versus the He gas pressure obtained by the reproducing the experimental data points of Brito Cruz [23] and our measurements [21] are depicted in Figure 3.18. Each dashed curve in this figure corresponds to a group of N_2 gas pressure having 0 – 20, 30, and 50 *torr* pressure variations. Except at the zero He gas pressure, where a very low output energy was reported by the authors of reference [23] and the appearance of sudden jumps in each of the curves that refer to their laser arc suppression, both experiments have shown clearly that, regardless of type of preionizers, the laser output energy drops drastically at the high He gas pressure.

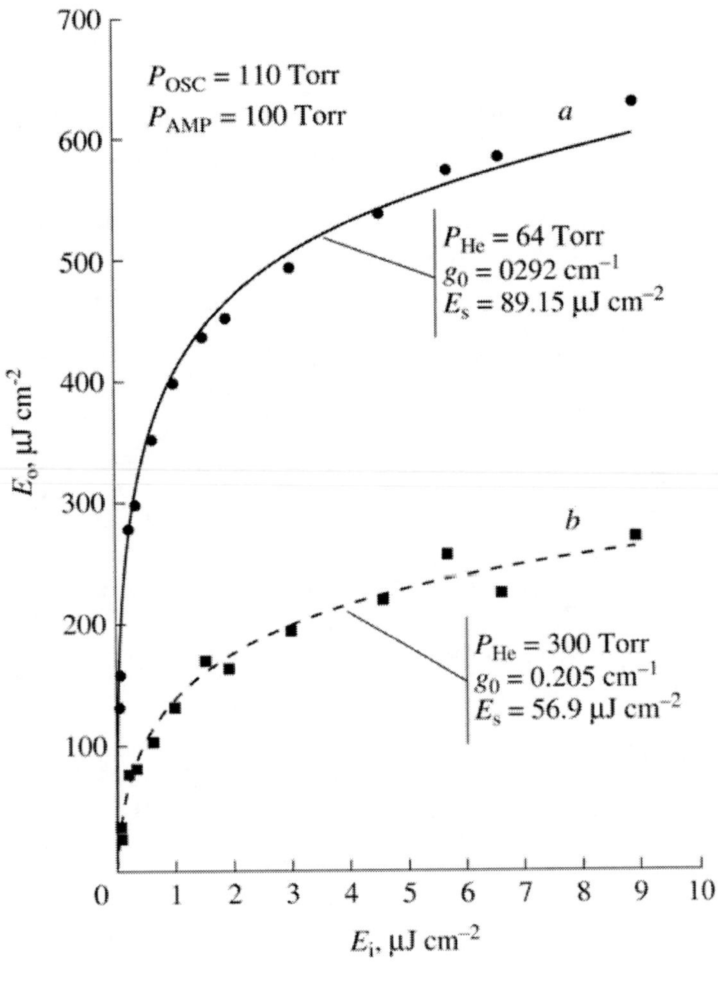

Figure 3.16. Plots of measured output energy density versus input energy density at $p_{OSC} = 110\ torr$, $p_{AMP} = 100\ torr$ for: (a) 64 *torr* ; and (b) 300 *torr* of *He* buffer gas.

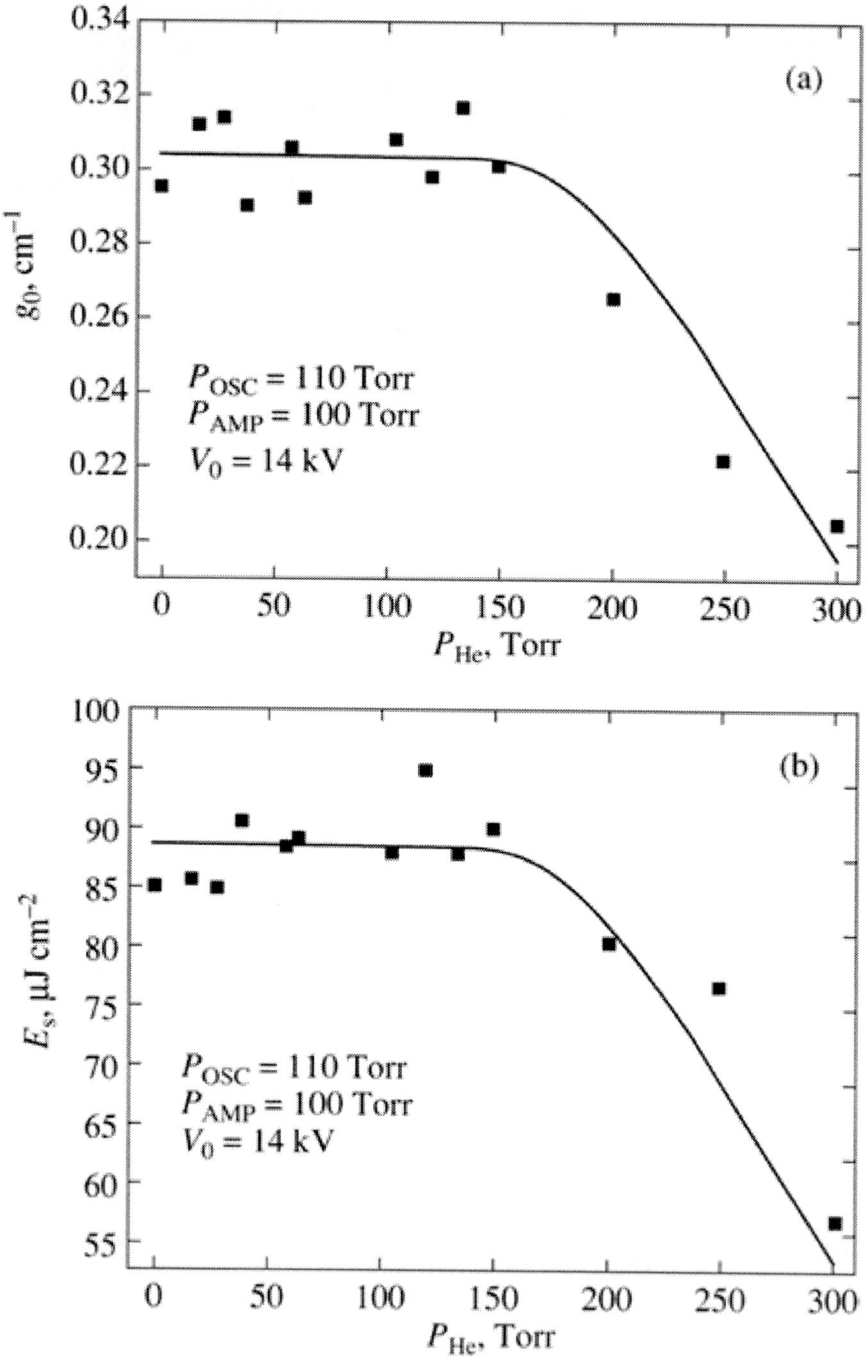

Figure 3.17. Plots of: (a) small-signal gain, g_0; and (b) saturation energy density, E_s, versus the pressure of *He* buffer gas. g_0 and E_s are almost constant in the range of 0-150 *torr* of *He*; $g_0 = 0.303 \pm 0.01\,cm^{-1}$, and $E_s = 88.4 \pm 3\,\mu J cm^{-2}$.

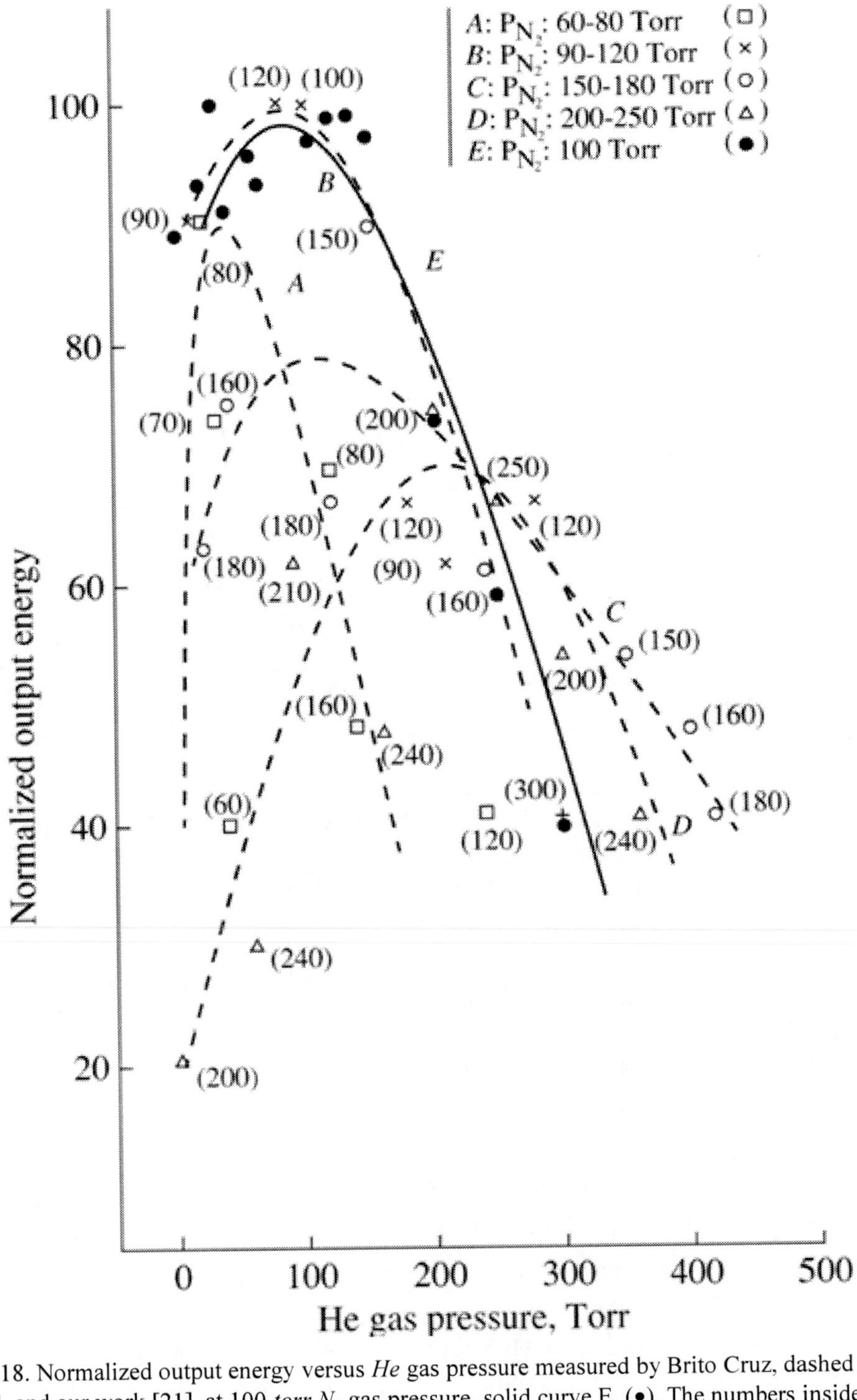

Figure 3.18. Normalized output energy versus *He* gas pressure measured by Brito Cruz, dashed curves A-D [23], and our work [21], at 100 *torr* N_2 gas pressure, solid curve E, (•). The numbers inside the parentheses refer to the laser N_2 gas pressures. The dashed-curves have been depicted for different N_2 gas pressures with N_2 pressure variations of 0-20, 30, and 50 *torr*: A (60-80 *torr*); B (90-120 *torr*); C (150-180 *torr*); and D (200-250 *torr*). A behavioral agreement was found between our work at $p_{N_2} = 100\,torr$ (curve E) [21] and Bito Cruz observation at $p_{N_2} = 90-120\,torr$ (curve B).

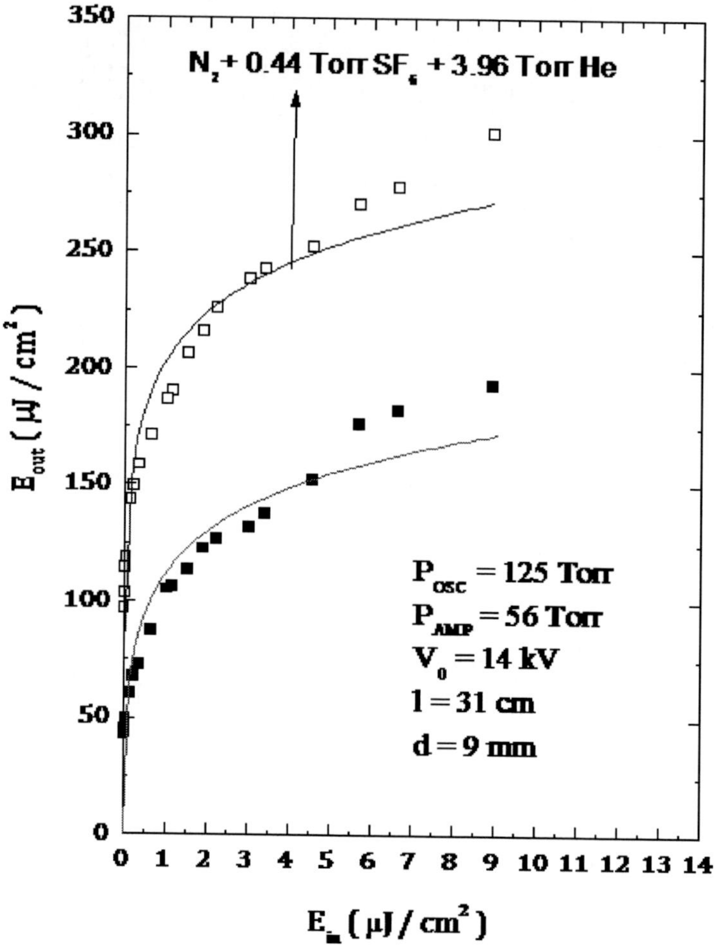

Figure 3.19. Plots of E_{out} vs. E_{in} . The upper curve shows the effect of a gas mixture of SF_6 + He added to N_2 entering the AMP section. The lower curve corresponds to the experiment with N_2 only.

The Effect of SF$_6$ on g_0- and E_s- Values

The effects of the different gases such as He, SF_6, BF_3, CF_4, Ar , and F_2 at low and high pressure N_2 – lasers on the laser output power have been observed and reported. Details of the observations have been reviewed by Kunabenchi *et al.* [24]. In Ref. [24], also many other subjects related to N_2 – lasers, including types of excitations, laser construction, pumping mechanism, etc., are reviewed. With a small amount of SF_6 added to N_2, we observed an enhancement for the output energy, and gain coefficient. At higher gas pressure, however, the laser operation encounters with arc formation, and for suppressing the influence of the arc in the discharge channel, addition of a small amount of He buffer gas to the gas mixture was found to be helpful for the measurements. In this experiment for operating the laser amplifier at higher gas pressure when SF_6 is added to the gas mixture we had to increase the AMP electrodes separation and $d_{AMP} = 9\, mm$ was used through out the measurements. Figure 3.19 shows the output energy density, E_o, versus input energy density, E_i, when a mixture of $He+SF_6$ was added to the flowing N_2 gas entering to the AMP.

The data points in this Figure should be compared with the corresponding data when pure N_2 was used for the measurement. For the low input energy we used E_{in} in logarithmic scale, and also used $G=10\log E_o/E_i$ (in dB). The corresponding plot is giving in Figure 3.20. The data points corresponding to low and high input energy densities have been omitted for the analysis, since the neglected data points are referring to ASE and pumping mechanisms, and these two mechanisms have not been considered in the rate equations for deducing the analytical expression given by Eq. (2.18).

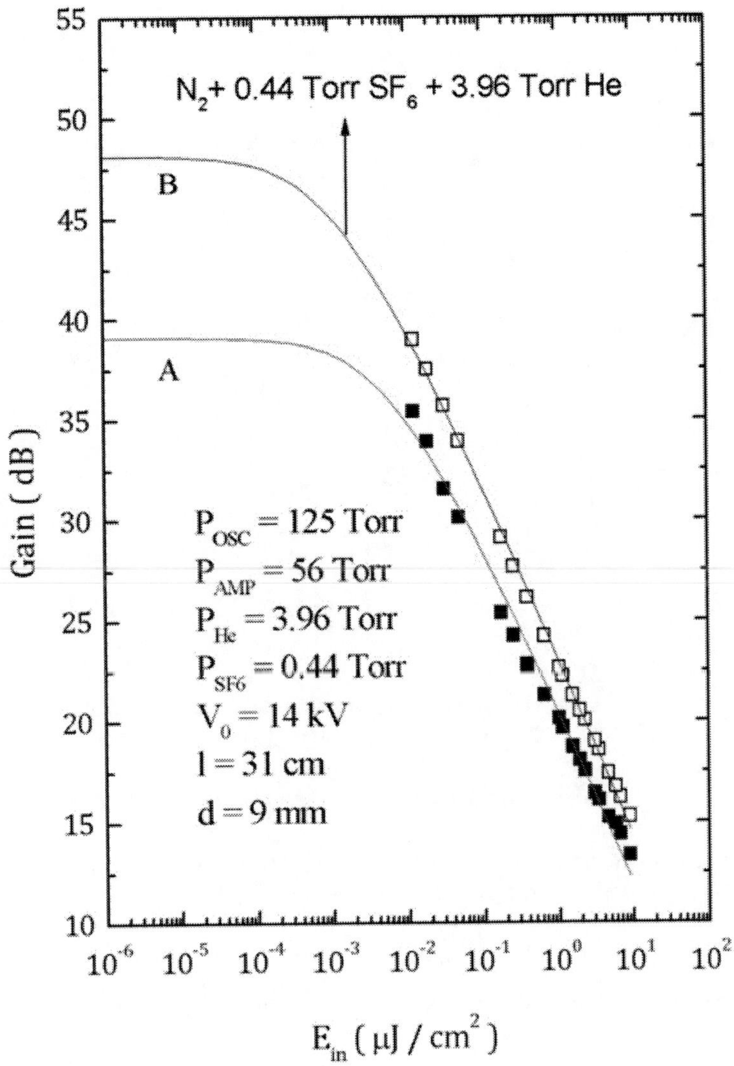

Figure 3.20. Plots of gain (in dB) versus E_{in} in logarithmic scale. The effect of adding $SF_6 + He$ to N_2 flowing to the AMP on the gain value is indicated in this figure: (A) pure N_2 ; (B) N_2+SF_6 + He.

The output energy density versus AMP total pressure for two different cases of using pure N_2, and adding a small amount of SF_6 (0.15 *torr*) to the AMP gas mixture is given in Figure 3.21.

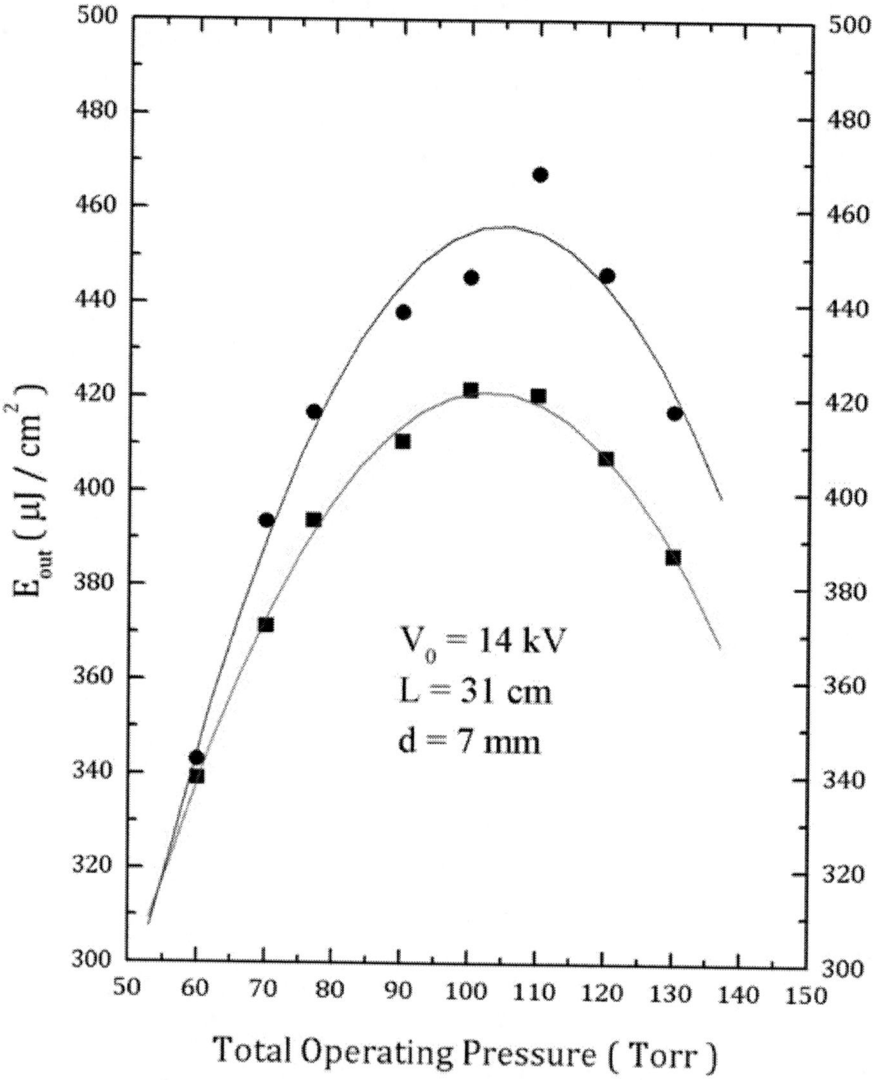

Figure 3.21. Plots of output energy density with respect to AMP total operating pressure: (A) pure N_2; (B) SF6 with pressure of 0.15 *torr* is added to the gas mixture.

If we define the fractional pressure as $x = p_{SF_6} / p_{N_2}$, we can also observe the E_o behavior with respect to the fractional pressure. In Figure 3.22 this plot is given. In figures 3.23 and 3.24 the behavior of the measured small – signal gain, and saturation energy density with respect to *x* for 3 different N_2 gas pressures of 56, 61, and 70 *torr* are shown. It is seen that by increasing the N_2 gas pressure the laser – parameters drop to lower values.

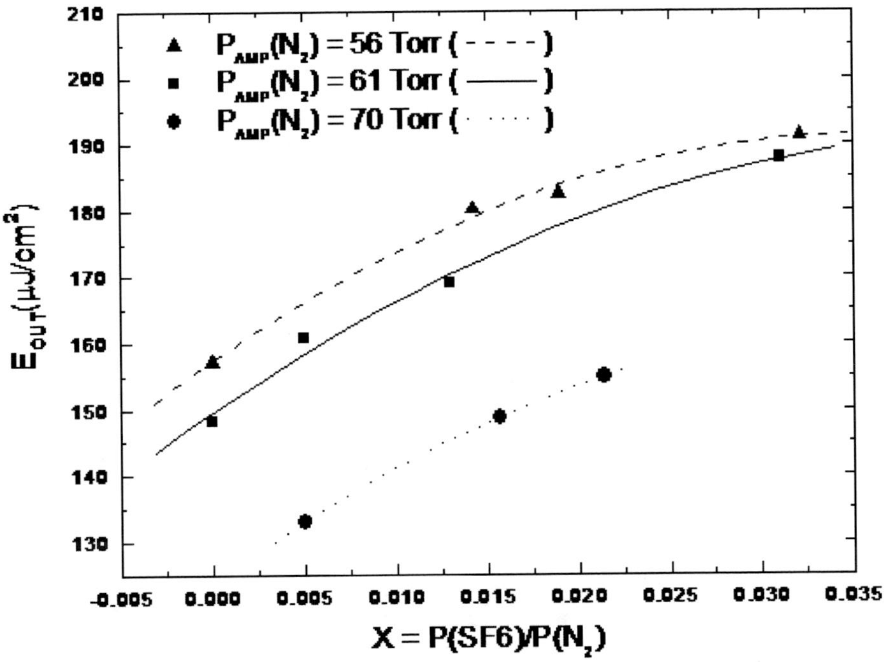

Figure 3.22. Plots of AMP output energy, E_{out}, versus $x = p_{SF_6} / p_{N_2}$ for different values of the AMP N_2 gas pressure, $p_{AMP}(N_2)$.

Finally, the time behavior of the laser output when a pure N_2 laser was used in the case when 3.0 *torr* SF_6 was added to the gas mixture is given in Figure 3.25. This Figure shows that we get a pulse broadening when a small amount of SF_6 is added to the AMP gas mixture.

Based on our observations, we realized that the output energy enhancement occurs when N_2 is at low pressure, and a small amount of SF_6 is added to the gas mixture.

From the theoretical point of view, the electron distribution function for simulated conditions of $SF_6 - N_2$ gas mixture has been calculated by Judd [25], where the results show that the electron distribution function, transport quantities, and electric excitation rate change a few percent in the presence of SF_6, for a fixed E/N. The output enhancement with the addition of SF_6 is attributed to an increase of the E/N - value in the gas discharge.

Polarization Dependence of g_0- and E_s- Values

There are a number reports on observations of the polarization – dependence of the small signal gain in different active media. In general, in an amplifier the selective depletion of a subset of the inverted population may lead to hole burning. This phenomenon is categorized as spectral and polarization hole burnings. In solid – state lasers, for example, the spectral hole burning arises from variations in the center frequency and line width of ions in different sites. Polarization hole burning, on the other hand, arises from anisotropy of the stimulated emission cross – section. In Nd: glass amplifiers it is observed that although the initial gain is isotropic, but after propagation of linearly polarized saturating pulse, the gain becomes polarization dependent. It is believed that this observation is a direct proof of polarization hole burning [26]. Observation of the spectral inhomogeneities of laser transitions upon exciting the ions at the laser wavelength, using a polarized beam is one approach to identify the polarization hole burning. Hall and Weber in their experiment observed polarized $^4F_{3/2} \rightarrow {}^4I_{9/2}$ and $^4F_{3/2} \rightarrow {}^4I_{11/2}$ fluorescence spectra in Nd^{3+} in LG-670 silicate glass upon exciting the sample by $1.064\ \mu m$ laser radiation.

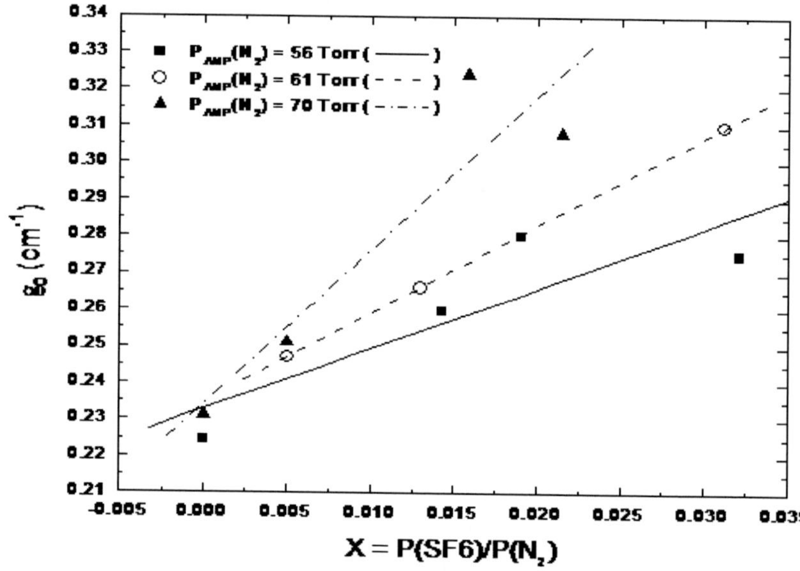

Figure 3.23. Small signal gain, g_0, versus $x = p_{SF_6} / p_{N_2}$ for $p_{AMP}(N_2) = 56, 61$, and 70 *torr*.

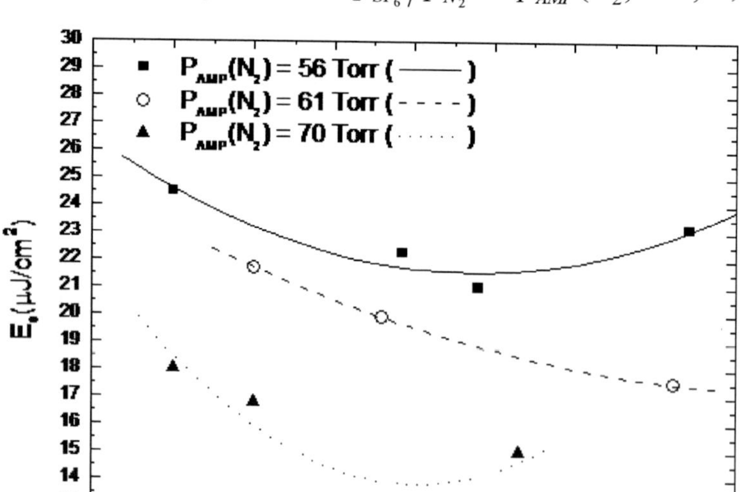

Figure 3.24. Saturation energy density, E_s, versus $x = p_{SF_6} / p_{N_2}$ for $p_{AMP}(N_2) = 56, 61$, and 70 *torr*.

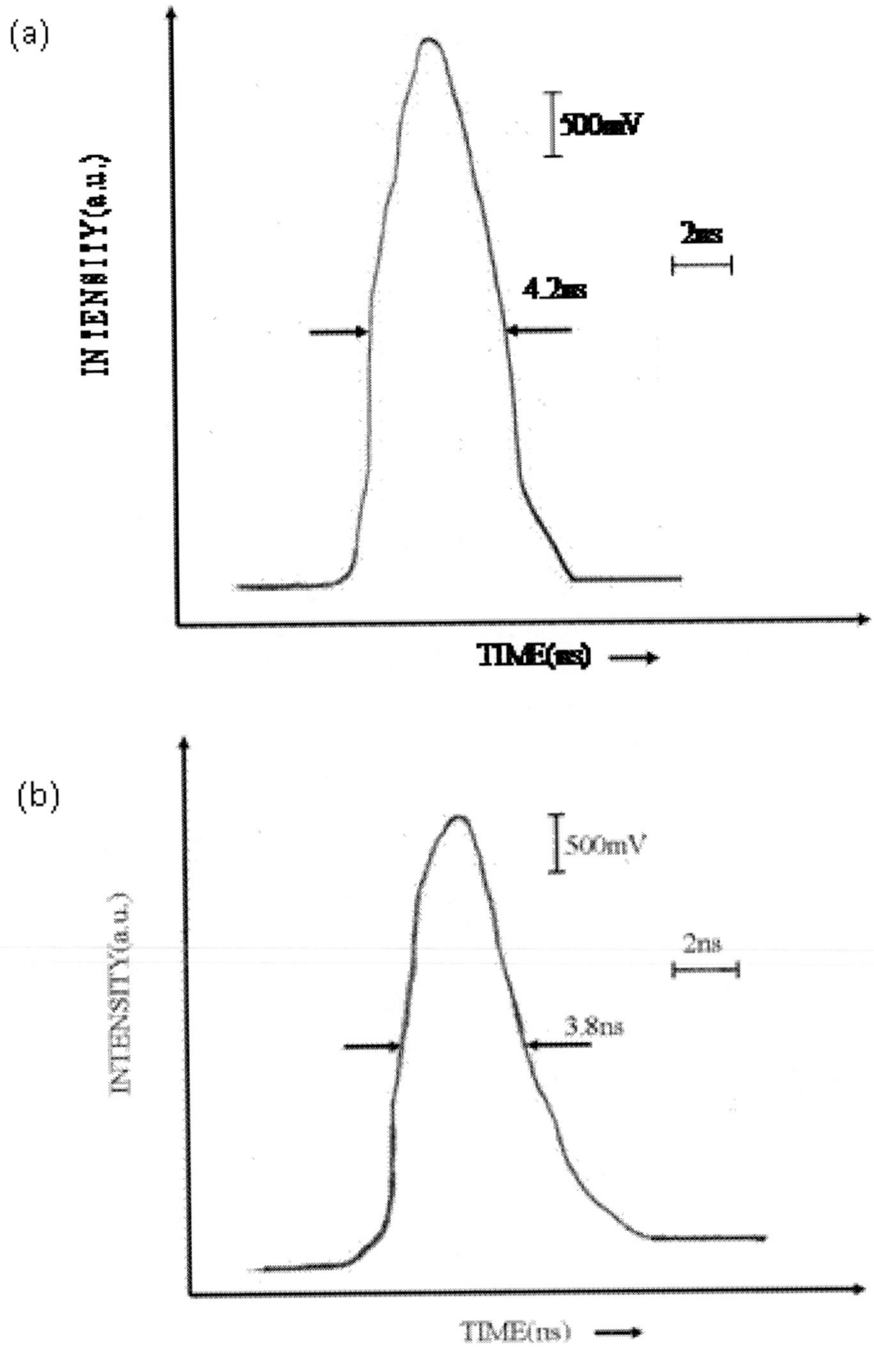

Figure 3.25. Time behavior of the output pulse from the AMP: (a) shows a pulse broadening of 4.2 *ns* FWHM using 3.0 *torr* SF_6; (b) with a pure N_2 the pulse width is 3.8 *ns*.

Other approach for identifying the polarization hole burning is to measure g_0 – parameter, using an OSC – AMP laser system as shown in [27]. The technique can provide information for laser glass compositions which is useful for reducing hole burning and energy extraction efficiency. In addition to solid materials, polarization effects have also been recognized in dye lasers, using an OSC-AMP laser system [28a], where in this respect, dye molecules often act as randomly oriented dipoles [28b].

The earliest report for the electron excitation of N_2 molecule goes back to 1969, where Burns *et al.* observed the polarization of (0,0) 337.1 *nm* band, and found a slight evidence for a polarization of about 1% in the light from the beam [29]. Base on our measurements using an OSC – AMP laser system, reported in Ref. [30], the polarization – dependent gain, g_0, and saturation energy density, E_s, in a TE N_2 laser amplifier at 337.1 *nm*, $C^3\Pi_u \rightarrow B^3\Pi_g$ laser transition were realized.

It was observed that for the AMP with the gap separation of 7 and 9 *mm*, where the pd_{AMP} value has its optimum value of 54 *torr cm*, and for *p*=77 and 60 *torr* AMP gas pressures, the gain – coefficient for the input beam with the polarization parallel to the discharge electrodes (P – polarized beam) is slightly higher than the case when the beam polarization is perpendicular to the discharge electrodes (S – polarized beam).

For the E_s- parameter, the reversed order is true. Also it was found that the saturation energy densities, E_s, for three states of polarization are linearly related to the output energy densities, E_o, having different slopes of 0.11, 0.14, and 0.17 for R (randomly), P- and S-polarization, respectively. This last observation is consistent with Martin and Milam report on the study of gain saturation in Nd: doped laser materials [31].

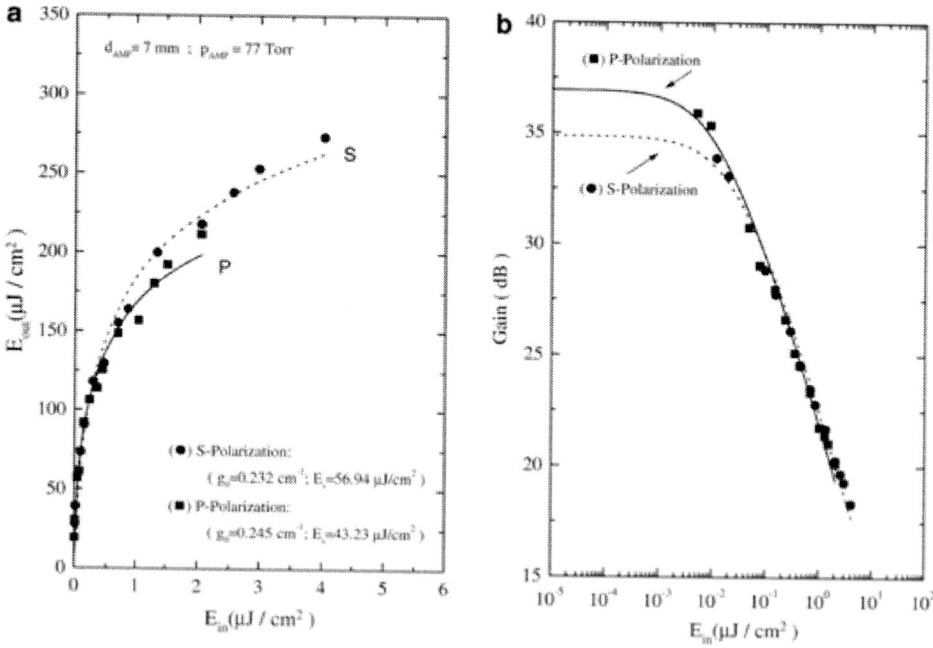

Figure 3.26. (a) Typical linear-linear plots of output-input energy densities for different states of the probe beam polarization; (b) Gain in dB (i. e., $10 \log E_{out}/E_{in}$) versus log of the input energy density. Data points are the same as those in (a). $p_{AMP} = 77\,torr$, and $p_{OSC} = 110\,torr$, $V_0 = 14\,kV$.

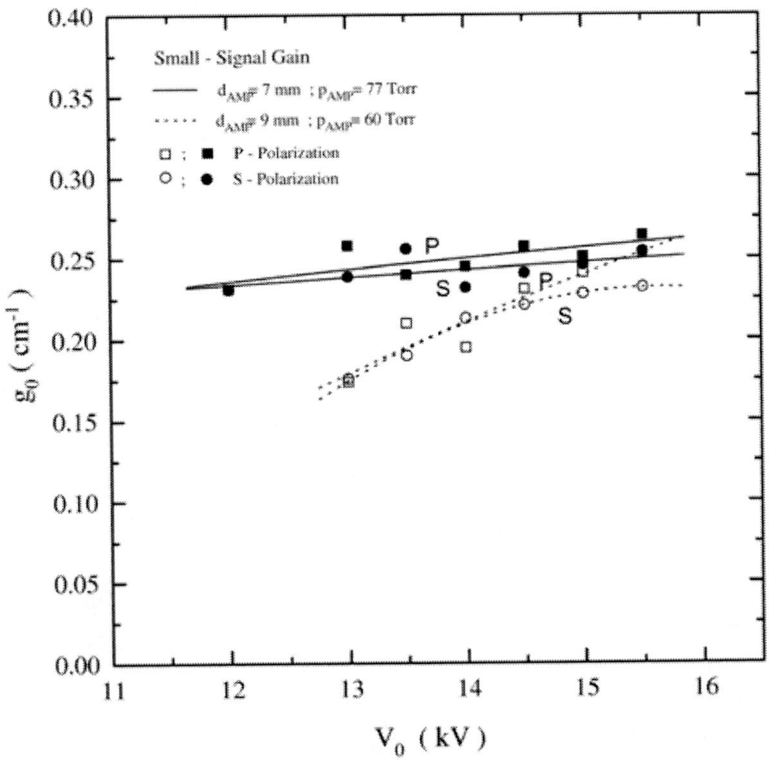

Figure 3.27. Plots of small signal gain, g_0, versus the input ignition voltage for P-, S-states of polarization. $p_{OSC} = 110\,torr$, and the AMP gas pressures are 77 and 60 *torr* for $d_{AMP} = 7$ and 9 *mm*, respectively.

For the gain measurement the same setup as introduced in the previous subsection was used. E_o and E_i were obtained using relations ε_o/A_o and ε_i/A_i, where ε_o and ε_i are the output and input energies, leaving the AMP and entering into the AMP, respectively. The cross – section areas, A_o and A_i, are 0.16 and 0.15 cm^2 (for $d_{AMP} = 7\,mm$); and 0.27 and 0.24 cm^2 (for $d_{AMP} = 9\,mm$), respectively. Typical results of the measurement (E_o versus E_i in linear – linear plots) for the P- and S- polarization of the probing beams entering into the AMP of $d_{AMP} = 7\,mm$, at $V_0 = 14\,kV$ are shown in Figure 3.26 (a). In Figure 3.26 (b) the gain-value, *G*, expressed in dB, i.e., $10\log(E_o/E_i)$, versus the logarithm of the input energy density, E_i, for the same data points presented in Figure 3.26(a) is given. These two figures are showing clearly that the output energy density for the S-polarized light is higher than that of the P-polarized light and the corresponding saturation energy density, i.e., the E_s-values, for these two beams are 56.94 and 43.23 μJcm^{-2}. The gain-values are in reversed order. They are 0.232 and 0.245 cm^{-1}, respectively, for two S- and P- beams. In these types of measurements we realized that to obtain reasonable distinctions between S and P data points, the system must be operated with pd_{AMP}-value $= 54\,torrcm$. This corresponds to

$d_{AMP} = 7-9\,mm$, where as we will see in section 4, with this condition the laser output energy, ε_o, is maximized. Operation of the AMP in other d_{AMP} arrangements, such as 4 and 11 *mm*, which correspond to pd_{AMP}-value of 66 and 44 *torr cm*, drive the laser in an operational condition where ε_o is not maximized, and distinction between S and P data points becomes difficult. Also we found that by increasing the laser operational voltage up to 15.5 *kV*, the separations between S-, and P-polarizations become more distinct. Figure 3.27 shows plots of g_0-values for $d_{AMP} = 7$ and $9\,mm$ versus input ignition voltage, V_0, for P- and S-polarized beams. For the AMP gap separation of $d_{AMP} = 4\,mm$, the results of measurement for S- and P- and R-polarizations are shown in Figure 3.28.

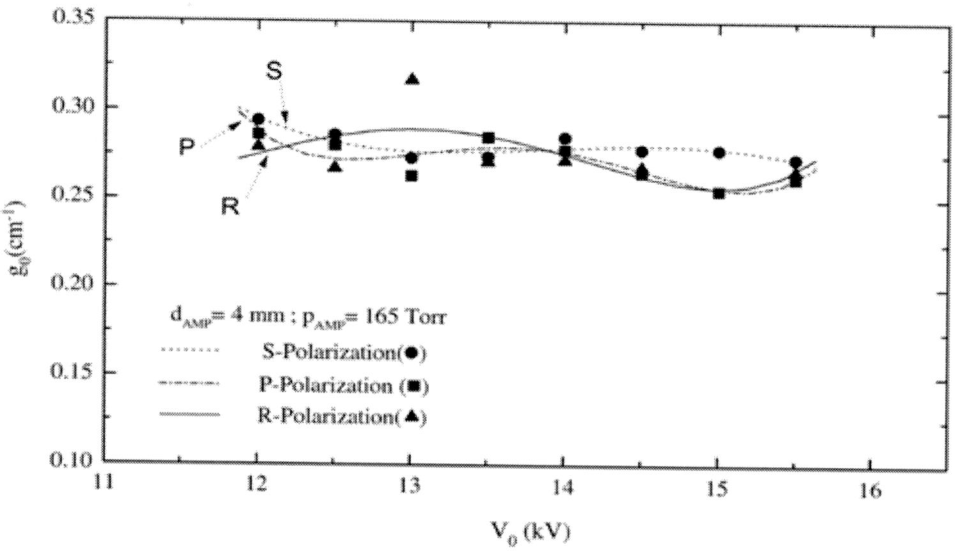

y

Figure 3.28. Plots of g_0 vs. V_0 for P-, S-, and R-states of polarization. The AMP gap separation is 4 *mm* and the gas pressure is 165 *torr*. The random oriented data for three states of polarization are evident in the measurement.

The randomly oriented data points for three different polarizations are evident in this figure. The plots of the saturation energy density for P- and S-polarized light versus the pumping voltage are shown in Figure 3.29. Again, it was realized that at the low-input operational voltage two curves, corresponding to S- and R-polarizations, tend to approach to a single-value. A similar behavior is also hold when we plot the measured output energy density, E_o, for the S-, P- and R-polarized light versus the input voltage, as shown in Figure 3.30. The data corresponding to the R-polarization are also introduced for the comparison. Clearly, E_o for $d_{AMP} = 4\,mm$ is higher than that for $d_{AMP} = 7\,mm$ and 9 *mm*.

The orders of R-, S- and P-polarization for these three cases, however, remain the same.

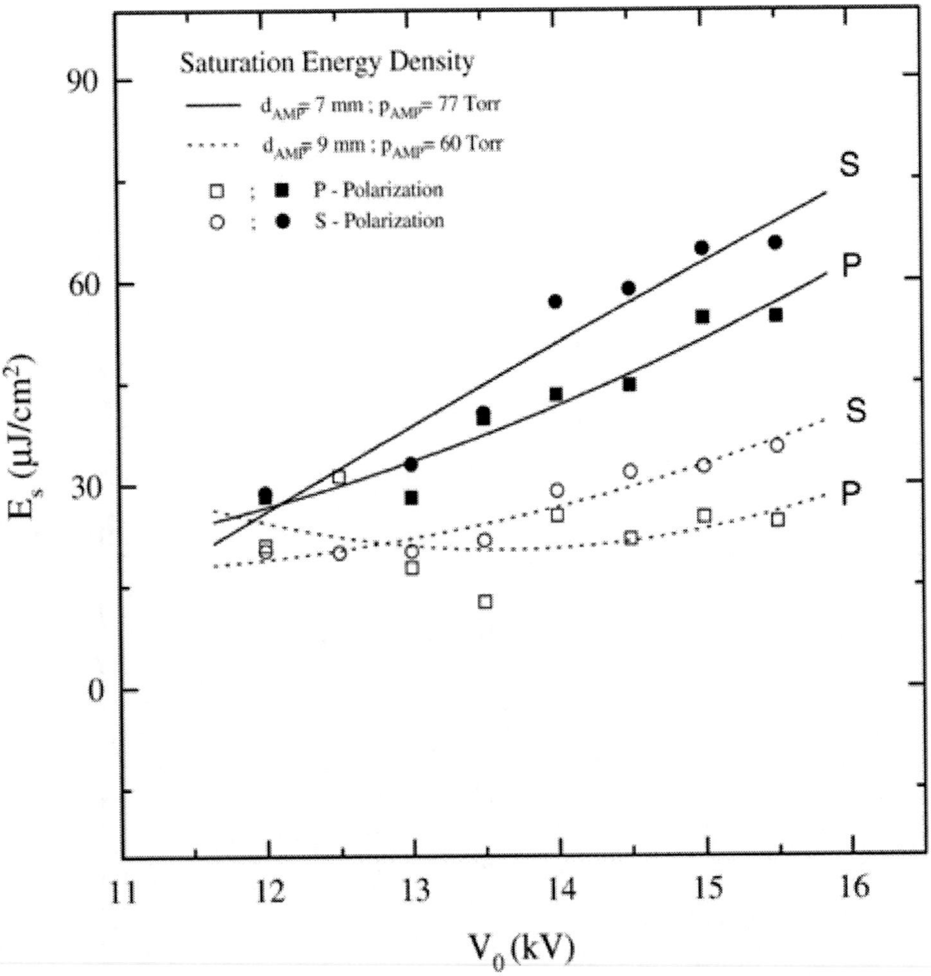

Figure 3.29. Plots of E_s for two different states of polarization versus input ignition voltage. $p_{OSC} = 110\,torr$, the AMP gas pressures are 77 and 60 *torr*, for $d_{AMP} = 7$ and 9 *mm*, respectively.

The similarities that appeared in figures 3.29 and 3.30 are indicating that a simple linear relation must exist between E_s and E_o quantities. In fact in Figure 3.31 we have shown that these two quantities are linearly related for three states of polarization. This behavior, on the study of Nd: doped laser materials have been observed and a linear relation as: $E_s = a_0 + b_1 E_{out}$ was introduced [31], where a_0 and b_1 are some constants and for our study on N_2-laser they are given in the inset of Figure 3.31. Our observation for $d_{AMP} = 9\,mm$ is indicating that E_o and E_s are in their lowest values, compared to the cases measured for $d_{AMP} = 4$ and $7\,mm$. This observation is consistent with the measurements for the longitudinally excited N_2-laser, appeared in section 3.3.

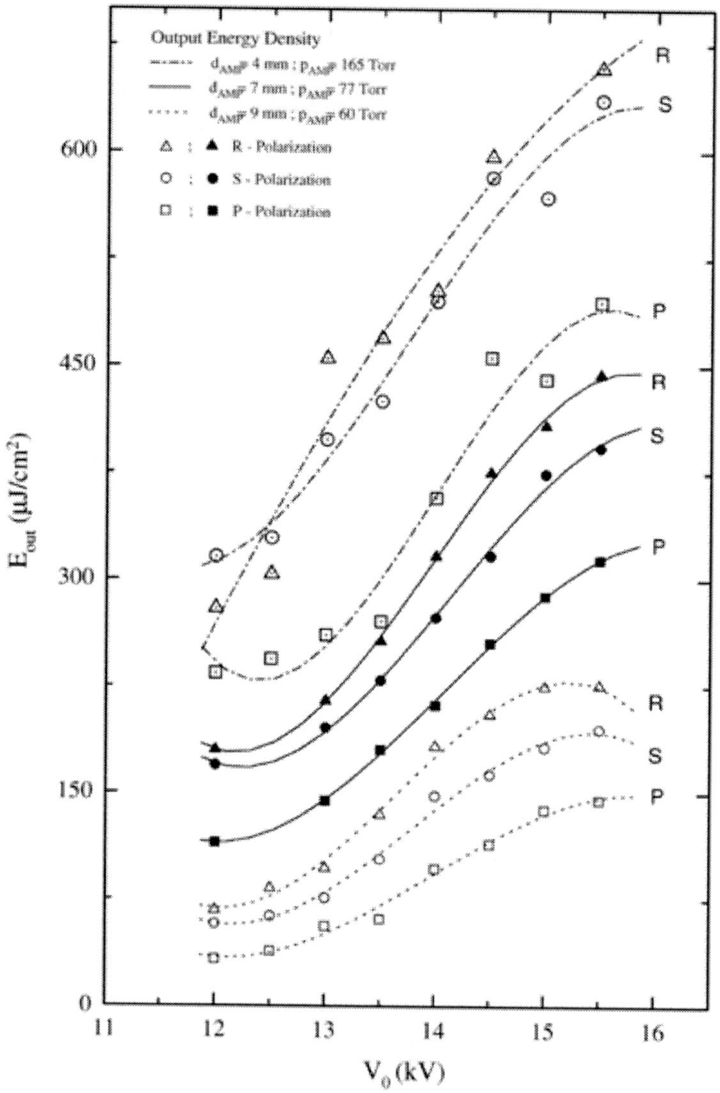

Figure 3.30. Plots of E_{out} versus V_0 for the three different gap separations. The profile for R-polarized light (randomly polarized beam) is given to be compared with the S- and P-polarized profiles.

3.3. Amplification in Longitudinally Excited (LE) Nitrogen Lasers

There are generally two types of gas discharge which are commonly used for the N_2-laser excitation for the $C^3\Pi_u \rightarrow B^3\Pi_g$ transition at 337.1 *nm*. They are transversal and longitudinal discharges. The first type that has been described in section 3.2 is usually used for extraction of high output powers. As it was shown earlier, the corresponding laser pulse-width is usually short of a few *n sec*, and can be varied upon operating the system at high gas pressures, perhaps up to atmospheric gas pressure [19]. Operation at a few atmospheric gas pressures from this device has also been reported [32]. The alternative approach is the longitudinal discharge scheme (LE) which is appropriate to be used at the low gas pressure operational regime. The advantage of longitudinal discharge devices is the discharge uniformity that can be easily achieved at gas pressures of a few *torr*. In addition, a low beam divergence and good circular mode structure of Gaussian type can be expected from the LE configurations. The output energy of lasers based on the longitudinally excited mechanism, however, is limited in the order of tens of μJ and that is mainly due to the higher impedance of the gas discharge compared to that of the transversal discharge. On the other hand, for improving the laser beam quality, enhancement of the laser output energy, and reducing the laser beam divergence designs based on oscillator-amplifier (OSC-AMP) N_2-laser, as described earlier, have also been proposed. In a laser system consists of an OSC-AMP in addition to achieving improvement in laser performances, it gives also a method of probing the laser output extracted from amplifier as the input energy to the amplifier changes.

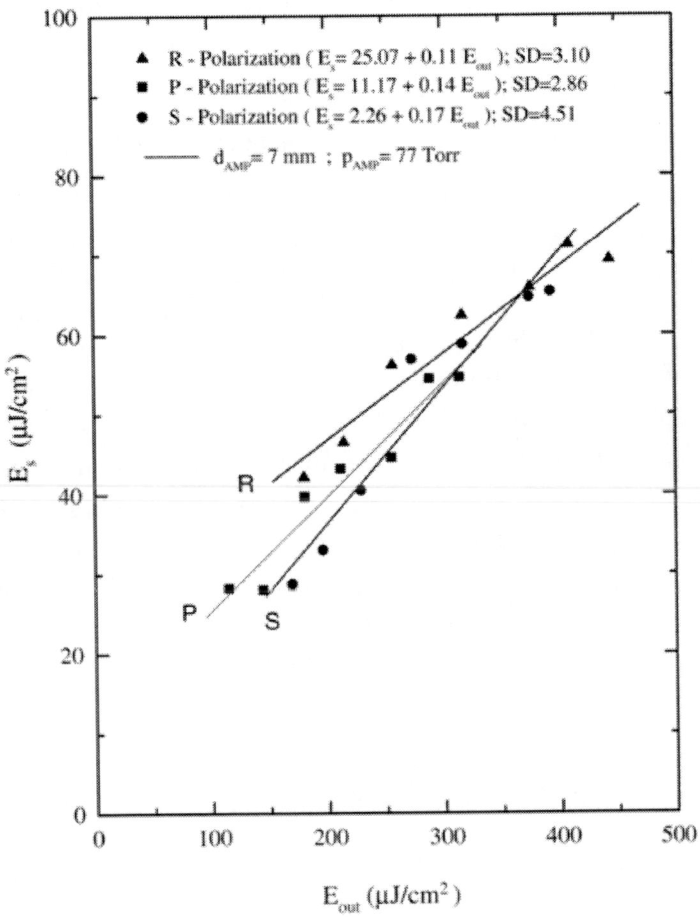

Figure 3.31. Experimental measurements of E_s versus E_{out} for three different states of polarization. The data points are fitted to $E_s = a_0 + b_1 E_{out}$. a_0 and b_1 for S-, P-, and R-polarizations are indicated in the inset of the figure. SD = saturated deviation.

In this section, based on our previously published article, the operational performances of a longitudinally excited OSC-AMP N_2-laser (LE-LE) using a fast rising Blumlein circuit, constructed from a commercial printed board will be described [33]. The introduced system is capable to be applied to other self-terminating gas lasers of similar designs and charging mechanisms. It can be easily constructed and the time-delay adjustment needed to be set between the OSC and AMP sections is simply accomplished by moving the AMP laser tube along the excitation length on the circuit board. The system, as it is the goal of this section, can also be used to measure laser parameters, g_0, and E_s, for laser tubes of different lengths and diameters. These observations and measurements will help understanding the gas excitations in tubes of different diameters, and can be further applied for theoretical investigations.

(a) LE N_2-Laser Design

A schematic diagram of the excitation circuit and the layout of the OSC-AMP system used for the experiment are shown in Figure 3.32. The $107 \times 107\, cm^2$ commercial printed circuit board is double-sided copper with a thickness of 1.5 *mm* fiberglass insulation. The board has been cut to produce two excitation circuit of $45 \times 43.5\, cm^2$ for the OSC; and $107 \times 62.5\, cm^2$ for the AMP section. A 25-*mm* copper strip is removed from sides of the printed board to prevent arcing. Also, 25 and 45-*mm* strips of copper are removed from central parts of the OSC, and AMP sections, respectively, to produce three capacitors of 2.4, 6.8, and 10.2 *nF* each. The OSC excitation section on the circuit board is in the shape of a double trapezoid with the shorter and longer sides of 34 and 38 *cm*, respectively. The laser tubes are made of Pyrex-glass with tungsten electrodes of ~ 1.5 *mm* in diameter sealed at both ends of the laser tube. The OSC part has a total length of $L_{OSC} = 20\ cm$, the excitation length of $l_{OSC} = 15.5\ cm$ and the inner bore diameter of $\phi = 2\, mm$. Various types of laser tubes of different lengths of $L_{AMP} = 20,\ 27,\ 31, \text{and}, 41\, cm$ with the excitation lengths of 15.5, 21, 25, and 35 *cm*, respectively; and 4 *mm* inner bore diameter are used for all the AMP laser tubes. Each laser tube has been ended to glass tubes of larger bore diameter of 10 *mm*, cut at Brewster angle and sealed with flat-plate fused-silica windows.

In order to minimize the overall circuit inductance, a single spark gap working at atmospheric N_2 gas pressure is directly soldered to the upper section of the circuit board. The gas pressures for the input of each laser tube are adjusted by two input values, one for the OSC and the other for the AMP. The gases are exhausted from the exit channels of the OSC and AMP tubes, and are controlled by two independent pressure gauges.

(b) Gain and Saturation Energy Density Measurements

Typical laser pulse shapes from the AMP output operating at different gas pressures, using a fast detecting system as described in section 3.2, are shown in Figure 3.33. At the zero-AMP gas pressure, it is only the OSC signal which is observed. By increasing the AMP gas pressure, the output signal grows gradually reaches at a maximum (case C in the figure) and decreases rapidly. Case (C) in the Figure which corresponds to 5.25 *torr* AMP gas pressure refers to completion of the laser pulse amplification.

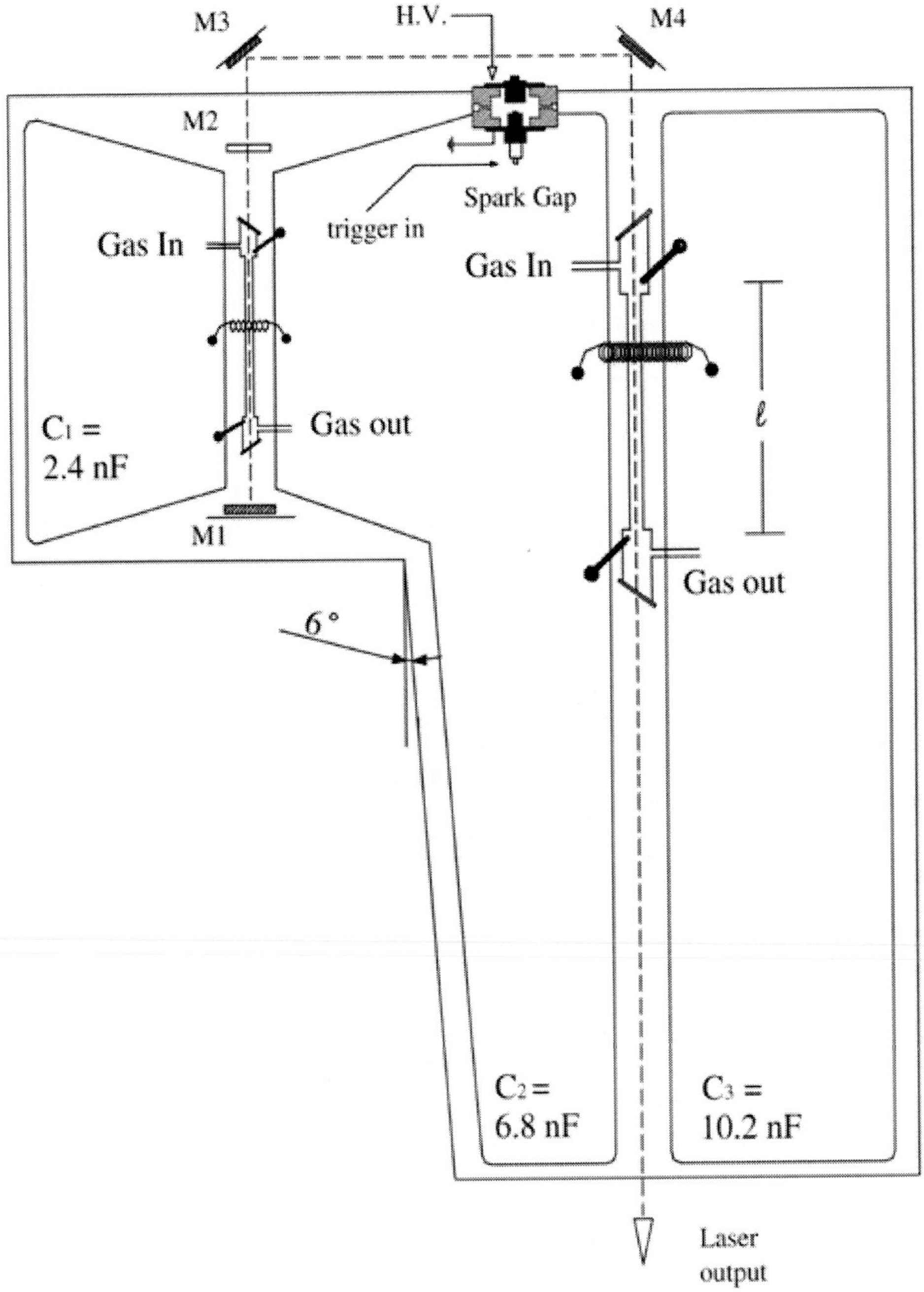

Figure 3.32. Top view schematic of the LE-LE N_2 laser.

Due to absence of the amplified spontaneous emission (ASE) in the AMP tube, in contrast to the TE or TEA types of N_2-lasers, we did not observe any double or distorted pulse shape at the AMP output. As the laser is operating at a low gas pressure, the laser pulse width is longer than that of TE/TEA types nitrogen lasers (i.e., typically, pulse width is ~ 10-20 ns).

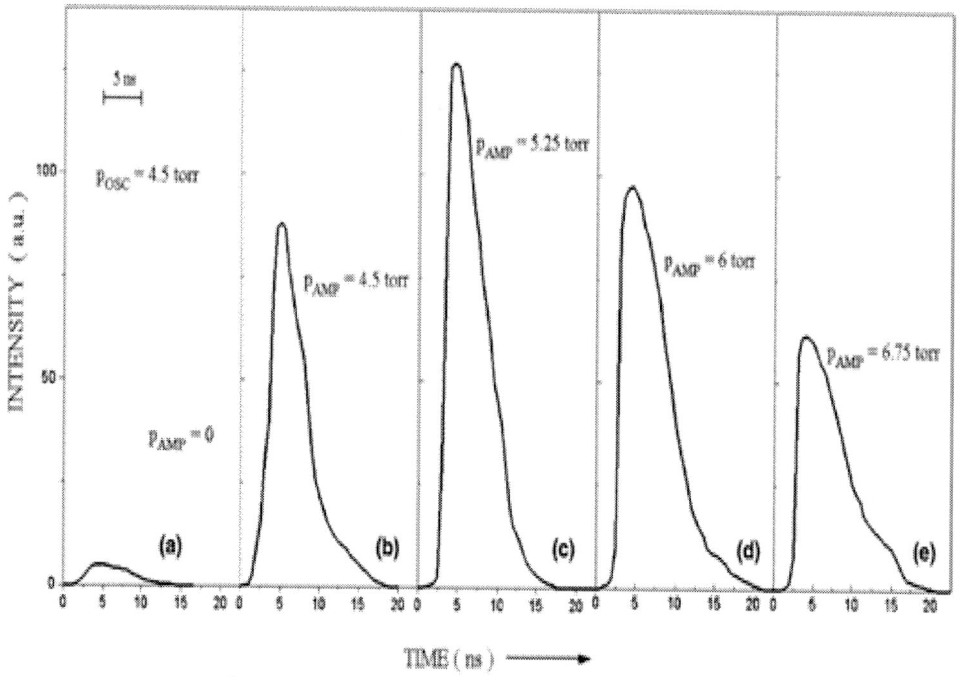

Figure 3.33. Typical laser output pulse shapes from the AMP tube obtained by an ITT vacuum photodiode. The excitation length of the AMP tube in this observation is 21 *cm*. The OSC gas pressure is 4.5 *torr*. The AMP gas pressures are: (a) zero, when the AMP is not working and the signal corresponds to the OSC, only; (b) 4.5 *torr*; (c) 5.25 *torr*; (d) 6 *torr*; and (e) 6.75 *torr*.

For the gain measurement it is necessary to measure the output power from the OSC and AMP sections. Also, for calculating the input and output energy densities, we need to measure the beam cross-sections from both OSC and each AMP. The output energy measurements from the OSC as well as from each AMP were obtained using a photodiode with a read-out micro-ammeter, calibrated at 337.1 *nm* wave length. The special intensity distribution of lasers outputs were determined by scanning a pinhole (0.22 *mm* diameter) across the laser beam and measuring the transmitted radiation. Figure 3.34 shows the spatial intensity distribution of the laser output from a typical AMP of the total length of $L_{AMP} = 31\,cm$, in vertical and horizontal directions. The solid line in each Figure corresponds to a Gaussian profile of the type $\varepsilon(r) = \varepsilon(0)\exp(-2r^2/w^2)$, where w is obtained by the method of least-squares fit of the experimental points to the Gaussian function. The output and input energy densities have been calculated using $E_o = \varepsilon_o/\pi w_o^2$ and $E_i = \varepsilon_i/\pi w_i^2$, where ε_o is the output energy extracted from the AMP, and ε_i is the input energy enters into the AMP. w_o and w_i are the beam radii at the AMP output and into the AMP, respectively. Figure 3.35 shows the LE-LE gain, G, expressed in dB, i.e., $10\log(E_o/E_i)$, versus the logarithm of the input energy density. The low-input energy region of the curve is evidenced in this figure. The solid line corresponds to Frantz-Nodvik equation, given by Eq. (2.18). g_0 and E_s were obtained by least-squares fit of the experimental points to Eq. (2.18).

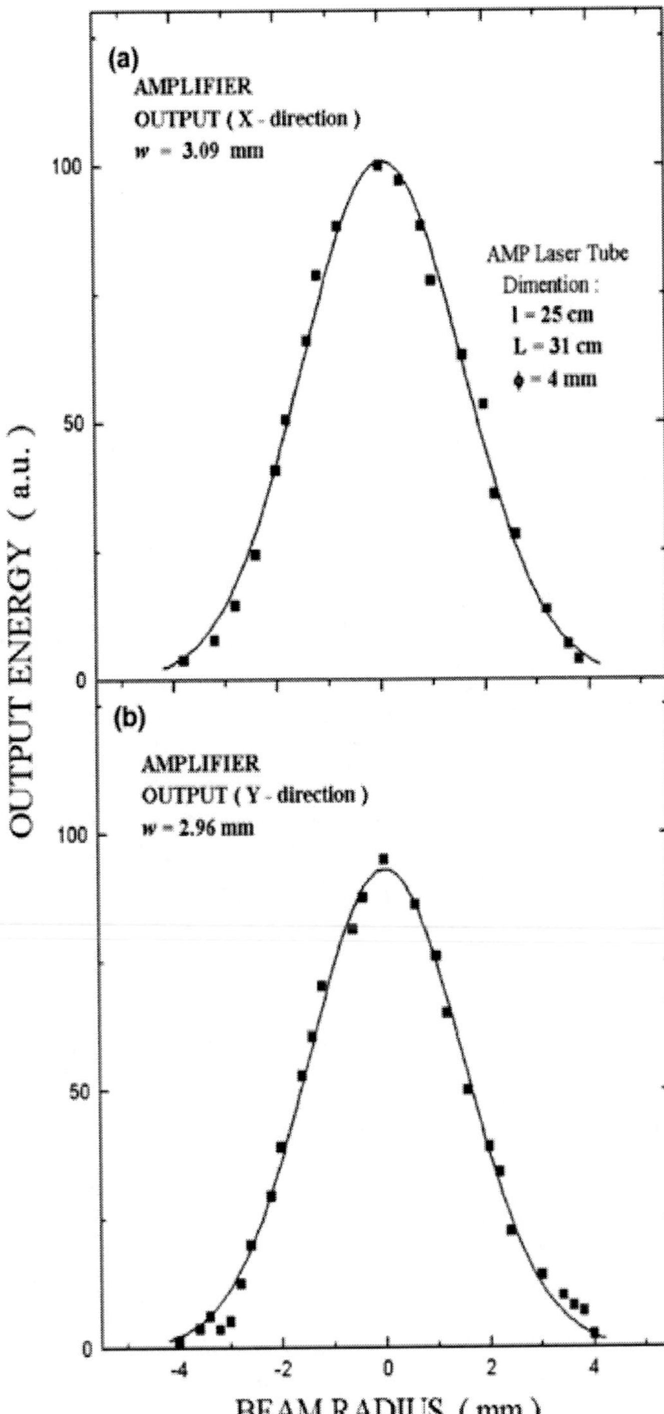

Figure 3.34. Typical spatial distributions of the laser output beam versus beam radius measured at a distance 168 *cm* from the amplifier output: (a) horizontal direction; (b) vertical direction. Solid lines refer to Gaussian profiles of $\varepsilon(r) = \varepsilon\ (0)\ \exp\ (-2r^2/w^2)$, where ε means output energy.

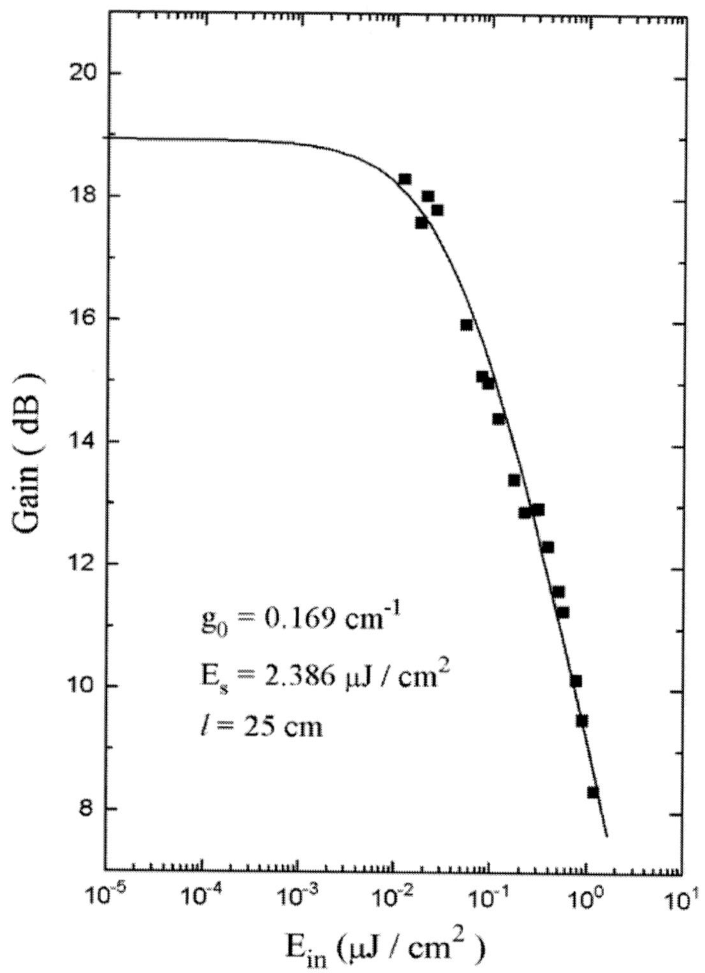

Figure 3.35. Gain in dB (i.e., 10 log E_0/E_i) versus log of the input energy density. The AMP excitation length is 25 *cm* and V_0 = 14 *kV*.

The plots of small signal gain, g_0, and saturation energy density, E_s, for AMP tubes of different excitation lengths operating at input voltage of 14 *kV*, are shown in Figure 3.36 (A) and Figure 3.36 (B), respectively. For a tube with an excitation length of 21 *cm*, the small signal gain gets its maximum value of 0.27 cm^{-1}. The saturation energy density for this length is $4.43\, \mu J\, cm^{-2}$ and the corresponding stored energy per unit volume is $E_{st} = g_0 E_s = 1.2\, \mu J\, cm^{-3}$. As it is indicated in this figure, g_0 grows rather slowly, reaches a maximum and drops to a lower value of ~ 0.06 cm^{-1} for the tube of 35 *cm* in length. As E_s is rather a small quantity in LE N_2-lasers, we are not expecting to obtain a high output power from this type of laser excitation, since we have already shown that $E_o \propto E_s$. For $l_{AMP} < 30\, cm$, Figure 3.36 shows that E_s has its peak value at 21 *cm*. Theiss, on measuring the laser output versus the LE N_2-laser excitation length reported an optimal discharge length of 28 *cm*, when the laser oscillator was operating at 60 *kV* input voltage with a peak power of

30 *kW* at ~ 2.5 *ns* pulse width [34]. Upon reproducing and interpolating Theiss' experimental data for the maximum peak powers, P_{max}, versus different tube lengths and also P_{max} versus the laser gas pressures to lower input voltages, close to our experimental conditions, we found that the operating gas pressure, optimal discharge length, and output power are 5 *torr*, 22 *cm*, and 82 *W*, respectively. The first two values are very much consistent with our present observation. According to Theiss' measurements, at the higher input voltages, the optimal laser excitation lengths shift to the laser tubes of longer lengths, and evidently by increasing the input voltage, the laser must be operating at higher gas pressure to keep E/p-value constant, where E is the electric field and p is the gas pressure. El-Osealy *et al.* in a laser tube of 5 *mm* inner diameter and length of 15 *cm* with pure N_2 gas pressure of 7 *torr* and input charging voltage of 24 *kV*, reported output energy of 0.45 *mJ*; and a gain value of 0.1 cm^{-1} was estimated for this system [35].

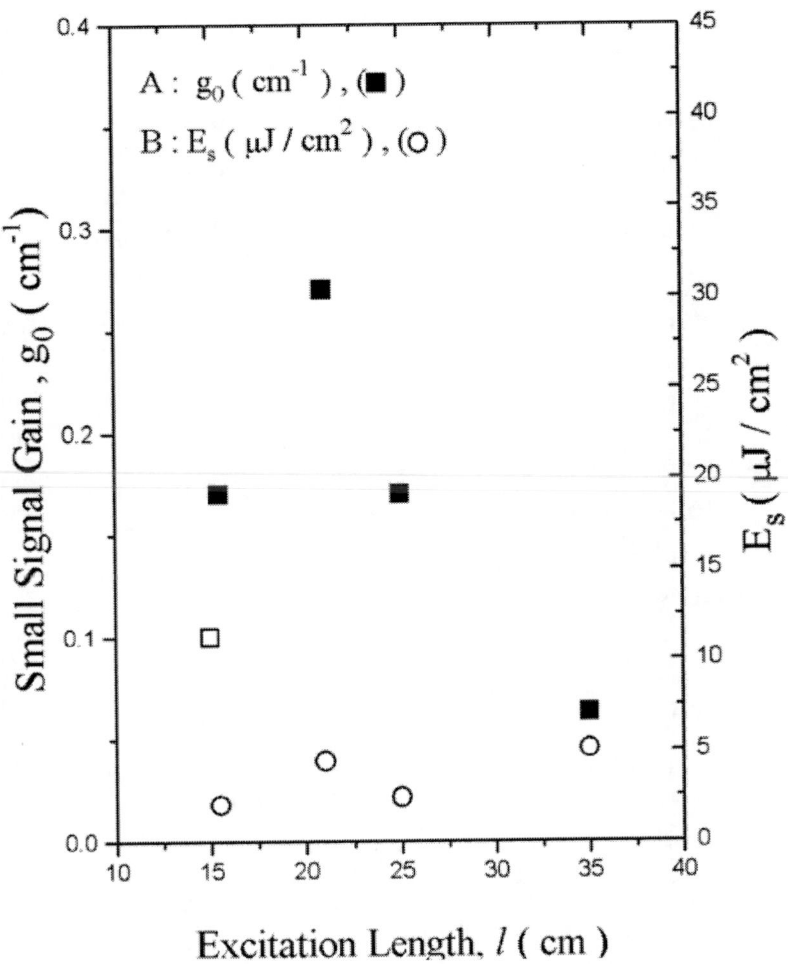

Figure 3.36. Plots of: (A) small signal gain, g_0, (■); (B) saturation energy density, E_s, (○) versus the excitation length of AMP laser tubes (l). The optimal gain value corresponds to l ~ 21 *cm*. The gain value of 0.1 cm^{-1} corresponds to [35] (□).

In Ref. [33], the N_2-laser parameters measured by different researches during the past few years have been summarized; also it includes the references corresponding to reported LE nitrogen lasers, and their power density specifications, along with the references corresponding to LE-nitrogen lasers, appeared in the literature. It should be cited that the experimental setup for an OSC-AMP laser system using a Blumlein circuit to be used for longitudinally excited gas lasers (i.e., LE-LE type) has been introduced by the author for the first time in the literature and it has shown its suitability to be used for g_0 and E_s measurements in any types of lasers. The advantage of the system is its potential for applying input voltages considerably lower than the reported designs. In addition, construction and operation of the laser system is simple.

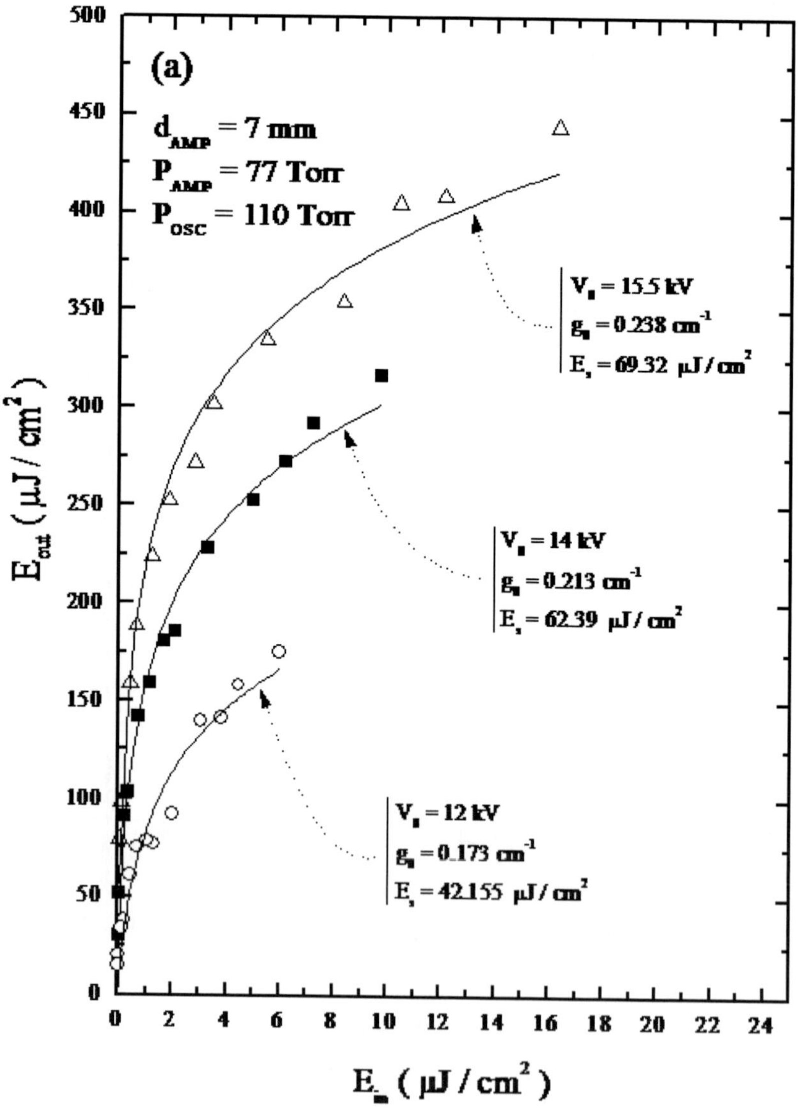

Figure 4.1. (Continued).

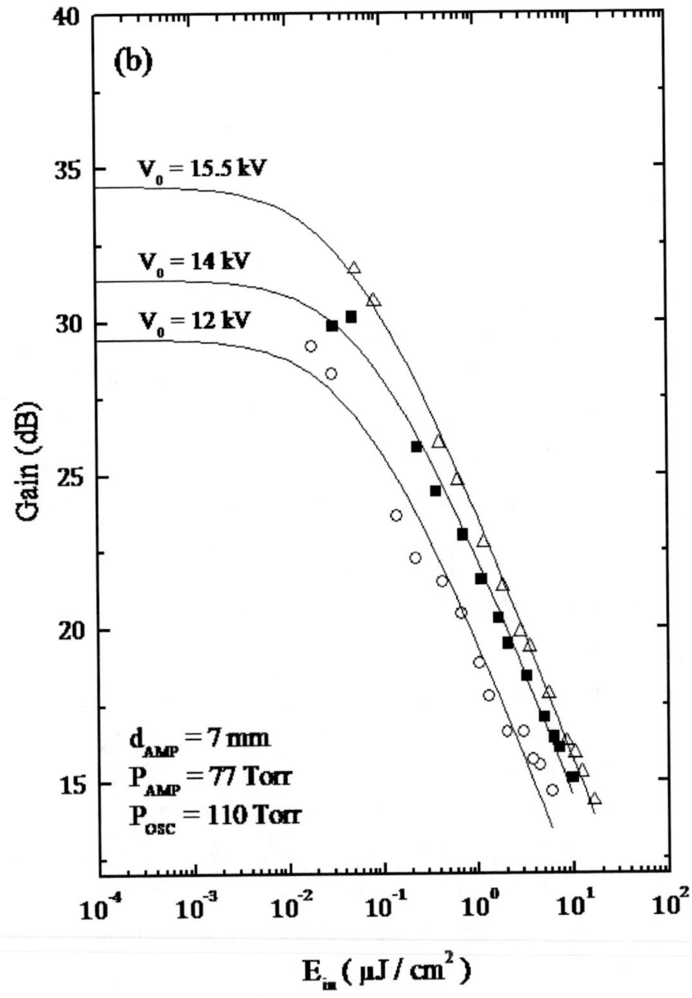

Figure 4.1. (a) Linear-linear plots of output-input energy densities in the TE-AMP for $V_0 = 12,\ 14$, and 15.5 kV ignition voltages. (b) Gain in dB vs. log of the input energy density. Data points are the same as those given in (a).

4. Laser Geometrical Dependence of g_0- and E_s- Values: General Treatment

In any laser system, some variations in the external operational parameters such as gas pressure, and ignition voltage will help optimizing the laser operation. But, this approach is rarely leads to understanding the details of laser behavior unless some other system modifications or variations on the internal parameters are applied.

For example, by changing some design-parameters and by keeping the laser-impedance unchanged, while operating the system in its optimized operational conditions will reveal valuable information which often lead to extending the physical understanding of the process involved in a given laser system. Pure mathematical approaches, which are usually implemented independent of the experimental observations, will not help the experimental researchers in realizations and understandings of the physical events. In this trend, therefore, mutual understandings between both groups of theoretical and experimental researchers are highly recognized.

4.1. Experimental Realization

The experimental measurements that constitute the contents of this section rely on our two published papers [36, 37], where in the first one, the effects of the change of amplifier gap separations, d_{AMP}, and in the second one, the effects of the change of electrodes length of the AMP on laser parameters, mainly g_0- and E_s-values are introduced. So, section 4.1 is arranged for the experimental approaches; and section 4.2 is organized for theoretical approaches based on our recent publications [38, 39]. A few examples for predicting gain-values for different laser designs are also given for justifying the theoretical treatment. The experimental arrangement is the same as that introduced in section 3.2 (b).For the first example, the electrodes length of the AMP were fixed at $l_{AMP} = 31\,cm$, but we arranged to do measurements with different electrodes gap separation, d_{AMP}. To obtain the AMP gas pressure, for a given AMP electrodes separation, d_{AMP}, and the OSC gas pressure of 110 *torr*, the AMP gas pressure was varied while the output energy from the AMP was monitored to observe a complete amplification. Then, the optimal AMP gas pressure was recorded. Typical results of the measurements (E_o vs. E_i in linear-linear plots) for three different input voltages of 12, 14, and 15.5 *kV* for the AMP gap separation of 7 *mm* are shown in Figure 4.1 (a). Figure 4.1 (b) shows the gain-values, G, expressed in decibel. The solid-lines in both figures correspond to Frantz and Nodvik equation. The corresponding g_0 and E_s values for the AMP gap separation of 7 *mm* at different input voltages (12 - 15.5 *kV*) are plotted in figures 4.2 (a) and 4.2 (b), respectively. Here, the abscissas are in E/p, or according to equation $kT_e(eV) = 0.11(E/p)^{0.8}$,where E is the electric field and p is the gas pressure, have been converted to the electron temperature. The numbers at each point refer to the operational voltages in kilo volts.

By reducing the AMP gap separation, or increasing the gap separation, it is required to increase or decrease the AMP gas pressure accordingly, to keep the electron temperature at the proper value. The results of g_0 and E_s measurements are shown in figures 4.3 (a) and 4.3 (b), respectively. These figures show clearly that laser parameters drop to lower values as the gap separation increases. These results can also explain the lower values for the E_s-parameter in LE N_2-lasers appeared in section 3.3. We will see in the next section that behavior of g_0 vs. d_{AMP} can be applied in an empirical equation of the type $g_0 = r + q/d_{AMP}$, where r and q are some constants that can be determined by least squares fitting program.

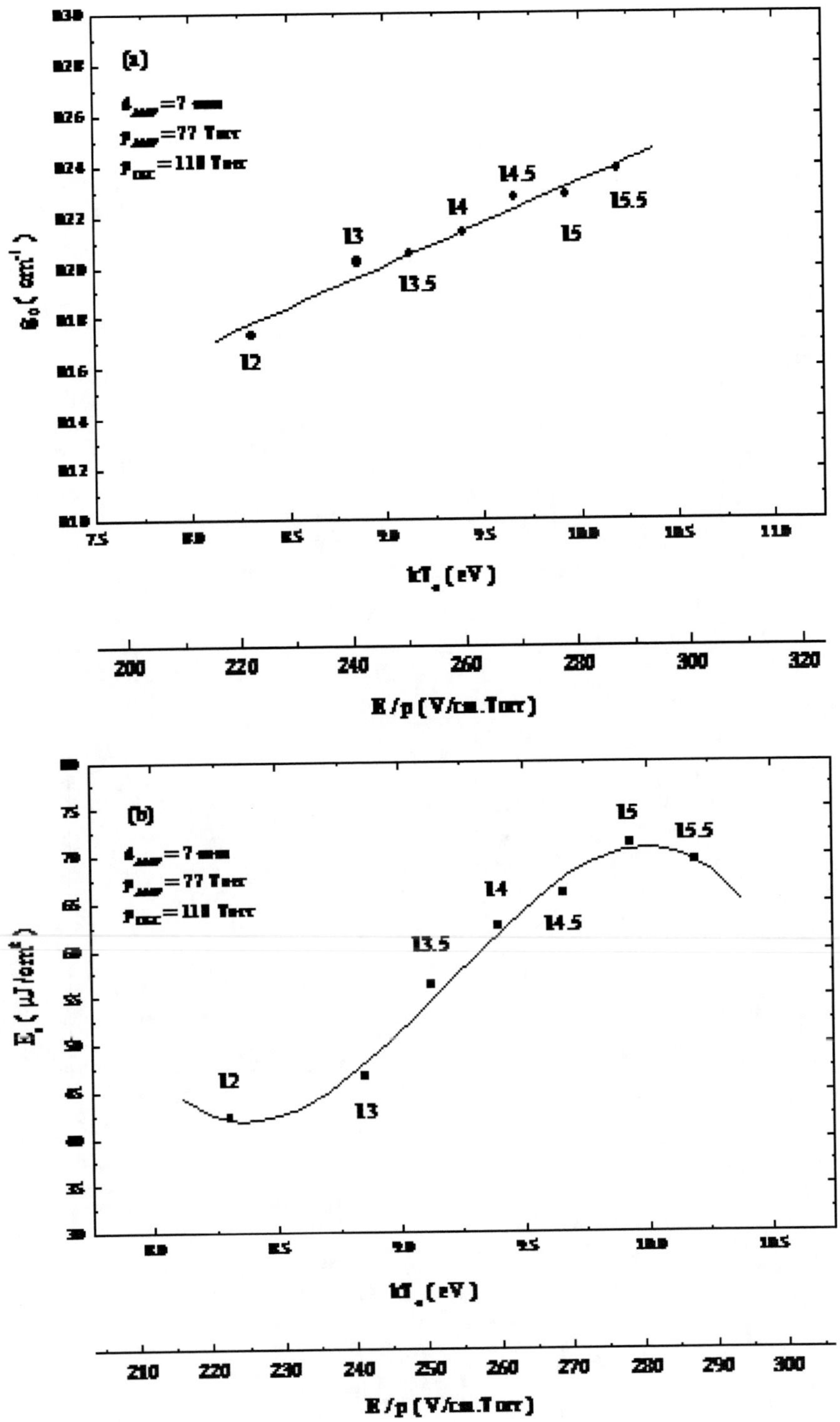

Figure 4.2. Plots of: (a) g_0 and (b) E_s vs. the electron temperature, T_e, or corresponding E/p-values, for d_{AMP} = 7 *mm* for different input voltages, as indicated by small numbers (in kilovolts) in both figures.

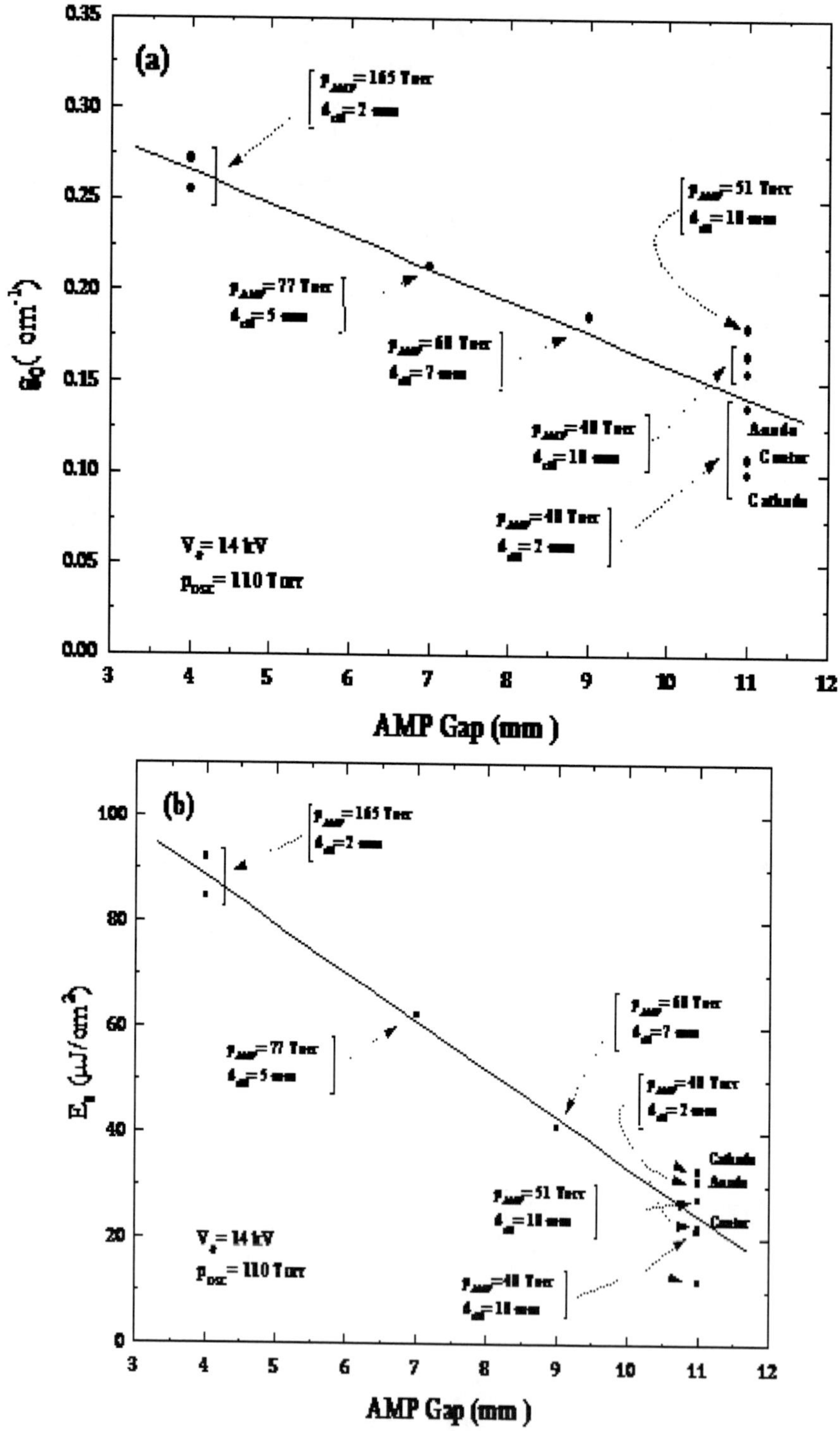

Figure 4.3. Plots of: (a) g_0 and (b) E_s parameters vs. the AMP gap separation, d_{AMP}. The corresponding g_0- and E_s-values for d_{AMP} = 11 *mm* at three distinct positions across the discharge width are shown in both figures.

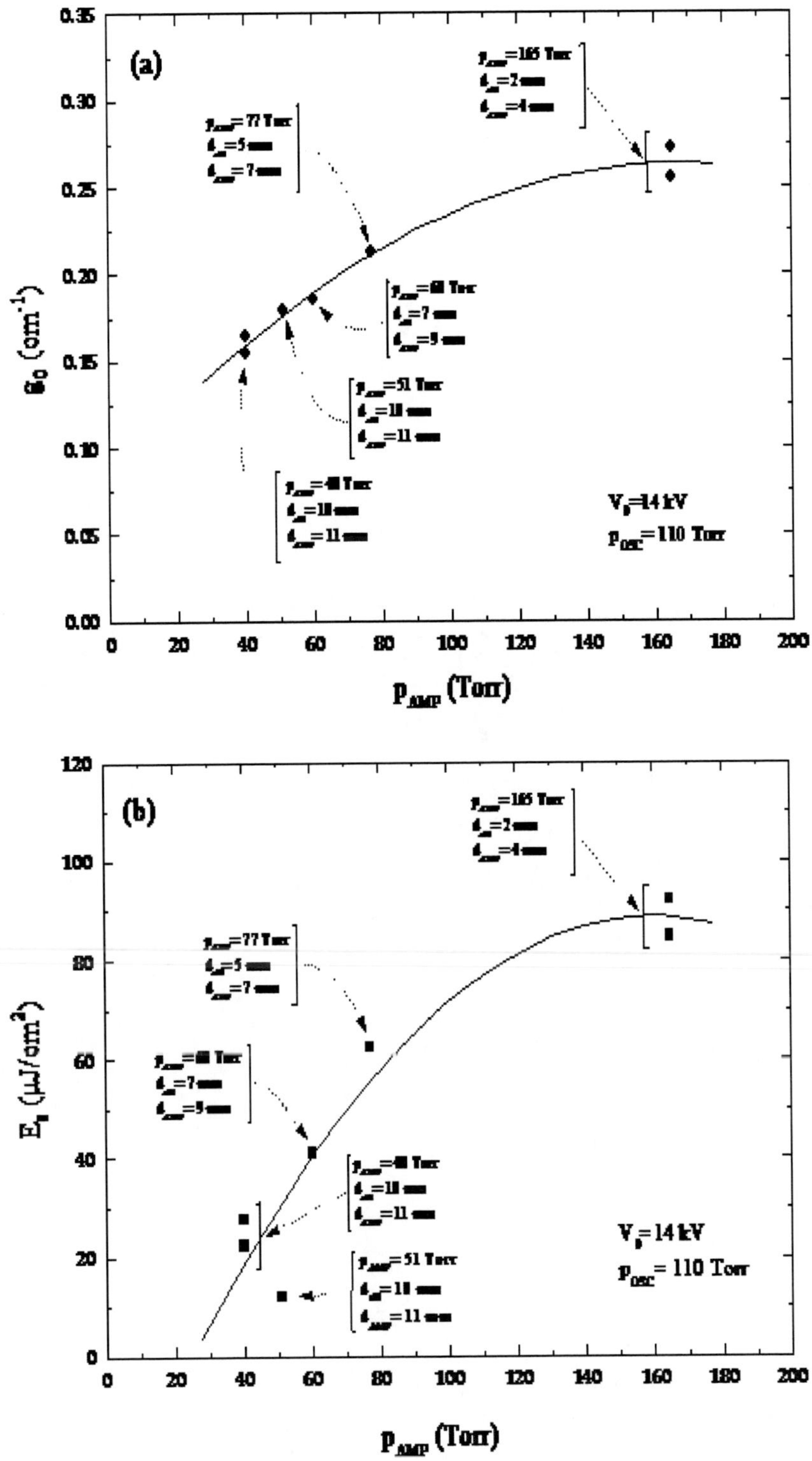

Figure 4.4. Plots of laser parameters: (a) g_0 and (b) E_s vs. the AMP gas pressure.

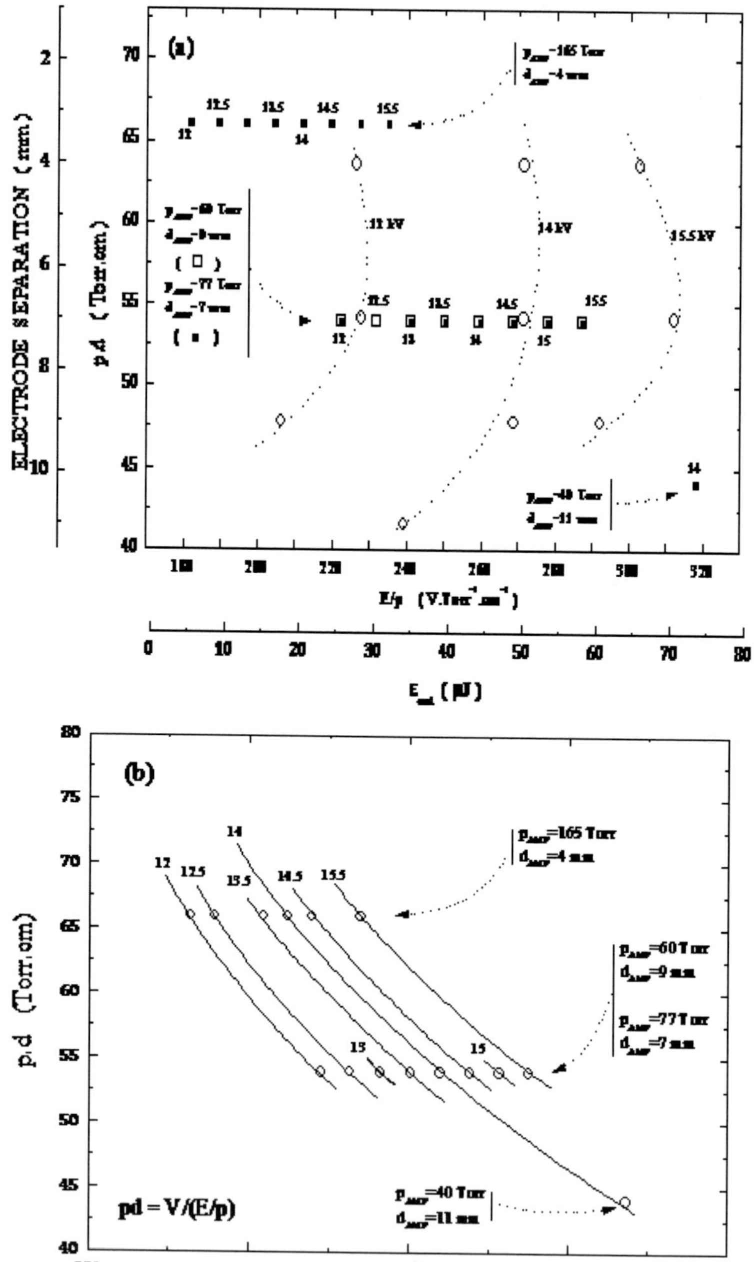

Figure 4.5. (a) Experimental measured values of *pd* versus *E*/*p* at different ignition voltages of 12-15.5 kV, shown as numbers (in kilovolts) for each experiment (■ and □); and the electrode gap separation vs. the output energy, ε_{out}, for different input voltages (○, and dashed lines), presented in different scales. It is seen that $(\varepsilon_{out})_{\max}$ corresponds to the range of d_{AMP}-values where its $p_{AMP}\,d_{AMP}$ parameter is ~54 *torr cm*. (b) The same *pd*-values versus *E*/*p*, corresponding to different input voltages taken as a parameter for each profile.

For the measurements, we realized that for each AMP gap separation, d_{AMP}, for a given p_{OSC}, there is an optimum AMP gas pressure, p_{AMP},where the OSC-AMP laser system is operating at its optimized operational condition. The dependencies of g_0 and E_s parameters on the operational N_2 gas pressure are shown in figures 4.4 (a) and 4.4 (b), where the plots of laser parameters versus the AMP gas pressure are given. The experimental conditions of the laser system are indicated in each figure, as well. These figures are showing that for a small d_{AMP}, it is required to operate the laser AMP at high gas pressure to have a proper T_e-value.

Also, the g_0 and E_s measurements at three different selected positions across the glow discharge, for $d_{AMP} = 11\,mm$, are indicating that the g_0-parameter has a higher value near the anode, and it is non-uniformly distributed across the gap separation. E_s-values, on the other hand, is more symmetrically distributed and has its minimum value at the center of the electrodes gap separation. This observation is consistent with that has been reported for the CO_2 laser [6]. It is also interesting to observe the behavior of the product of the AMP optimum operational gas pressure and the AMP electrode separation, $p_{AMP}d_{AMP}$, with respect to operational value of E/p_{AMP} at different ignition voltages. The curves corresponding to four gap separations of 4, 7, 9, and 11 *mm*, along with the curves representing the behavior of the output energy versus the electrodes gap separation, using independent scales, are given in Figure 4.5 (a). These data show that $p_{AMP}d_{AMP}$-values corresponding to the gap separation of 7 and 9 *mm* are overlapping exactly, though they belong to different experiments, different optimum gas pressures, and different operational voltages. The $p_{AMP}d_{AMP}$-values are 53.9 *torr cm* (for $d_{AMP} = 7\,mm$, $p_{AMP} = 77\,torr$) and 54.0 *torr cm* (for $d_{AMP} = 9\,mm$, $p_{AMP} = 60\,torr$), when the corresponding data were fitted to straight lines. By accepting an optimum value of $(pd)^{opt}_{AMP} = 54\,torr\ cm$ for the present laser system, it can be further realized that the gap separation corresponding to this optimum value of $(pd)_{opt}$ is located at the optimum range of the AMP output energy (ε_{out}), and also it is located at the optimum range of the average output power of the AMP when it is working as an OSC [20]. If we further write $pd = V_0/(E/p)$, and by rearranging our experimental data points, we can obtain different sets of profiles as indicated in Figure 4.5 (b), where for each profile the ignition voltage is considered to be a constant value. The data points given in this Figure were fitted to functions of the form $y = V_0/x$, where $V_0 = 12 - 15.5\,kV$. With these profiles the effect of input voltage on the $p_{AMP}d_{AMP}$-behavior can also be observed more explicitly. Thus, it seams that for any transversely excited N_2-laser we can define an optimum value for the product of the pressure and gap separation, similar to that known for longitudinally excited gas lasers (e.g., for He-Ne laser we have $pD = 3.6 - 4\,torrmm$, where D is the capillary diameter [40]). Operation of the laser in other d_{AMP}-values such as 4 and 10 *mm*, where they are not located in the range of laser optimum output energy, leads to $p_{AMP}d_{AMP}$-values of 66 and 44 *torr cm*, respectively. In other words, there is only a certain range for the laser gap separations where the $p_{AMP}d_{AMP}$-value remain constant, and for an optimum gas pressure the system will be operating on its maximum output power, whereas other gap separation will derive the system to work in lower output power, and consequently different $p_{AMP}d_{AMP}$ -values, which are quite different from that of the optimum $p_{AMP}d_{AMP}$ -value i.e., $(pd)_{opt} = 54.0\ torrcm$.

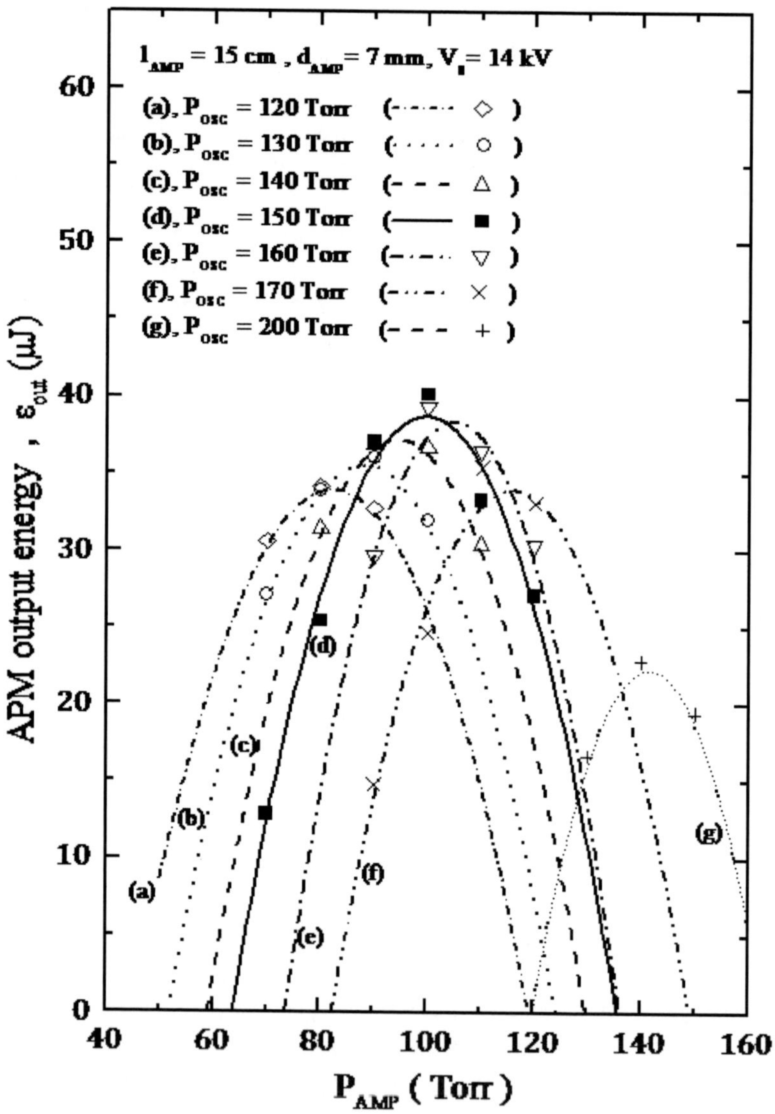

Figure 4.6. Typical plots of the AMP output energy, ε_{out}, versus AMP gas pressure for different OSC gas pressures to obtain appropriate operational gas pressures condition for the AMP maximum output energy: (a–g) $p_{osc} = 120 - 200\ torr$. $l_{AMP} = 15\,cm$, and $d_{AMP} = 7\ mm$. Curve (d) with $p_{osc} = 150\ torr$, and $p_{AMP} = 100\ torr$ corresponds to an optimized operational pressure where ε_{out} is maximum.

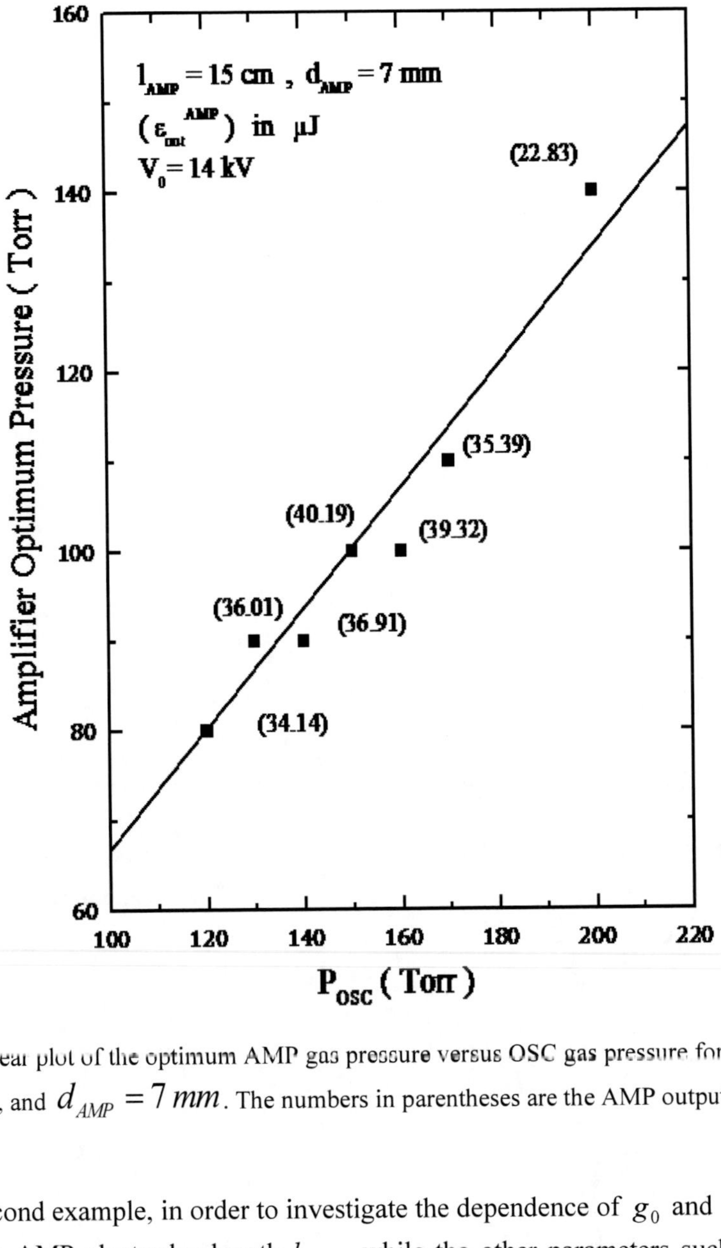

Figure 4.7. A linear plot of the optimum AMP gas pressure versus OSC gas pressure for $l_{AMP} = 15 cm$, and $d_{AMP} = 7\ mm$. The numbers in parentheses are the AMP output energies in μJ.

For the second example, in order to investigate the dependence of g_0 and E_s-parameters with respect to AMP electrodes length, l_{AMP}, while the other parameters such as electrodes gap separation, d_{AMP}, input voltage, and laser impedance remain constant, various electrodes, ranged from 5 to 31 *cm*, were used in the amplifier. The laser was operating under the optimized gas pressure and $V_0 = 14 kV$ was kept constant throughout the experiment. The experimental results show that g_0-parameter follows a profile of the type $m + n/l_{AMP}$, where *m* and *n* are some constants [37]. For the measurements, it was necessary to find the optimum AMP gas pressure for different l_{AMP}, at a given p_{OSC}. For this purpose at a fixed OSC gas pressure, the AMP gas pressure was varied, while the output energy from the AMP section was monitored. Typical results for the AMP output energy measurements for $l_{AMP} = 15 cm$

and OSC gas pressure, which were selected from 120 to 200 *torr*, corresponding to 7 different experiments are given in figures 4.6 (a-g). As it is seen in this Figure for $p_{OSC} = 150 torr$, for example, by varying the AMP gas pressure, the extracted output energy maximized only for $p_{AMP} = 100 torr$. So, $p_{OSC} = 150 torr$ and $p_{AMP} = 100 torr$ are the operational gas pressure for the OSC-AMP gas pressure at 14 *kV*, and $l_{AMP} = 15 cm$ (peak of curve (d) in this figure). Clearly, for other selected OSC gas pressures, the proper desynchronization time condition is obtained at different p_{AMP}, but the output from the AMP is not necessary maximized.

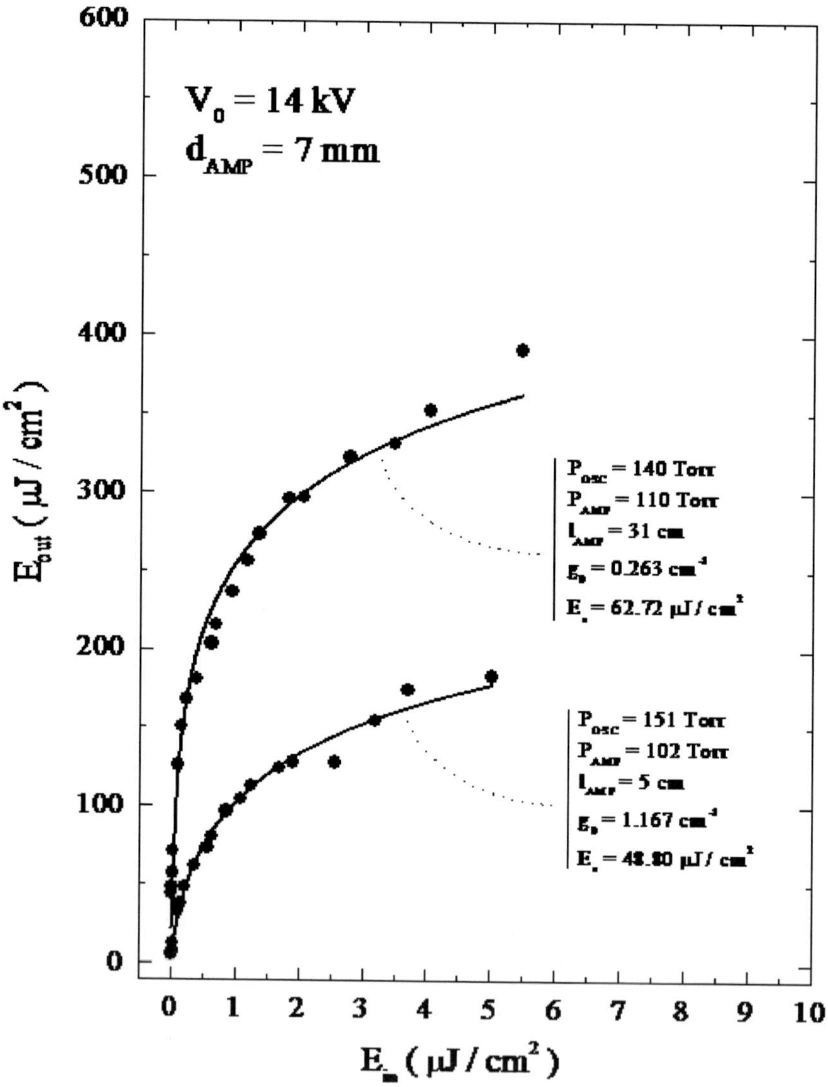

Figure 4.8. Typical plots of measured output energy density versus input energy density for $l_{AMP} = 31 cm$ (upper curve) and $l_{AMP} = 5 cm$ (lower curve). The operational gas pressures are given in the inset of the figure.

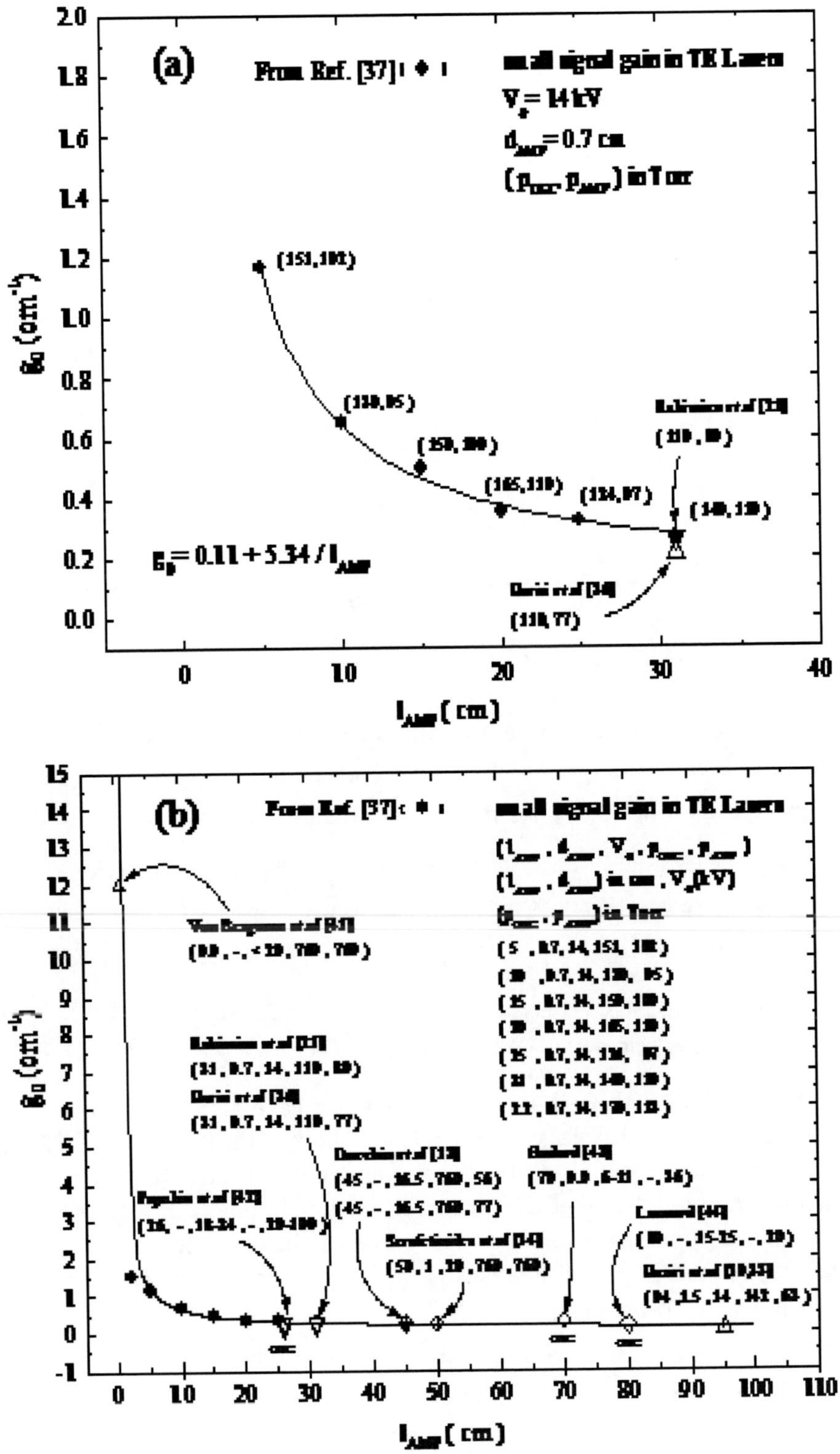

Figure 4.9. (Continued)

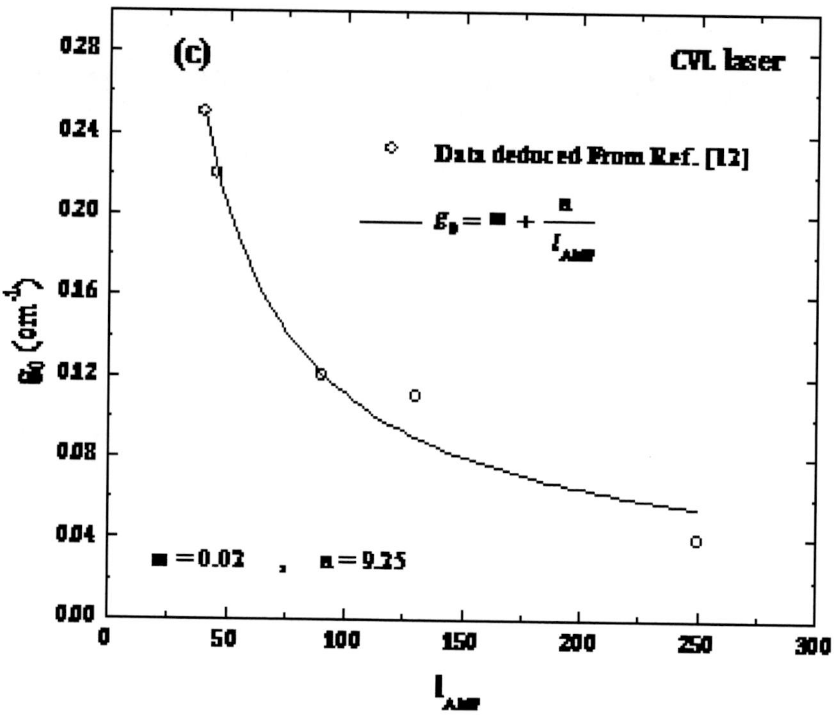

Figure 4.9. Plots of: (a) gain coefficient versus AMP electrode length obtained for $l_{AMP} = 5$ to 31 cm [37]. The numbers in parentheses are the OSC and AMP gas pressures. Data points are fitted to a function of the type $g_0 = m + n/l_{AMP}$, where m and n are 0.11 and 5.34, respectively; (b) All the g_0 −values using either a single OSC or OSC-AMP laser systems that are introduced in the literature have been included to show their values and their positions with respect to the proposed function. The numbers in parentheses are: l_{AMP}, d_{AMP}, V_0, p_{OSC}, and p_{AMP}, respectively; (c) the functional dependence of g_0 vs. l_{AMP} in CVL lasers. The data points are tabulated in Ref. [12]. l_{AMP} is the length (distance between electrodes) of CVL amplifiers.

By further use of our data in this experiment, we have also observed that for a given AMP electrodes length, there exists a linear relation between the OSC and AMP gas pressure, where at these pressures the OSC-AMP is operating with the maximized AMP output energy. These maxima are those that appeared as curves' maximum points in Figure 4.7. The output energies extracted from the AMP are also given in the parentheses in this figure. The procedure used for the gain measurements is the same as introduced earlier. Typical results for the measurements for two extreme cases of $l_{AMP} = 31\,cm$, and $5\,cm$ for given OSC and AMP gas pressures, in linear-linear plots of E_o versus E_i are shown in Figure 4.8. Distinct evidence of the experimental results for different electrodes length is evident in this figure.

Figure 4.9 (a) gives the behavior of g_0-parameter with respect to l_{AMP} corresponding to our measurements at 14 kV [37]. Data points are fitted to a function of the type $g_0 = m + n/l_{AMP}$, where m and n are constants, and determined by least-squares fit of the experimental points to the proposed function. This behavior will be proved in the theoretical section of this review. Other reported g_0-values, as given in Ref. [33], for either applying OSC-AMP laser systems or using single oscillators are introduced in Figure 4.9 (b). The same functional dependence of g_0 versus l_{AMP}, as given for Figure 4.9 (a) was used to show the positions of all the reported g_0-values with respect to the proposed function.

One particular example that appeared in Figure 4.9 (b) is referring to an experiment given by von Bergmann, and Hasson [41], where their system was operating under a very fast excitation condition (~ 2 *ns* current pulse duration) with a low impedance (1-6 Ω) and low inductance (~ 1 *nH*). They reported the gain coefficient as high as $12\,cm^{-1}$ (for the current density of $J_c \cong 30\ kA/cm^2$, and $Z_c = 1.3\,\Omega$) and as low as $3-4\,cm^{-1}$ (for the current density of $J_c \cong 2.8\ kA/cm^2$, and $Z_c = 6\,\Omega$). The shortest AMP electrodes length of $l_{AMP} = 0.9\,cm$, and the highest current density of $30\ kA/cm^2$ in an OSC-AMP system belong to this work.

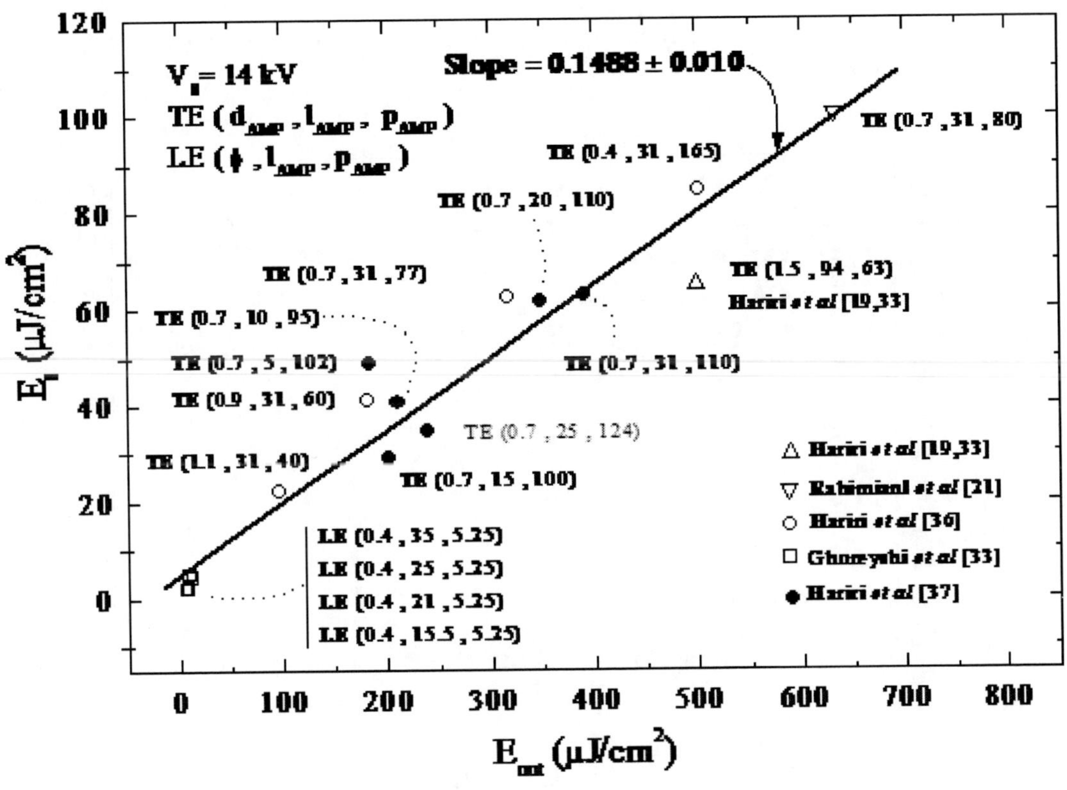

Figure 4.10. Plot of saturation energy density, E_s, versus output energy density, E_{out} for TE and LE N_2-laser measurements operating at a constant input voltage of 14 *kV*. Existence of a linear relation between these two quantities is evident in this figure. ϕ is the AMP laser tube diameter for LE lasers.

Their laser design and measurements stand at two extreme values of short-l_{AMP} and high-J_c, which are valuable measurements for theoretical investigations. Papakin and Sonin used a single oscillator to evaluate gain parameter in the range of 20-100 *torr* gas pressure, and $V = 18-24\,kV$, they obtained the gain values within a range of 0.14 - 0.25 cm^{-1} reaching at a maximum at $E/p = 110-170\,Vcm^{-1}torr^{-1}$ [42]. Other references shown in this graph are the gain-values reported by Docchio et al. [13], Serafetinides and Tsikrikas [14], Godard [43], and Leonard [44].

It is interesting to see if similar behaviors on the geometrical configurations for other types of lasers exist. For this purpose, the data points corresponding to the measured g_0-values for CVL lasers, tabulated in Ref. [12], have been used, and the profile for l_{AMP} dependency of the g_0-parameter is deduced, as shown in Figure 4.9 (c), where l_{AMP} is the length of CVL-AMP. The data appeared in the literature have been fitted to $g_0 = m + n/l_{AMP}$, where m and n are some constants and by least squares-fitting of the data points to the introduced function have been evaluated and introduced in the inset of Figure 4.9 (c).

These g_0-profiles as appeared in N_2 and CVL lasers are indicating that laser geometrical configuration is an important parameter and deserves to be studied for better understanding the lasers behavior.

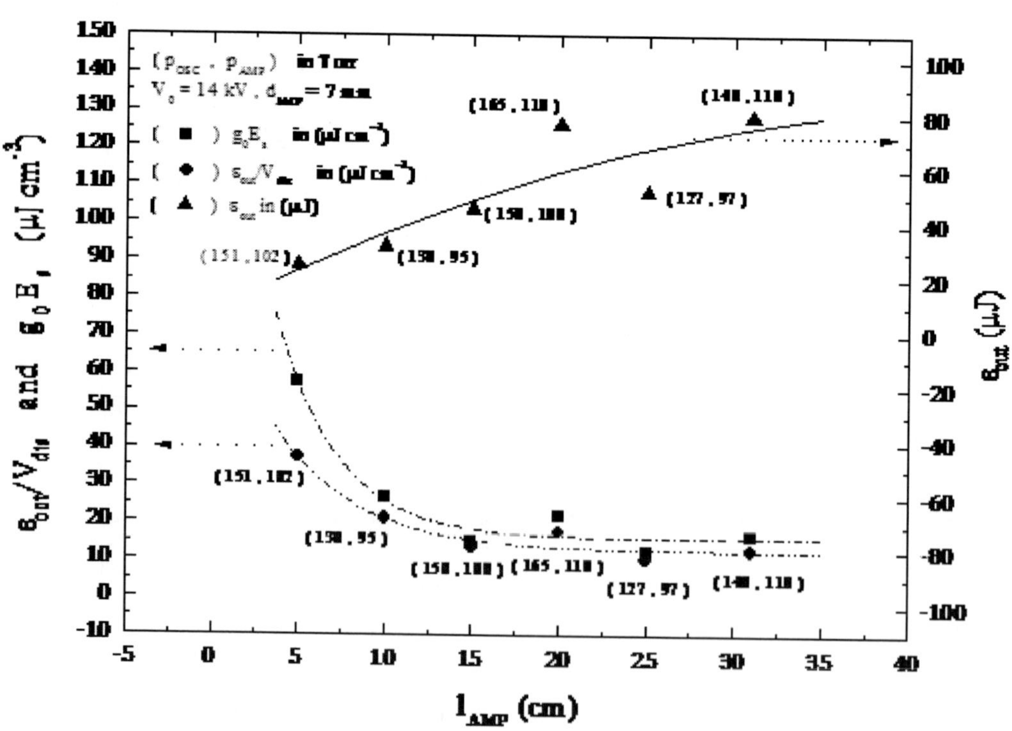

Figure 4. 11. Plots of the output energy density ($g_0 E_s$, E_{out}/V_{dis}), and output energy, ε_{out}, extracted from AMP versus l_{AMP}. Data points correspond to $l_{AMP} = 5-31\,cm$.

The result of measurements for TE and LE-lasers show that at a constant input voltage, i.e., $V_0 = 14\,kV$, the saturation energy density and output energy density are linearly correlated [37]. The plot corresponding to this behavior is shown in Figure 4.10. The laser specifications including the geometrical parameters, d_{AMP}, l_{AMP}, ϕ; gas pressure, p_{AMP}, and the references are indicated in the inset of the figure. This observation which is relating E_s and E_o linearly with a slope ~ 0.15 in N_2-lasers, was initially reported by Martin and Milam in solid-state lasers [31] and it seams it is worth to be study in other types of lasers. The measured values of g_0 and E_s parameters as given in [37], and their products, ($g_0 E_s$), which is giving the energy density for different electrodes lengths along with the extracted output energy per unit discharge volume, given by ε_o / V_{dis}, where V_{dis} is the corresponding discharge volume, are plotted in Figure 4.11. The sameness of these two plots, in particular in the range of $l_{AMP} > 10\,cm$ is evident in this figure. Also, the plot of output energy, ε_o, versus l_{AMP} in Figure 4.11 shows that ε_o for the given range of l_{AMP} increases linearly with respect to l_{AMP} and then reduces to a saturation level.

One important realization for g_0 and E_s parameters is the observation of their behavior when the ignition operating voltage of the laser system changes. In this respect, measurements of E_o, g_0, and E_s-values were carried out for $l_{AMP} = 5$, 15 and 25 *cm* (figures 4.12). Typical results of the measurements for $l_{AMP} = 15\,cm$ are given in Figure 4.12 (a). It is seen in this Figure that all these quantities have similar responses to the input voltage, they increase and approach to saturation levels when the input voltage increases; and also they are following the same behavior as V increases. Comparison of the measured g_0-values for three lengths of $l_{AMP} = 5$, 15 and 25 *cm* are given in Figure 4.12 (b). The plot of $g_0 l_{AMP}$-product versus the input voltage is also shown in this figure, showing that $g_0 l_{AMP}$ increases very slightly by increasing the input voltage.

Based on our observation of E_s- values versus E_o, given in Figure 4.10 with the slope of 0.1488 at 14 *kV*, and two sets of E_s-measurements for $l_{AMP} = 25$ and 15 *cm*, for different voltages from 12 to 15 *kV*, and the fact that E_s / E_o is a constant ratio at a constant voltage, and also by deducing this ratio from Refs. [19, 33, 13, 14] at their reported operating voltages, it was realized that E_s / E_o ratio is following an exponentially decaying function with respect to the input operating voltage, as shown in Figure 4.13. The plot is able to predict E_s-value of any N_2-laser system if the output energy, ε_o, output beam cross-section, and operational voltage are available. In [37] a few examples corresponding to g_0, E_s and ε_o predictions using the given empirical equations are given. For predicting the output energy, it is sufficient to get $g_0 E_s V_{dis}$ product.

At last, to observe the effect of l_{AMP} on the beam divergence, we introduce the plot of AMP beam divergence for the horizontal direction, versus the length of active medium, as shown in Figure 4.14. The data points are corresponding to $l_{AMP} = 5 - 31\,cm$ [37], and those reported for $l_{AMP} = 31$ and 94 *cm* [19, 33, 21]. It is seen that for lasers of short channels the beam divergence reaches at high values, but it drops to a constant value for long amplifying media.

Although an experiment was also performed for $l_{AMP} = 2.2\,cm$, but because of the failure to obtain laser oscillation within the amplifying medium using appropriate feedback mirrors, and observing a beam divergence which was comparable to that of the OSC section, so all the measured laser parameters including the beam divergence for $l_{AMP} = 2.2\,cm$ were not considered to be reported here. Up to this stage of observations, and measurements, our data points seems to be sufficient enough to develop a model based on measurements and mathematical approaches in order to predict N_2-lasers gain coefficient. Thus, in the next section this approach will be introduced to finalizing this review.

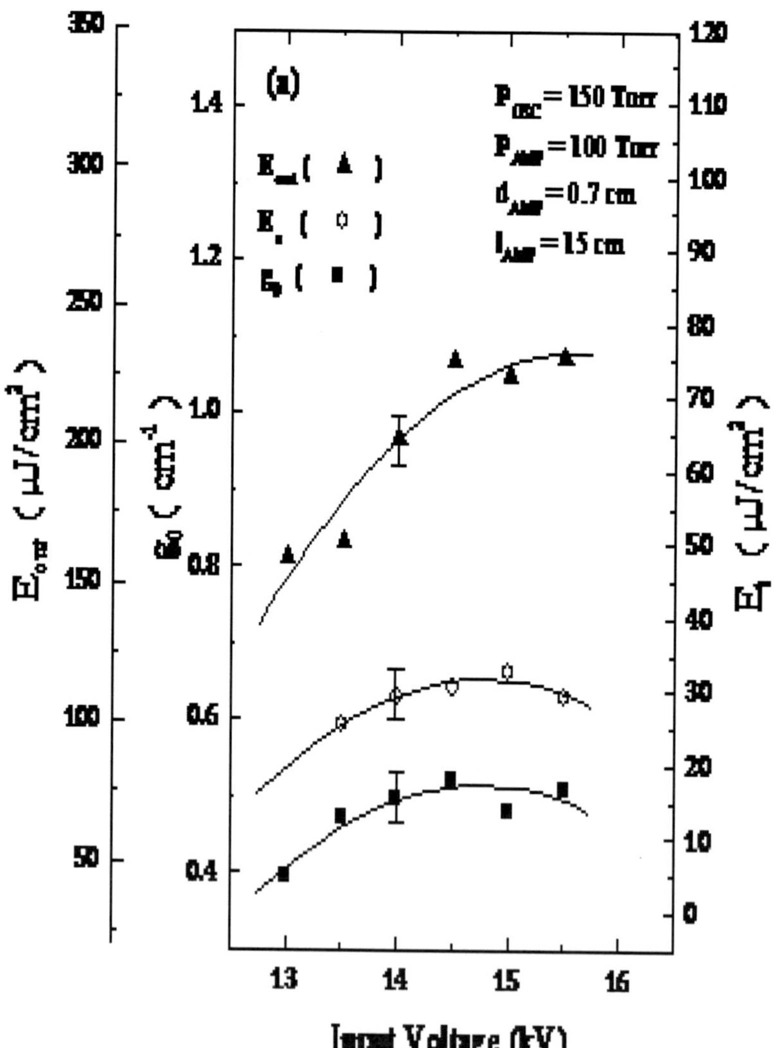

Figure 4.12. (Continued).

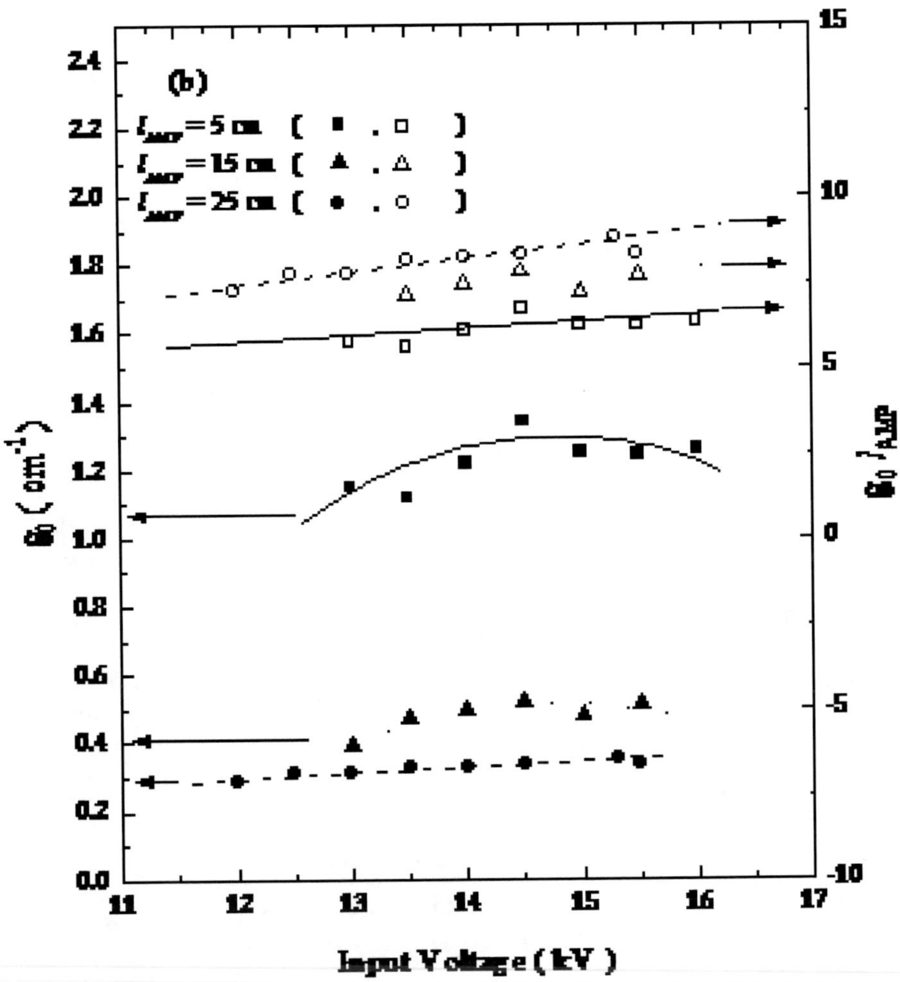

Figure 4.12. Plot of: (a) output energy density, E_{out}, gain coefficient, g_0, and saturation energy density, E_s, versus input voltage. These quantities behave similarly as the input voltages increase; (b) results of g_0 −measurements for $l_{AMP} = 5, 15, \text{and} 25\ cm$, and $g_0 l_{AMP}$ behavior with respect to the input voltage.

4.2. Theoretical Approaches

This section is dealing with theoretical considerations related to introducing a model, where it becomes possible to predict empirical equations obtained through the experimental measurements. We recall that in the previous subsection we found that $g_0 = m + n/l_{AMP}$ and $g_0 = r + q/d_{AMP}$ are the relations that have been verified experimentally. l_{AMP} and d_{AMP} are AMP electrodes length and gap separation, respectively; *m*, *n*, *r*, and *q* are some constants that can be determined through the gain measurement.

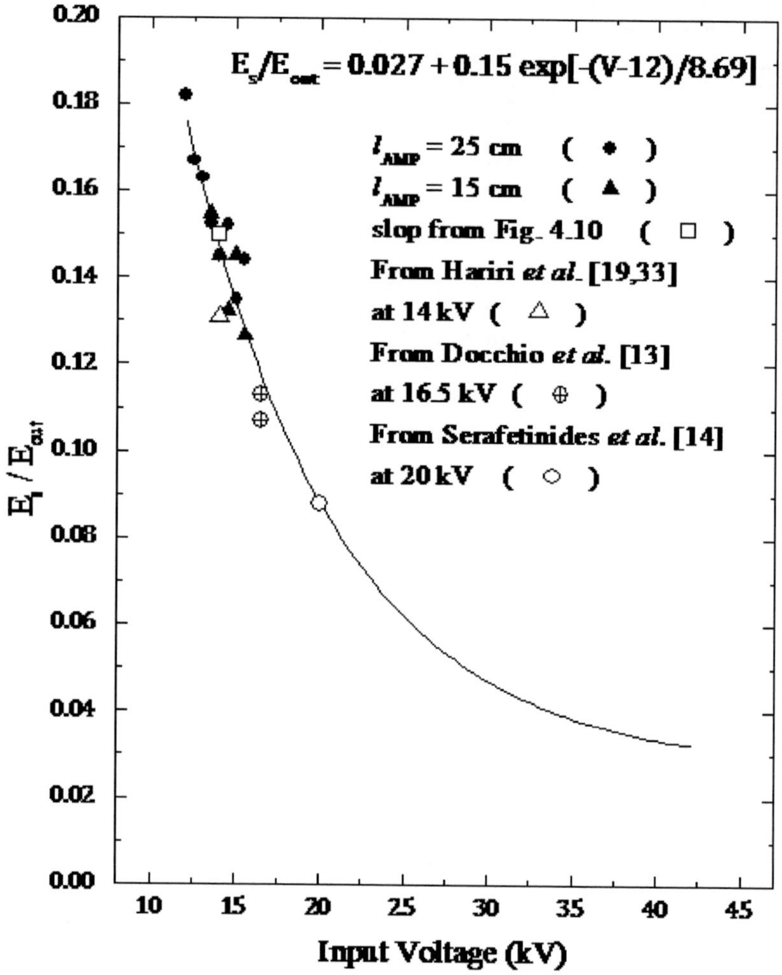

Figure 4.13. Plot of E_s/E_{out} versus input voltage obtained from the references indicated in the inset of the figure. The E_s/E_{out} ratio is fitted to an exponentially decaying function of the input operating voltage.

For a single oscillator, rate equations along with the known macroscopic properties of nitrogen discharges, electron impact ionization for the N_2 molecules proved that they can be used for the theoretical predictions [45]. The method is to determine electron density, n_e, from the discharge current, and electron temperature, T_e, from the applied electric field and gas pressure. Upon using rate equations we can calculate numerically the output energy and gain values corresponding to different l_{AMP} and d_{AMP} laser geometrical parameters.

For an OSC-AMP laser system, in contrast to a single laser oscillator, for analyzing the laser system some modifications to the relevant equations, both experimentally and theoretically are needed to simplify the analysis. Therefore, this section is divided into two parts (a and b) where they are dealing with the electric circuit behavior, and rate equations, respectively.

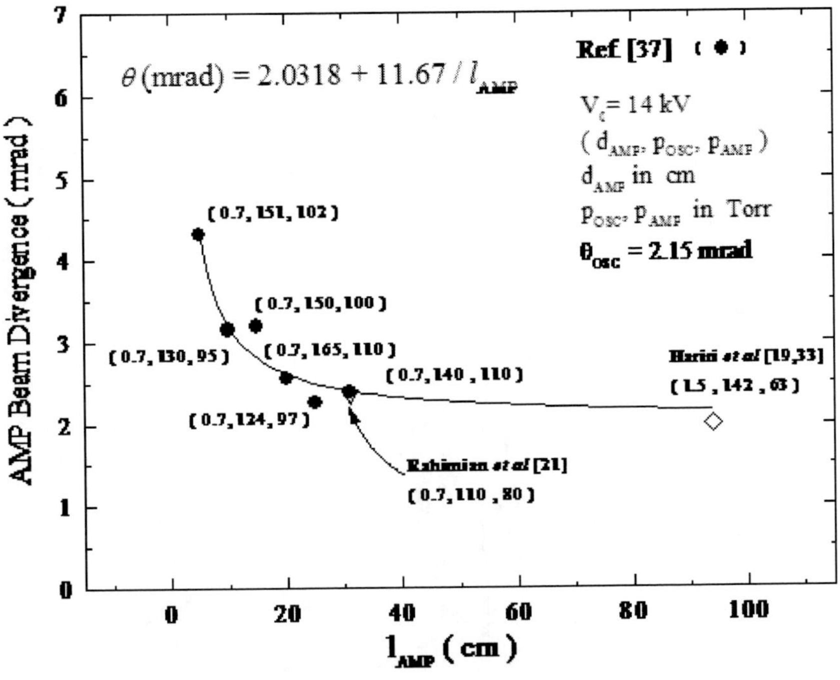

Figure 4.14. Plot of AMP beam divergence (horizontal direction) versus the length of amplifying medium, l_{AMP} . Data points correspond to our measurements, given by the references in the inset of the figure.

(a) Electric Circuit Behaviour

Generally, analysis of an OSC-AMP discharge circuit meets with difficulties because the inductance and resistance of different sections, i.e. spark gap, OSC and AMP discharge channels, cannot be measured directly. However, these quantities can be evaluated using a theoretical model, and by comparing the observed voltage waveforms of different laser sections with those calculated based on an equivalent electric - circuit.

The differential equations that describe the current and voltage waveforms at different sections of the OSC-AMP N_2-laser system are as follows, where use is made from Figure 4.15 (a) and its equivalent circuit diagram as appeared in Figure 4.15 (b). A three - dimensional arrangement for the OSC-AMP equivalent circuit has been applied.

$$L_1 \frac{dI_1}{dt} + R_1 I_1 + \frac{1}{C_1} \int (I_1 - I_2 - I_3)\, dt = 0 \tag{4.1}$$

$$L_2 \frac{dI_2}{dt} + R_2 I_2 + \frac{1}{C_2} \int I_2 dt - \frac{1}{C_1} \int (I_1 - I_2 - I_3)\, dt = 0 \tag{4.2}$$

$$L_3 \frac{dI_3}{dt} + R_3 I_3 + \frac{1}{C_3} \int I_3 dt - \frac{1}{C_1} \int (I_1 - I_2 - I_3)\, dt = 0 \tag{4.3}$$

For the analysis it is assumed that the resistance and inductance of the gas discharges, i.e. the spark gap (SG), OSC, and AMP sections are constants. In the above equations, as well as in Figure 4.15 (b), S_1, R_1 and L_1 are referring to SG switch, resistance and inductance of the SG section, S_2, R_2 and L_2 refer to the AMP switch, and its electric parameters, and S_3, R_3 and L_3 refer to the OSC section and its corresponding electric parameters.

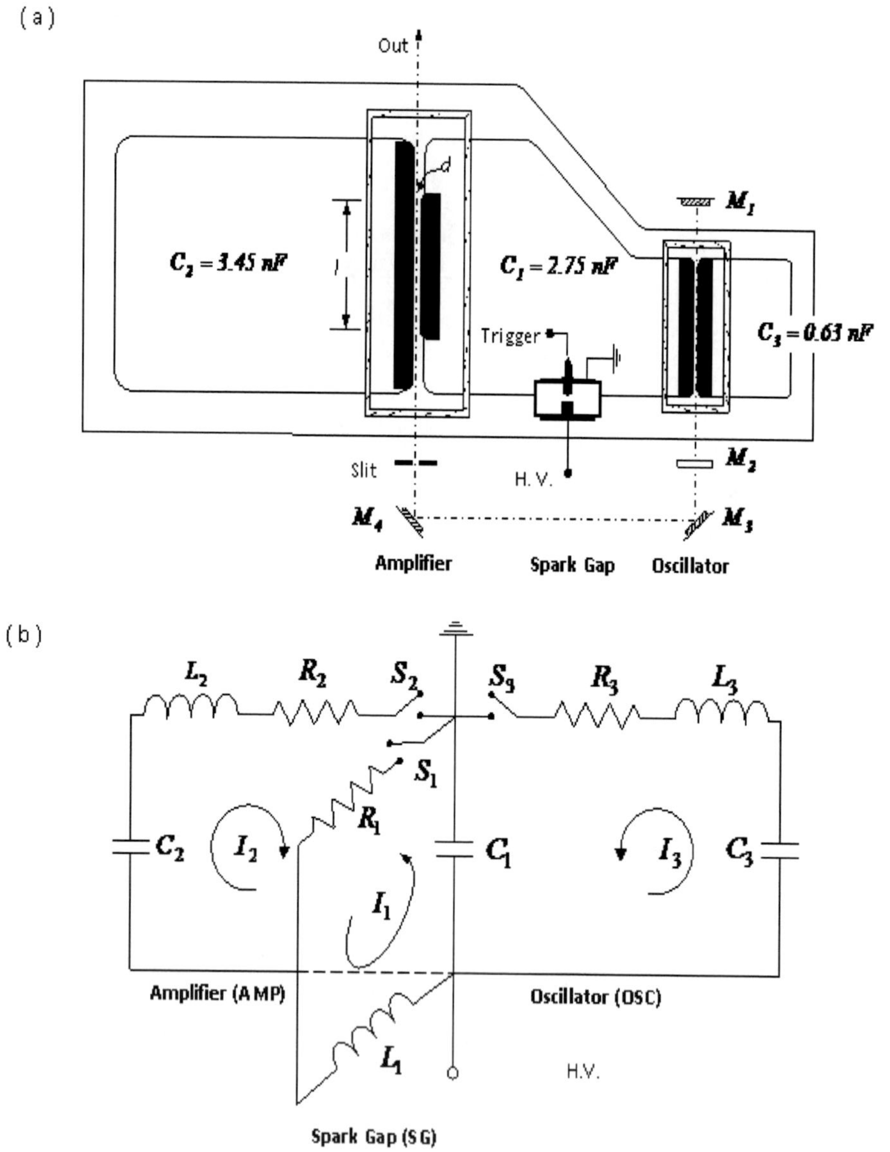

Figure 4.15. (a) Schematic diagram of the OSC-AMP N_2 -laser system; and (b) the equivalent circuit for the laser where it is depicted in a three-dimension, showing that the SG section is in share with OSC and AMP sections.

The time derivatives of equations (4.1)-(4.3) give rise to the following equations, where the notations of $D = d/dt$ and $D^2 = d^2/dt^2$ have been used.

$$\left(L_1 D^2 + R_1 D + \frac{1}{C_1}\right) I_1 - \left(\frac{1}{C_1}\right) I_2 - \left(\frac{1}{C_1}\right) I_3 = 0 \tag{4.4}$$

$$-\left(\frac{1}{C_1}\right) I_1 + \left(L_2 D^2 + R_2 D + \frac{1}{C_2} + \frac{1}{C_1}\right) I_2 + \left(\frac{1}{C_1}\right) I_3 = 0 \tag{4.5}$$

$$-\left(\frac{1}{C_1}\right) I_1 + \left(\frac{1}{C_1}\right) I_2 + \left(L_3 D^2 + R_3 D + \frac{1}{C_3} + \frac{1}{C_1}\right) I_3 = 0 \tag{4.6}$$

Eqs. (4.4)-(4.6) are coupled together and cannot be easily solved analytically. But the approach to the numerical solution is possible if R_i, L_i parameters $(i = 1-3)$ are evaluated experimentally.

When loops 2 and 3 are open, loop 1 (SG section) is equivalent to a RLC circuit. This condition is obtained by letting I_2 and I_3 in Eq. (4.4) equal to zero, i.e. we have

$$L_1 \frac{d^2 I_1}{dt^2} + R_1 \frac{dI_1}{dt} + \frac{1}{C_1} I_1 = 0 \tag{4.7a}$$

In terms of voltage differential equation if we let $V_1 = \frac{1}{C_1}\int I_1 dt$, then Eq. (4.7a) becomes

$$L_1 C_1 \frac{d^2 V_1}{dt^2} + R_1 C_1 \frac{dV_1}{dt} + V_1 = 0 \tag{4.7b}$$

The solution to the above equation has been obtained by introducing the initial condition of $V_1(0) = V_0$ and the slope at $t = 0$, equal to some constant, β, to fit the open-circuit waveform. So, we have $(dV_1/dt)_{t=0} = \beta$, and V_1 is given by

$$V_1 = V_0 e^{-\lambda_0 t}\left(\cos\omega_0 t + \frac{\beta + \lambda_0 V_0}{\omega_0 V_0} \cdot \sin\omega_0 t\right) \tag{4.8}$$

where

$$\omega_0 = \left(\frac{1}{L_1 C_1} - \lambda_0^2\right)^{1/2} \tag{4.9a}$$

and

$$\lambda_0 = \frac{R_1}{2L_1} \tag{4.9b}$$

As $\lambda_0 << \omega_0$, in Eq. (4.9a) we can let $\omega_0 \approx (L_1 C_1)^{-1/2}$.

By introducing $(\beta + \lambda_0 V_0)/V_0 = \sin z_0$ and $\omega_0 = \cos z_0$, Eq. (4.8) becomes

$$V_1 = \frac{V_0}{\omega_0} e^{-\lambda_0 t} \cos(\omega_0 t - z_0) \tag{4.10}$$

where, $\tan z_0 = (\beta + \lambda_0 V_0)/\omega_0 V_0$ has been defined. For $\beta = 0$, we get the same solution as given in [46].

It is worth mentioning that in fact the right-hand side of Eq. (4.7b) is equal to an arbitrary constant. It is known, however, that the solution of a differential equation is sum of the homogeneous equation, i.e. Eq. (4.7b), and the particular solution, where in this case it is a constant. In addition, if we let $t \to \infty$ we must also have $V_1 \to 0$, and consequently, this arbitrary constant has to be taken equal to zero. With this argument, it is seen that the solution which is given by Eq. (4.8) or (4.10) satisfies the limiting condition of $V_1 = 0$ when $t \to \infty$. This limiting condition can be observed upon monitoring the voltage waveforms at the longer time scales of a few tens of microseconds.

Referring to Eq. (4.8), although at $t = 0$ and for $\beta = 0.7\ kVns^{-1}$, we get a correct overlap of the experimental voltage waveform to Eq. (4.8), but the best fit which is consistent with the overall experimental observation occurs if we let $\beta = 1.94\ kVns^{-1}$. Thus, upon observing the waveform at the ends of C_1, and by fitting the V_1-time behavior of the waveform to Eq. (4.8) or (4.10), the required parameters of R_1 and L_1 with the best fitting condition can be obtained.

At the next step, we need to evaluate R_2 and L_2 parameters. For an OSC-open circuit, we have $I_3 = 0$. Equations (4.4) and (4.5) then give

$$L_1 \frac{d^2 I_1}{dt^2} + R_1 \frac{dI_1}{dt} + \frac{1}{C_1} I_1 - \frac{1}{C_1} I_2 = 0 \tag{4.11a}$$

$$L_2 \frac{d^2 I_2}{dt^2} + R_2 \frac{dI_2}{dt} + \frac{1}{C_s} I_2 - \frac{1}{C_1} I_1 = 0 \tag{4.11b}$$

where, in Eq. (4.11b) $C_s^{-1} = C_1^{-1} + C_2^{-1}$ has been defined. These are the circuit equations for a Blumlein circuit used for a N_2-laser oscillator. The circuit has been studied in detail by

Papadopoulos et al. [46]. The corresponding solution to the voltage at the ends of C_2 and C_1 are given, respectively by

$$V_2(t) = \frac{V_{2+}}{\cos z_+} e^{-\lambda_+ t} \cos(\omega_+ t + y_+ - z_+ + \Phi_+) + \frac{V_{2-}}{\cos z_-} e^{-\lambda_- t} \cos(\omega_- t + y_- - z_- + \Phi_-) \tag{4.12}$$

$$V_1(t) = \frac{V_{1+}}{\cos z_+} e^{-\lambda_+ t} \cos(\omega_+ t - z_+ + \Phi_+) + \frac{V_{1-}}{\cos z_-} e^{-\lambda_- t} \cos(\omega_- t - z_- + \Phi_-) - \frac{C_2}{C_1} V_2(t) \tag{4.13}$$

where, $k_\pm q_\pm / C_2$ and $q_\pm / C_1$ as given in [46] have been replaced by $V_{2\pm}$ and $V_{1\pm}$ for simplicity. Also, $\tan z_\pm = \lambda_\pm / \omega_\pm$ and $k_\pm$, $y_\pm$ are obtained by the equations given in [46].

The quantities $q_\pm$ and $\Phi_\pm$ can be calculated from the initial conditions of the circuit and voltage of C_1 and C_2. Also for $\omega_\pm$ and $\lambda_\pm$, we have

$$\omega_+^2 + \omega_-^2 = \omega_1^2 + \omega_2^2 \tag{4.14 a}$$

$$\lambda_+ + \lambda_- = \lambda_1 + \lambda_2 \tag{4.14 b}$$

$\omega_{1,2}$ and $\lambda_{1,2}$ are introduced to simplify the relations. $\omega_1^2 = 1/L_1C_1$, $\omega_2^2 = 1/L_2C_s$, $\lambda_1 = R_1/2L_1$ and $\lambda_2 = R_2/2L_2$ are also defined. ω_1 and λ_1 are the OSC and AMP open circuit parameters (i.e. $\omega_1 = \omega_0$ and $\lambda_1 = \lambda_0$) as it was discussed previously. $\omega_\pm$ and $\lambda_\pm$ are the circuit parameters that can be evaluated by observing the waveforms V_2 and V_1, fitting by equation (4.12) or (4.13), and subsequently using Eqs. (4.14a) and (4.14b) for determining ω_2 and λ_2. Thus, by the waveforms observation the required circuit parameters $R_{1,2}$ and $L_{1,2}$ are determined.

A typical result of the voltage waveform for $l_{AMP} = 31\,cm$, where both OSC and AMP are open-circuit (i.e., OSC and AMP gas pressure are at 1 *atm*), is shown in Figure 4.16. In this condition the circuit corresponding to the SG section works as a RLC damped sinusoidal oscillation. The solid line indicated in Figure 4.16 is referring to the observed waveform corresponding to the SG section, while the dashed line corresponds to the theoretical model given by Eq. (4.8) or (4.10). ω_0 and λ_0 are obtained by the method of least-squares fit of the observed waveform to Eq. (4.8). By determining ω_0 and λ_0, $R_1 = 1.15\,\Omega$ and $L_1 = 43.4\,nH$ are evaluated using Eqs. (4.9a) and (4.9b). As it is shown in this figure, at $t = 0$, $V_0 = 14\,kV$, and at $t \sim 10\,ns$ V_1 raises to $\sim 20\,kV$. By operating the AMP at $110\,torr\,N_2$ gas pressure, where for $l_{AMP} = 31\,cm$ at this pressure the gain value was measured, and keeping the OSC in an open-circuit condition, further voltage waveform measurements on the C_2 and C_1 sides can be made. Typical voltage waveforms of V_2 and V_1 for the C_2 and C_1 sides, for $l_{AMP} = 20\,cm$ and $5\,cm$, and gas pressure of 110 and 102 *torr* are shown in figures 4.17, and 4.18, respectively. Again, as explained earlier, by fitting the observed waveforms to Eqs. (4.12) and (4.13) and by looking for the consistent results, $\omega_\pm$ and $\lambda_\pm$ can be determined. Consequently, the AMP inductance, L_2, for different AMP electrode lengths are determined, using Eq. (4.14a) for obtaining ω_2^2, and then the relation $\omega_2^2 = 1/L_2C_s$. Keeping in mind that ω_1 which is needed to be used in (4.14a) is obtained from The OSC-AMP-open circuit condition (i.e. when we let $\omega_1 = \omega_0$). Figure 4.19 shows that L_2 obtained by this method reduces slightly when l_{AMP} increases. For this measurement the experiment was repeated several times, and the error bars in Figure 4.19 are the maximum error appeared in the experiment and by the use of curve fitting program. On an average, $L_2 \approx 12.2\,nH$ can be adopted to be used in the numerical calculations. In this work, however, for each l_{AMP}, the corresponding measured L_2 was used for the calculation.

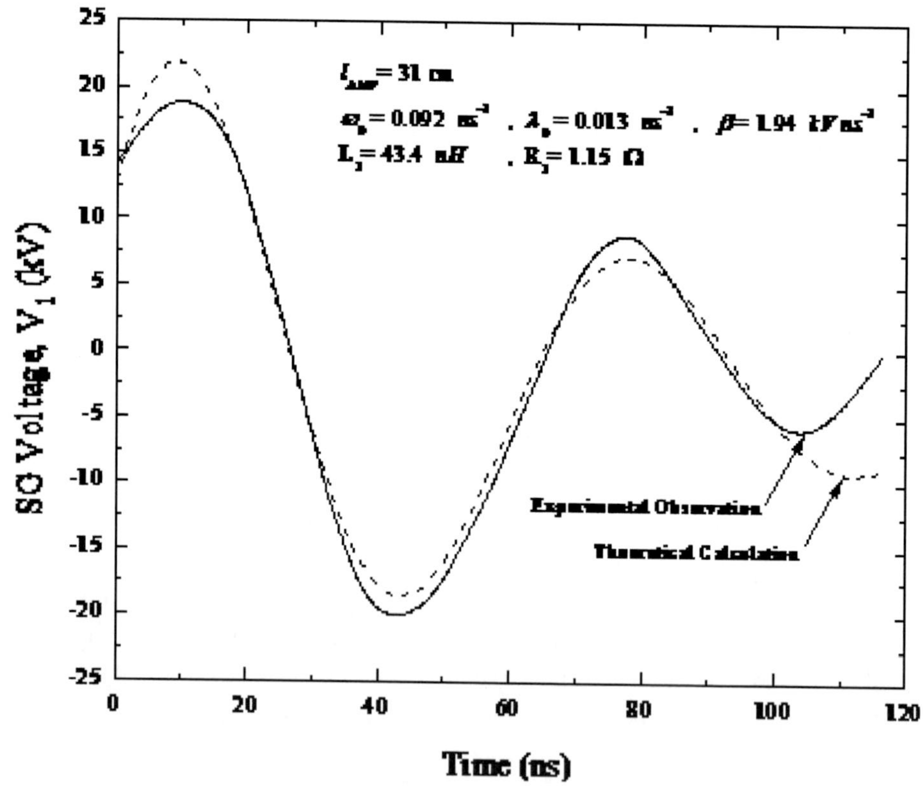

Figure 4.16. A typical experimental voltage waveform corresponding to the OSC- and AMP-open-circuit case, i.e., $I_2 = I_3 = 0$ (solid line). The dashed line corresponds to the RLC damped sinusoidal oscillation, given by Eq. (4.8) or (4.10).

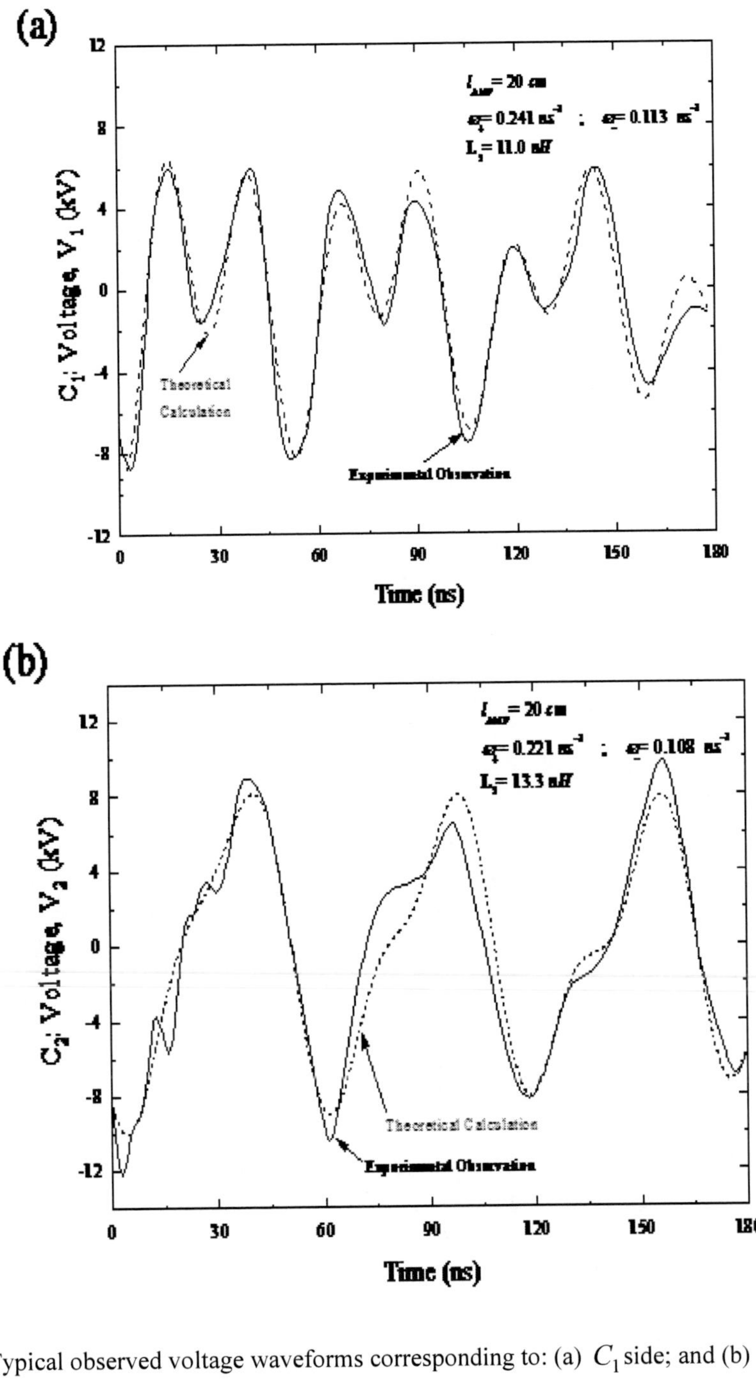

Figure 4.17. Typical observed voltage waveforms corresponding to: (a) C_1 side; and (b) C_2 side of Figure 4.15. AMP is operating at 110 *torr*. The system is in OSC-open-circuit condition, i.e., $I_3 = 0$, $l_{AMP} = 20cm$ (solid line). The dashed lines are given by Eqs. (4.12) and (4.13), respectively.

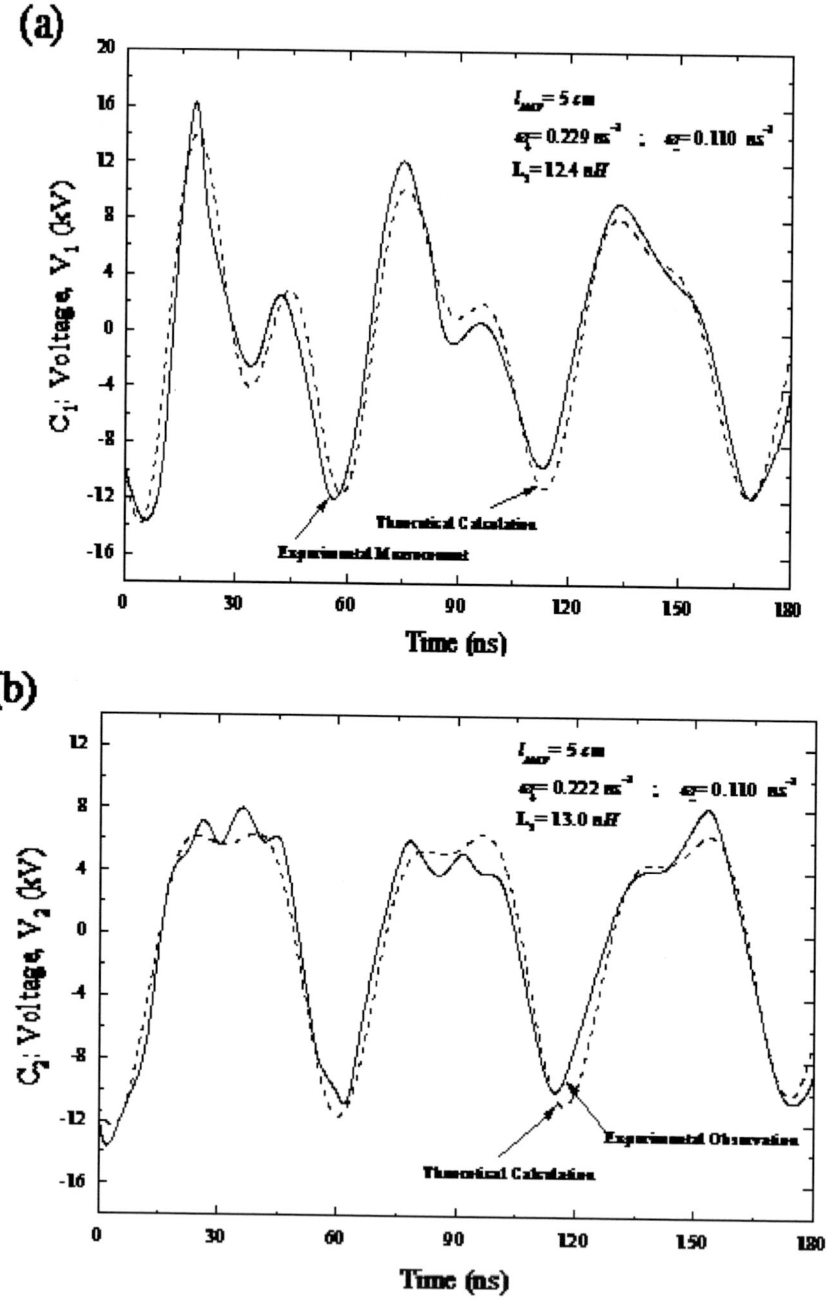

Figure 4.18. The same as Figure 4.17 for $l_{AMP} = 5cm$, and $p_{AMP} = 102torr$.

In a Blumlein driven laser which can represented by a transmission line circuit, the total resistance of the AMP section, R_2, given by $R_2 = R_g + Z_0$, where R_g is the resistance of the laser discharge and Z_0 is the characteristic impedance of the transmission line [45]. For a uniform electric field, E, between the AMP electrodes, R_g is given by

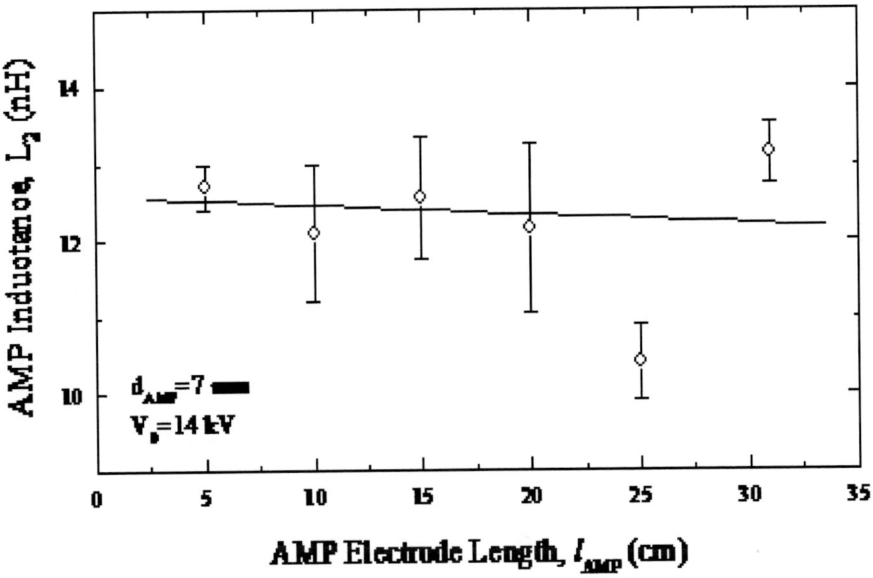

Figure 4.19. The behavior of AMP inductance, L_2, versus l_{AMP}.

$$R_g = \frac{d_{AMP}}{s\sigma l_{AMP}} \tag{4.15}$$

where σ is the conductivity of the gas, and s is the thickness of the discharge. σ as a function of time is given by

$$\sigma = \frac{e n_e \mathrm{v}_d}{E} \tag{4.16}$$

where n_e is the electron density in the laser channel and obeys the Townsend equation

$$\frac{d n_e}{d t} = \alpha \mathrm{v}_d n_e \tag{4.17}$$

$\mathrm{v}_d = 2.9\times10^5 (E/p)$, in $cm s^{-1}$, is the drift velocity. $\alpha = 1.4\times10^{-8}(E/p)^{3.7}\cdot p$ is the Townsend coefficient given in cm^{-1}, and p is the gas pressure in *torr*. Z_0 in our case was obtained from [47], and was found to be 0.74Ω.

The required electric field between the electrodes of the AMP is given by

$$E = \frac{1}{d_{AMP}}\left(R_g I_2 + \eta L_2 \frac{d I_2}{d t} \right) \tag{4.18}$$

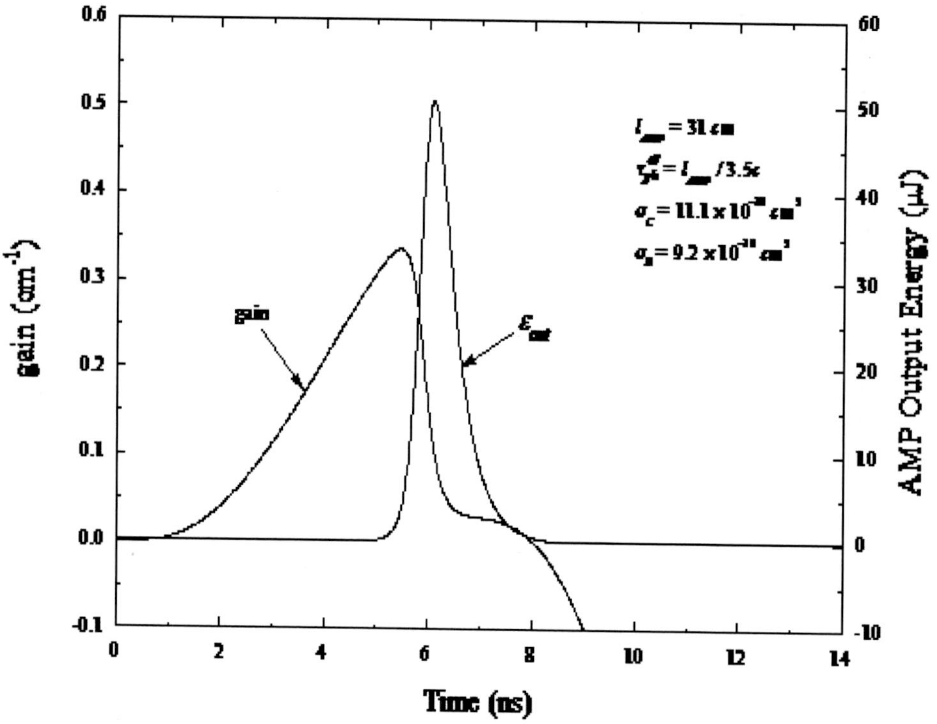

Figure 4.20. Time behaviors of the calculated gain coefficient, g_0, and output energy, ε_{out}. The peak values for g_0 and ε_{out} for $l_{AMP} = 31cm$ and $\tau_{ph}^{eff} = l_{AMP}/3.5c$ are $0.34\,cm^{-1}$ and $50.48\,\mu J$ respectively.

I_2 in Eq. (4.18) can be obtained from Eq. (4.2), by letting $I_3 = 0$. With the assumption that I_1 in Eq. (4.2) gets the same values of open-OSC and AMP circuit [45], we can write

$$L_2 \frac{dI_2}{dt} + R_2 I_2 = V_{eff} \tag{4.19a}$$

where

$$V_{eff} = \frac{V_0}{\omega_0} e^{-\lambda_0 t} \cos(\omega_0 t - z_0) - \frac{1}{C_s} \int_0^t I_2 dt. \tag{4.19b}$$

η in the second term of Eq. (4.18) is the ratio of d_{AMP} to the total perimeter length of the current flow in the laser channel. η in our case is ~ 0.04.

As the excited states corresponding to the $N_2(C^3\Pi_u) \rightarrow N_2(B^3\Pi_g)$ laser transition are close to the ionization limit, the relation

$$kT_e = 0.11(E/p)^{0.80} \text{ eV} \qquad (E/p \text{ in } \text{V}/cm \cdot torr) \tag{4.20}$$

which was obtained for the ionization cross section of N_2 molecule is expected to give approximately the electron temperature, T_e, of the electron excitation of the $N_2(C)$ and $N_2(B)$ states [45].

(b) Rate Equations

Following the rate equations given by Fitzsimmons et al. [45], the required equations for the $C^3\Pi_u$ and $B^3\Pi_g$ states along with the equation for the photon density are given by

$$\frac{dN_C}{dt} = n_e N_{gas} \int_0^\infty g(T_e, \text{v}) \sigma_C(\text{v}) 4\pi \text{v}^3 d\text{v} - \sigma_{stim} n_{ph} c (N_C - N_B) - \frac{N_C}{\tau_C} \tag{4.21a}$$

$$\frac{dN_B}{dt} = n_e N_{gas} \int_0^\infty g(T_e, \text{v}) \sigma_B(\text{v}) 4\pi \text{v}^3 d\text{v} + \sigma_{stim} n_{ph} c (N_C - N_B) - \frac{N_B}{\tau_B} + \frac{N_C}{\tau_C} \tag{4.21b}$$

$$\frac{\partial n_{ph}}{\partial t} = \sigma_{stim} n_{ph} c (N_C - N_B) - \frac{n_{ph}}{\tau_{ph}} + \gamma \frac{N_C}{\tau_C} \tag{4.21c}$$

For the small signal gain and the output energy extracted from the gain medium, we have

$$g_0 = \sigma_{stim}(N_C - N_B) \tag{4.22a}$$

and

$$\varepsilon_{out} = n_{ph} h \nu V_{dis} \tag{4.22b}$$

where in Eqs. (4.21) and (4.22), N_{gas} is the ground-state gas density; σ_C and σ_B are the electron impact cross sections of the $N_2(C)$ and $N_2(B)$ states, respectively; $g(T_e, \text{v})$ is the normalized Maxwell-Boltzmann distribution; σ_{stim} is the stimulated emission cross section and is given by $\lambda^2 / 8\pi\Delta\nu\tau_C \approx 4.5\times10^{-15}\ cm^2$ for $\Delta\nu = 278 GHz$ [42]; τ_C and τ_B (40 *ns* and 6 μs, respectively) are the spontaneous lifetime of the C and B laser levels; τ_{ph} represents the mean cavity lifetime; γ is the geometrical factor for spontaneous emission; and $V_{dis} = s d_{AMP} l_{AMP}$ is the discharge volume.

For the simulation, we have used $N_2(C_{\text{v}=0})$ cross sections introduced by Burns [29] and other researchers as appear in Burns' paper. They are (11.8, 10.8 and 8.12 $\times10^{-18}\ cm^2$), and the effect of each was evaluated in our gain coefficient and the output energy simulation study. For $N_2(B)$ cross section, we applied the value $\sigma_B = 6.35\times10^{-18}\ cm^2$ given by

Fitzsimmons et al. [45] and $\sigma_C = 8.12\times10^{-18}\,cm^2$. The results are indicating that our numerical calculations are in good agreement with the experiment. However, as the recent work on the σ_C measurement given by Y. Tohyama and T. Nagata [48] and Shemansky et al. [49] are showing that σ_C is slightly higher i.e. $\sigma_C = 11.1\times10^{-18}\,cm^2$, we decided to apply other values for the σ_B. The values of $\sigma_B \approx 9-10\times10^{-18}\,cm^2$ along with $\sigma_C \approx 11-12\times10^{-18}\,cm^2$ turn out to be consistent with our g_0 and ε_{out} observations. Thus, if we accept $\sigma_C = 11.1\times10^{-18}\,cm^2$, the selected value for σ_B will be $9.2\times10^{-18}\,cm^2$ which is very close to the value of $\sigma_B = 8.9\times10^{-18}\,cm^2$ at 14 kV given by Tabata et al. [50].

For a traveling wave excitation, we can write $\tau_{ph} = l_{AMP}/c$. This relation has already been introduced by Chang and Teii [51 – 52].

The numerical solution to the rate equations (4.21a)-(4.21c), and using Eqs. (4.22a) and (4.22b) for g_0 and ε_{out}, along with the required equation (4.17) which is needed for the electron density, n_e, leads to the time behavior of the laser parameters as shown in Figure 4.20. The peak values of g_0 and ε_{out} are those that have been accepted in the calculation. As in an OSC-AMP laser system the laser pulse- width is determined by the OSC section, we are not expecting to get a proper result for the laser pulse- width, and for this purpose the OSC section has to be analyzed numerically.

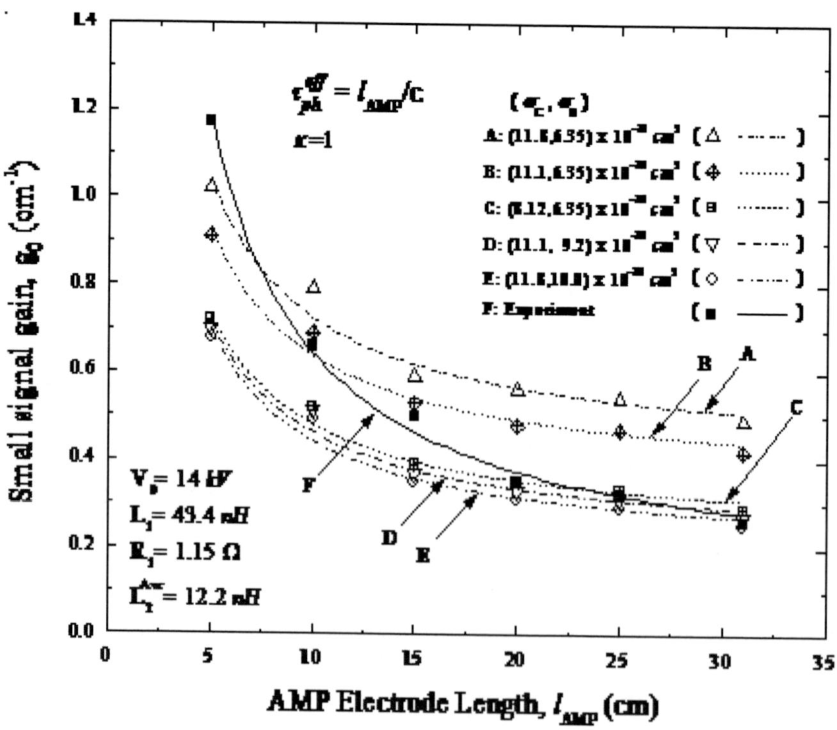

(a)

Figure 4.21. (Continued).

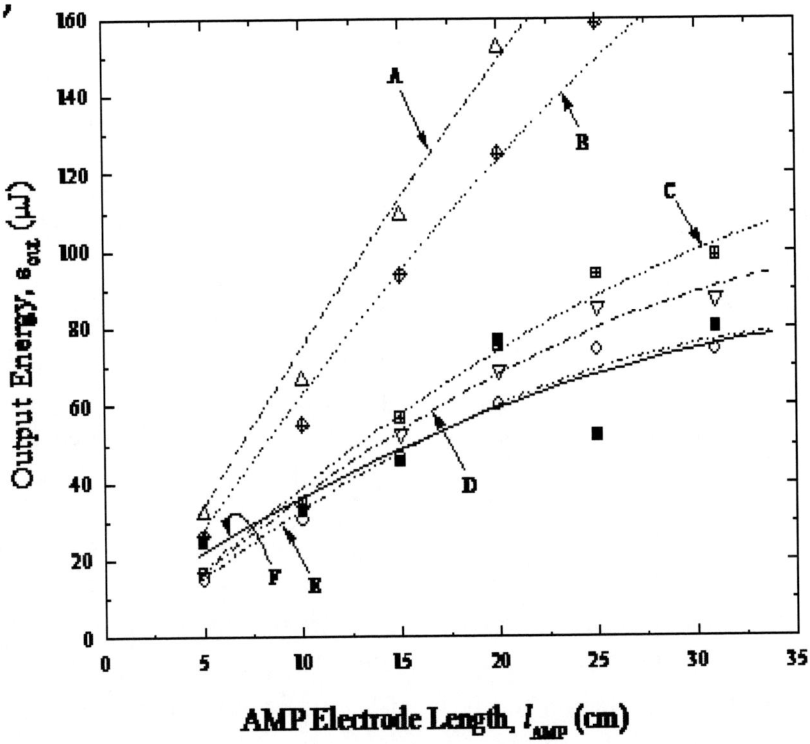

(b)

Figure 4.21. Plots of: (a) g_0, and (b) ε_{out} versus AMP electrode length. In both figures the mean cavity lifetime is taken to be $\tau_{ph}^{eff} = l_{AMP}/c$, i.e. $\kappa = 1$. Curves D and E appear to agree with the experiment (F) for large l_{AMP} (i.e. $l_{AMP} \sim 30\,cm$). Capital letters of A, B ... are referring to a pair of σ_C and σ_B cross-sections used in the calculation.

Referring to figures. 4.21 for both g_0 and ε_{out}, we realized that by adopting $\tau_{ph} = l_{AMP}/c$, the numerical results are rather excellent for long AMP length of ~ 30 *cm*. In particular the calculation gives appropriate pair of impact - excitation cross-sections of σ_C and σ_B. But for short l_{AMP} a strong deviation from the experiment exists for all the selected pair of cross-sections. For simplifying an OSC-AMP calculation, and by applying a model base on a single OSC, without considering any mirrors reflectivity and cavity losses, τ_{ph} has to be modified. The approach for this modification relies on our experimental observations, where we realized that for short electrode lengths τ_{ph} needs to be somehow smaller than l_{AMP}/c to get a better agreement with the experimental measurements. Our observations are indicating that for this purpose if we introduce the effective cavity lifetime $\tau_{ph}^{eff} = \kappa l_{AMP}/c$, for $\kappa = some\ constant$, a better agreement, for example for $\kappa = 1/2$, is realized with our experimental observations. Still, a better agreement exists when $\kappa = 1/3$ is used. For $\kappa = 1/4$, however, a deviation from experiment is observed. Upon trying a number between these two values i.e. $\kappa = 2/7$, a rather excellent agreement with the experiment is obtained. Thus, in graphs of g_0 and ε_{out} versus l_{AMP}, we applied five different τ_{ph}^{eff} in our calculation; they are corresponding to $\kappa = 1$, 1/2, 1/3, 2/7, and 1/4 and have shown the difference between the results in the g_0 and ε_{out} theoretical evaluation. Generally, it was realized that for $\kappa = 2/7$ or $\tau_{ph}^{eff} = l_{AMP}/3.5c$ the results for g_0 and ε_{out} are more consistent with the experiment for the short-length electrodes say $l_{AMP} < 5\,cm$. For longer electrode lengths i.e. $l_{AMP} \cong 5-10\,cm$, $\kappa = 1/3$; and for $l_{AMP} \cong 15-20\,cm$, $\kappa = 1/2$ shown to present appropriate selections. For $l_{AMP} > 25\,cm$ κ tends to be close to unity.

These observations are indicating that κ is a positive number and is less than 1, and for amplifiers of long electrode lengths of 30-50 cm, which are usually used in any laser laboratory, we can apply $\tau_{ph}^{eff} = l_{AMP}/c$, i.e. $\kappa = 1$. For amplifiers of short electrode lengths, however, the effective cavity lifetime of $\tau_{ph}^{eff} = \kappa l_{AMP}/c$ with an appropriate κ-parameter has to applied.

The results of the numerical calculations using different electron impact excitation cross sections of σ_C and σ_B for the gain-values, and output energies, corresponding to different AMP electrode lengths l_{AMP} and for $\kappa = 1$, 1/2, 1/3, 2/7, 1/4 are shown in figures 4.21-4.25. All σ_C in this calculation have been adopted from the literature. For the σ_B cross section the value of $6.35\times10^{-18}\,cm^2$ was taken from Ref. [45]. Other σ_B values that appeared in our analysis are (8.7, 9.0, 9.2, and 10)$\times10^{-18}\,cm^2$. They are chosen to present the best results close to our experimental measurements. In figures 4.21a and 4.21b the average photon lifetime $\tau_{ph}^{eff} = l_{AMP}/c$ was used in the calculation. It is seen in this figure, that for curve D which corresponds to $\sigma_C = 11.1\times10^{-18}\,cm^2$ and $\sigma_B = 9.2\times10^{-18}\,cm^2$, the calculated g_0 parameter has been deviated from the experimental g_0-measurements, in particular for short l_{AMP}. However, an excellent agreement exists for $l_{AMP} \sim 25-30\,cm$. For $\kappa = 1/2$ or $\tau_{ph}^{eff} = l_{AMP}/2c$, as shown in Figure 4.22, the agreement occurs for $l_{AMP} \sim 15-20\,cm$. In Figure 4.23 with $\kappa = 1/3$ the occurrence of the agreement is in short ranges of $l_{AMP} \sim 5-10\,cm$, while for $\kappa = 1/3.5$ the agreement occurs for $l_{AMP} < 5\,cm$.

Thus, the numerical calculations for different $\kappa-$parameters give correct answers for certain values of l_{AMP}. This is also true for other laser dimension, d_{AMP}. Therefore, we can find intersections of the experimental measurement with the profiles corresponding to the numerical calculations for each κ, to find the dependence of the $\kappa-$parameter with respect to l_{AMP}. Here, we have examined that the theoretical prediction for the gain value is also following the same empirical equation of the form $g_0 = m + n/l_{AMP}$. It is seen from numerical calculation that as κ decreases from 1 to 2/7, their corresponding calculated m, and n values reach to those of the experimental measurements. We further plot g_0 versus l_{AMP}

for different proposed effective photon lifetimes and the results are shown in Figure 4.26. Curves D in Figures 4.21 to 4.25 which are corresponding to $\sigma_C = 11.1\times10^{-18}\ cm^2$, and $\sigma_B = 9.2\times10^{-18}\ cm^2$, are used for the plots appeared in Figure 4.26 experiment for short l_{AMP} occurs.

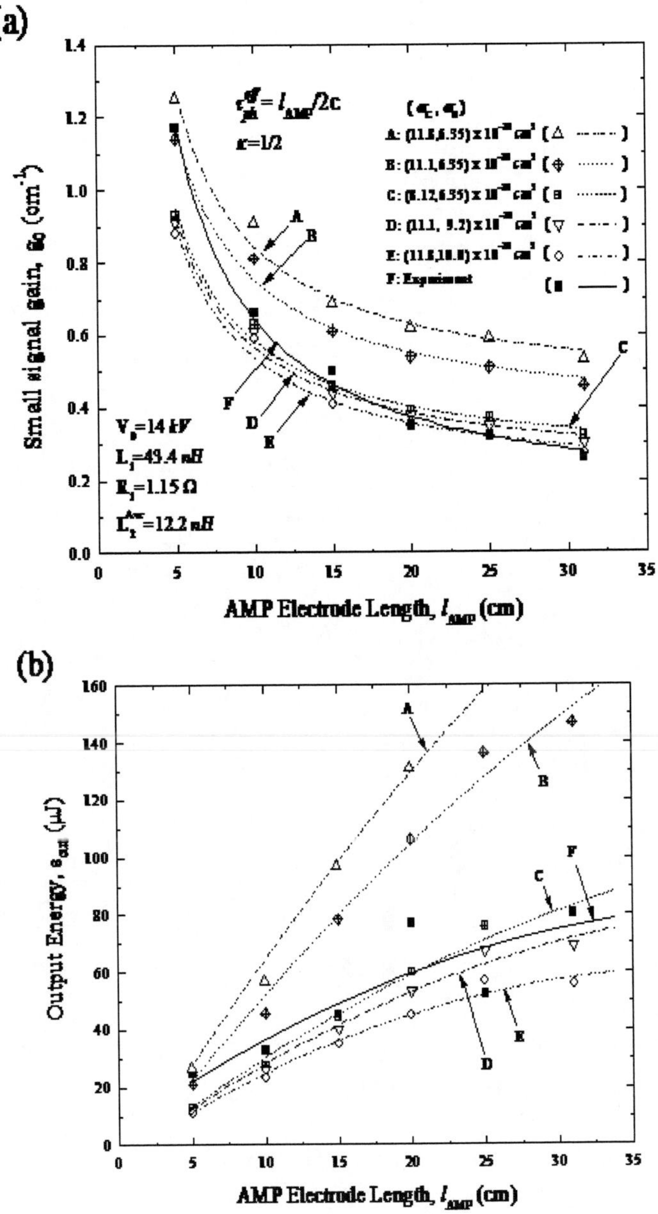

Figure 4.22. The same as Figure 4.21, but with $\tau_{ph}^{eff} = l_{AMP}/2c$ or $(\kappa = 1/2)$. Curve D in both figures appears to agree with the experiment (F) for $l_{AMP} \sim 15-20\,cm$.

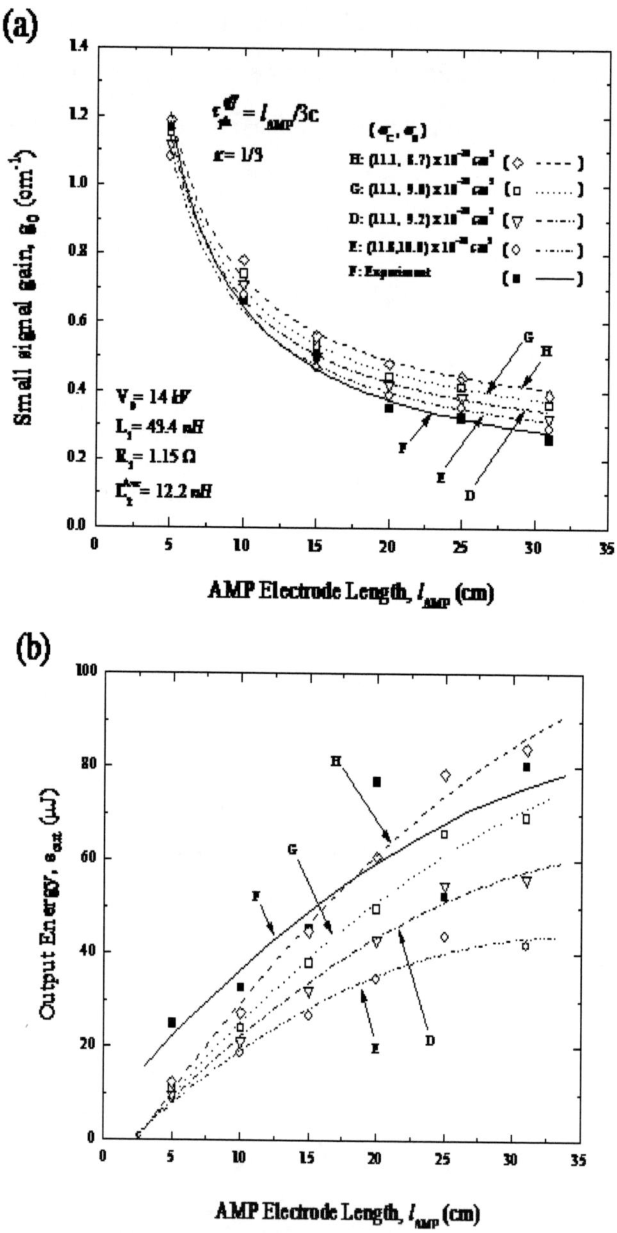

Figure 4.23. The same as Figure 4.21, but with $\tau_{ph}^{eff} = l_{AMP}/3c$ or $(\kappa = 1/3)$. Curve D in both figures appears to agree with the experiment (F) for $l_{AMP} \sim 5-10\,cm$.

(c) Analytical Approach (One Dimensional Treatment)

For a simplified two-level laser system, the rate equation for the photon density in an amplifying medium is given by

$$\frac{\partial n_{ph}}{\partial t} + c\frac{\partial n_{ph}}{\partial z} = cg_0 n_{ph} - \frac{n_{ph}}{\tau_{ph}} \tag{4.23}$$

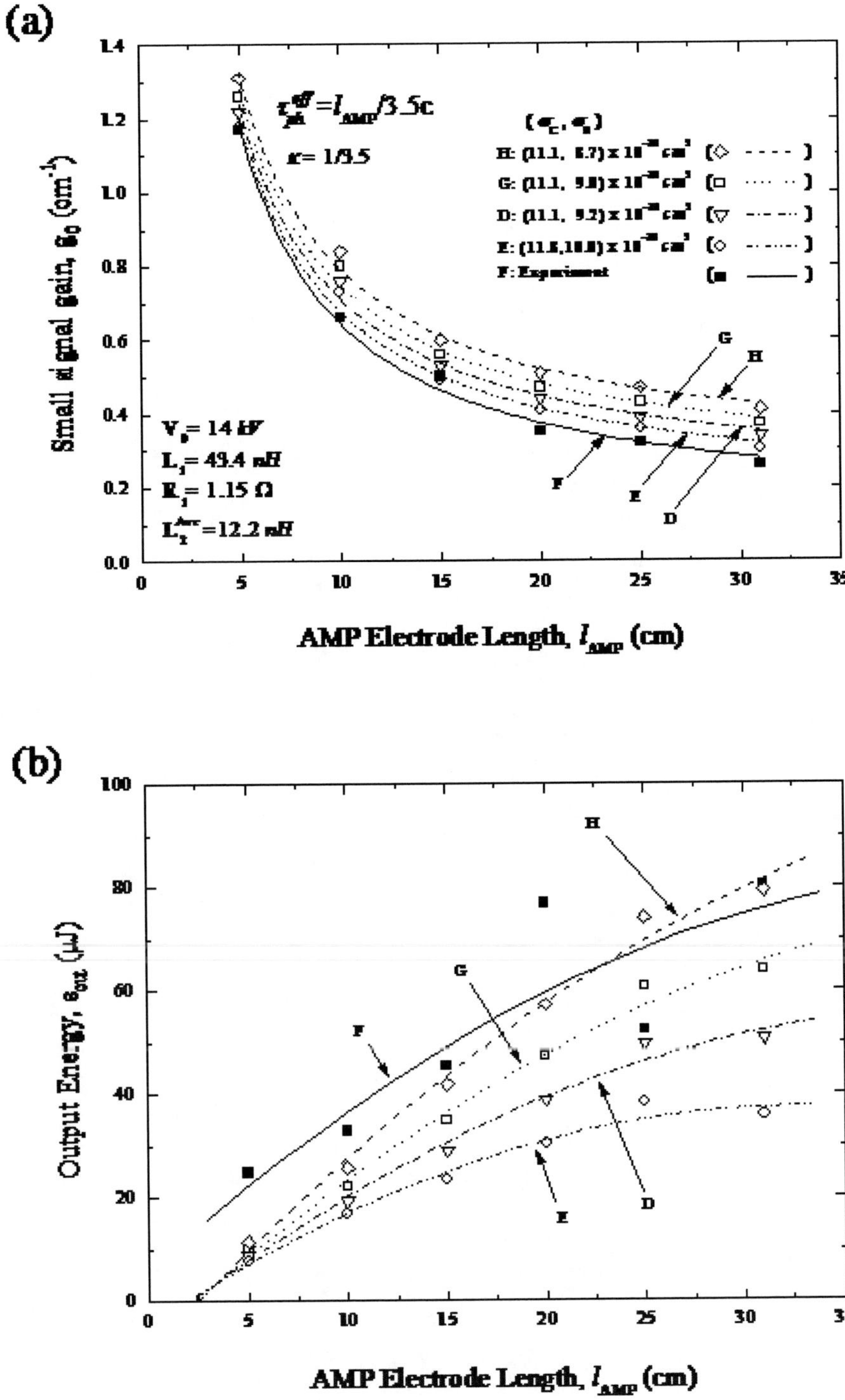

Figure 4.24. The same as Figure 4.21, except for selection of $(\kappa = 2/7)$ where the agreement with the experiment for short electrode lengths is excellent.

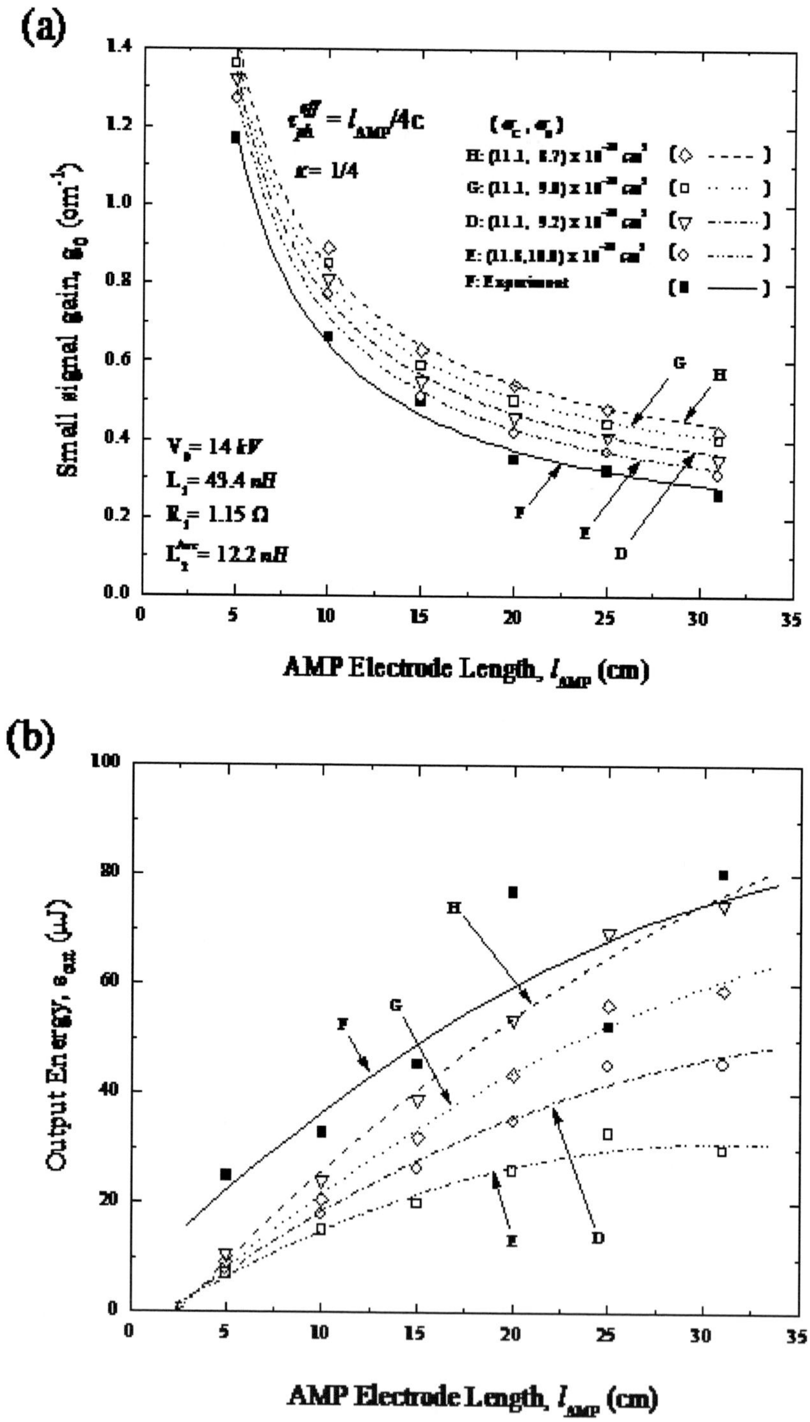

Figure 4.25. The same as Figure 4.21, but for $\kappa = 1/4$. Deviation from experiment for short l_{AMP} is evident in this figure.

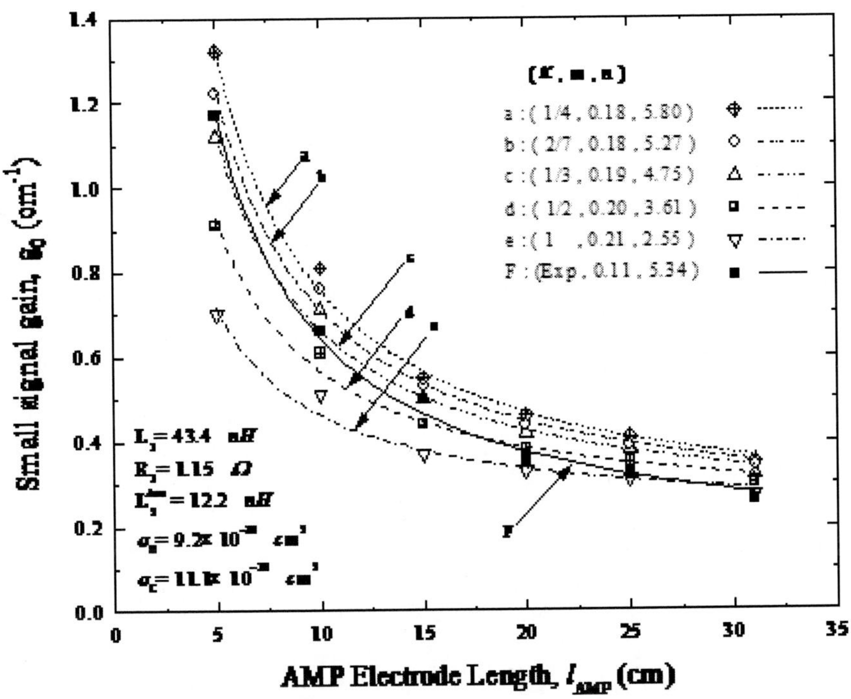

Figure 4.26. g_0 versus l_{AMP} for the mean cavity lifetimes ($\tau_{ph}^{eff} = \kappa l_{AMP}/c$) for $\kappa = 1$, 1/2, 1/3, 2/7, and 1/4 as discussed in the text. For $\kappa = 1/4$ deviation from

where $n_{ph} \equiv n_{ph}(z,t)$ is the photon density corresponding to photons traveling in the z-direction, τ_{ph} is the photon lifetime in the amplifier and for the traveling wave amplification in the z-direction, we have $\tau_{ph} = l_{AMP}/c$.

By introducing the equivalent OSC system, we have assume that photon density inside the AMP section can be written as

$$n_{ph}(z,t) = n_{ph}(0,t)\, e^{g_0(z)\,z} \tag{4.24}$$

In the theory of pulse amplification the above equation refers to the photon density growth rate which is given by the exponential law in the limit of infinitesimal pulse - width [1]. By substituting Eq. (4.24) in Eq. (4.23), and by considering that g_0 is also a function of z, or the length of the amplifying medium, ($z \equiv l_{AMP}$), the cavity effective lifetime for the introduced OSC will be obtained, i.e., we have

$$\frac{1}{\tau_{ph}^{eff}} = c\left\{\frac{1}{l_{AMP}} + \left|\frac{\partial g_0}{\partial l_{AMP}} l_{AMP}\right|\right\} \tag{4.25}$$

As the time behavior of g_0 is accompanied by a rise and a decay, the presence of the absolute value introduced in Eq. (4.25) is indicating that the positive value of the slope in the profile of g_0 vs. *t*, (i.e. the rise portion of the g_0-profile) is in our concern. Furthermore in the equivalent OSC system at the lasing threshold we can relate $g_0^{\text{th}}\, l_{AMP}$ to the power loss, given by $\gamma_L = \gamma_L^{\max} e^{-bl_{AMP}}$, where $\gamma_L^{\max}$ is the maximum power loss and *b* is a constant. γ_L can be determined by the simulation and experiment. Upon introducing $g_0^{\text{th}} = \gamma_L^{\max} e^{-bl_{AMP}} / l_{AMP}$, and calculating $\left| l_{AMP}\, \partial g_0^{th} / \partial l_{AMP} \right|$, and by the assumption that g_0 in Eq. (4.25) has the same value as g_0^{th}, then Eq. (4.25) leads to the following equation

$$\tau_{ph}^{eff} = \frac{l_{AMP}}{c(1+\gamma_L + b\gamma_L l_{AMP})} \tag{4.25a}$$

for small *b*, Eq. (4.25a) reduces to

$$\tau_{ph}^{eff} = \frac{l_{AMP}}{c(1+\gamma_L)} \tag{4.25b}$$

where this last equation is introduced in [38]. Thus, for the effective cavity lifetime of the OSC, we can write

$$\tau_{ph}^{eff} = \kappa \frac{l_{AMP}}{c} \tag{4.26}$$

where

$$\kappa = \frac{1}{(1+\gamma_L + b\gamma_L l_{AMP})} \tag{4.27}$$

In Figure 4.27, we plot κ and γ_L versus l_{AMP}. For $\kappa(l_{AMP})$ the points indicated in the figure are obtained from curves D in figuers 4.21 – 4.25, or equivalently Figure 4.26. No intersection of $g_0^{\exp}$ and g_0^{theo} occurs for $\kappa = 1/4$. For the power loss γ_L in this figure, first the functional dependence of $\gamma_L = \gamma_L^{\max} e^{-bl_{AMP}}$ is defined [38], then by the use of Eq. (4.27) and by knowing the l_{AMP} dependency of κ, the γ_L parameter, i.e. $\gamma_L^{\max}$ and *b*, are obtained.

We can further proceed to obtain the l_{AMP} dependency of the gain-value. The condition for the sustained oscillation requires that $\partial n_{ph} / \partial t 2\ 0$. From the rate equations we obtain $c\tau_{ph}^{eff} g_0 2\ 1$, where it implies that g_0 must be written as

$$g_0^{exact} = m' + \frac{1}{c\tau_{ph}^{eff}} = m' + \frac{1}{l_{AMP}} + \gamma_L^{\max}\left(b + \frac{1}{l_{AMP}}\right)e^{-bl_{AMP}} \tag{4.28}$$

m' is a constant and it is introduced to remove the inequality condition, and for ($c\tau_{ph}^{eff}$) term in Eq. (4.28), Eq. (4.25a) has been used. By expanding $\exp(-bl_{AMP})$ in Eq. (4.28), and adding two terms inside the parenthesis, we get

$$g_0^{exact} = m' + \frac{1}{c\tau_{ph}^{eff}} = m' + \frac{1+\gamma_L^{\max}}{l_{AMP}} + b\gamma_L^{\max}\sum_{n=1}^{\infty}\frac{n}{(n+1)!}\left(-bl_{AMP}\right)^n \tag{4.29a}$$

By neglecting the third term in Eq. (4.29a) due to possessing a small contribution due to higher powers of (bl_{AMP}), we obtain an approximate analytical function for the g_0, i.e.,

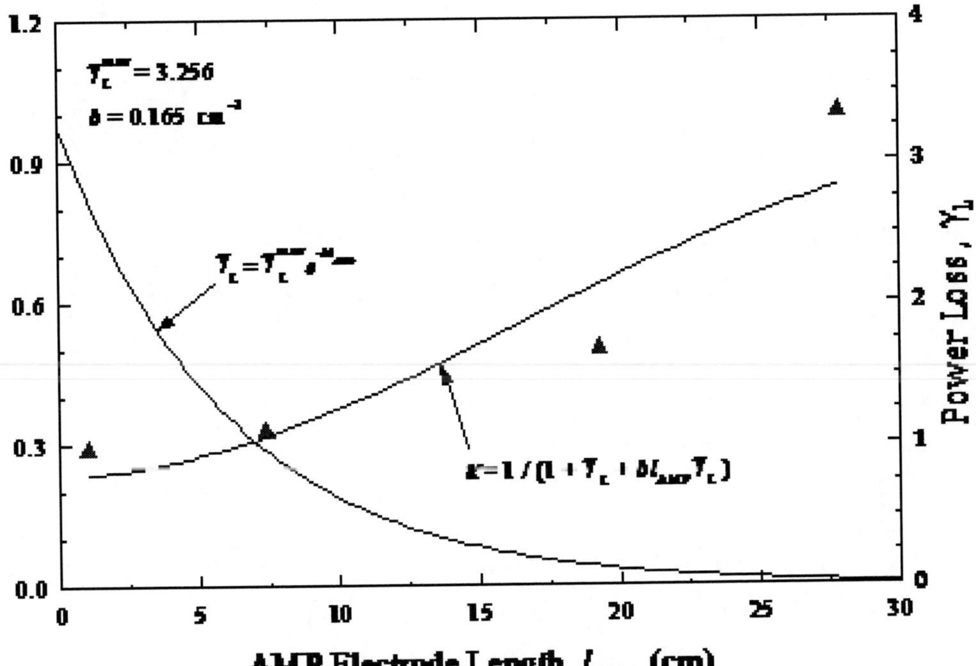

Figure 4.27. Plot of κ versus l_{AMP}. The triangles are referring to points where the profiles corresponding to the g_0-calculation corresponding to different κ-parameter, g_0^{cal}, and the g_0-experimental measurement, $g_0^{\exp}$, at certain values of AMP electrodes length intersect. $\gamma_L = \gamma_L^{\max}\exp(-bl_{AMP})$ is introduced as power losses. $\gamma_L^{\max} = 3.256$, and $b = 0.165$ obtained by the least squares fit of the indicated points to equation $\kappa = (1 + \gamma_L + b\gamma_L l_{AMP})^{-1}$.

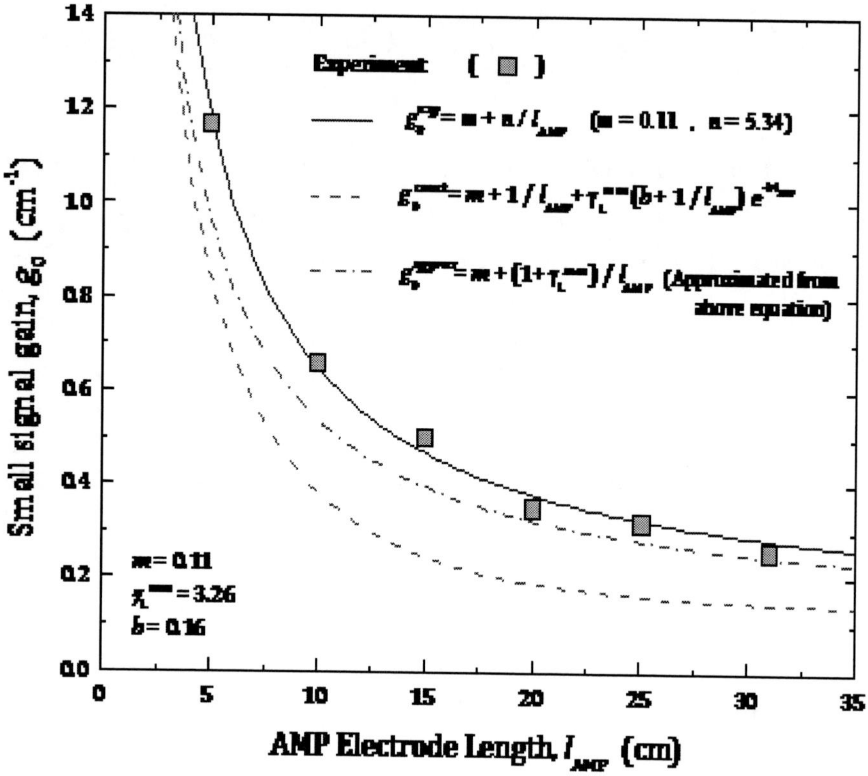

Figure 4.28. Plots of Eqs. (4.28) and (4.29b) corresponding to g_0^{exact} and g_0^{approx} with $m=0.11$, $\gamma_L^{\max}=3.256$, and $b=0.165$ versus l_{AMP}. The experimental observations are also indicated in this figure.

$$g_0^{approx} = m' + \frac{1+\gamma_L^{\max}}{l_{AMP}} \tag{4.29b}$$

If we compare Eq. (4.29b) with our empirical equation of the form $g_0^{\exp} = m + n/l_{AMP}$, the comparison shows that $m = m'$, and $n = 1+\gamma_L^{\max}$. With $\gamma_L^{\max} = 3.26$ from Figure 4.28., we get

$n = 1+\gamma_L^{\max} = 4.26$, which is close to our measurement of $n = 5.34$, and this can be compared with our previous report of $n = 1+\gamma_L^{\max} = 3.82$, when b in Eq. (4.27) was neglected [38]. The minimum value of κ occurs for the condition when $l_{AMP} \to 0$, then from Eq. (4.27) we get

$$\kappa^{\min} = \frac{1}{1+\gamma_L^{\max}} = \frac{1}{n} \tag{4.30}$$

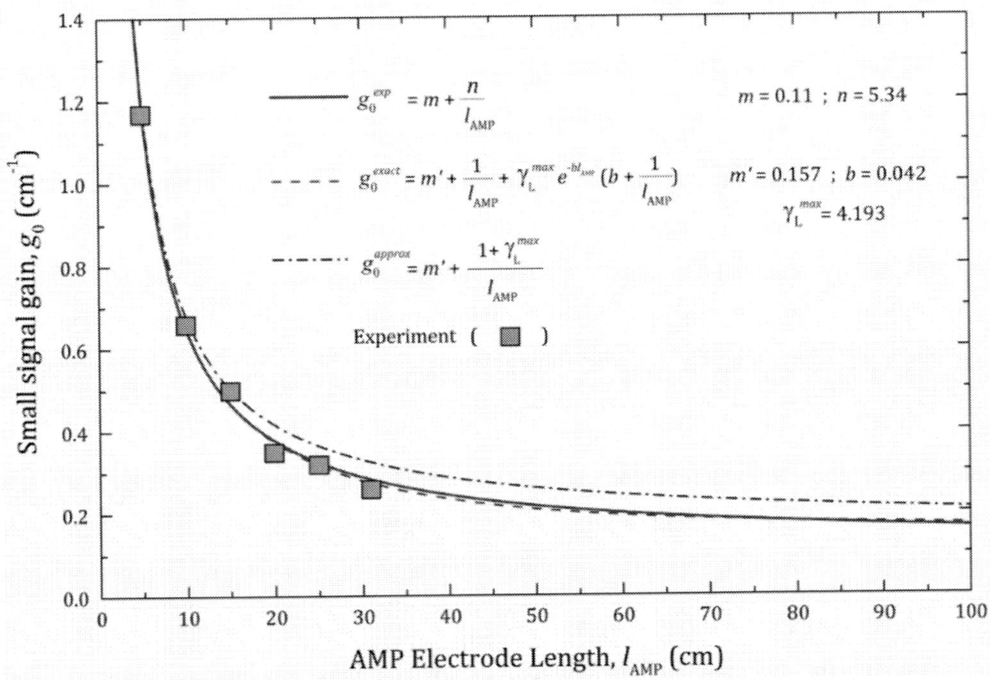

Figure 29. Plots of Eqs. (4.28) and (4.29b) corresponding to g_0^{exact} and g_0^{approx} versus l_{AMP} to obtain m', $\gamma_L^{\max}$, and b through the least squares fit to the experimental measurements as indicated with square symbols in this figure.

where for $\gamma_L^{\max} = 3.26$, we get $\kappa^{\min} = 1/4.26$. Our observation shows that for $\kappa = 1/4$, the calculated g_0-profile and the g_0-measurement do not intersect and $\kappa = 1/4$ has been considered as the lower limit for the κ-parameter [38].

With the introduced value for $\gamma_L^{\max} = 3.26$, and $b = 0.16$ deduced as given in Figure 4.27, we draw curves corresponding to g_0^{exact} and g_0^{approx} given by Eqs. (4.28) and (4.29b), respectively. The results are given in Figure 4.28. and the profiles are comparable with the experimental observation. In fact it is seen that the g_0^{approx}-profile, given analytically by Eq. (4.29b), is closer to the experiment, as compared to the g_0^{exact} -profile indicated by Eq. (4.28). For these profiles, m which is still an unknown quantity, was adopted from the experiment of $m = 0.11$

Our observations are indicating that $\gamma_L^{\max}$ and b parameters are two important factors and have to be determined more accurately. Perhaps, by least-squares fitting of the measured g_0-values to Eqs. (4.28) and (4.29b). The results corresponding to the least-squares fittings are given in Figure 4.29. As it could be expected in this case, the g_0^{exact} is overlapped almost completely with the experiment. In this approach $m' = 0.157$, $b = 0.042$, and $\gamma_L^{\max} = 4.193$ are obtained, where consequently we get $n = 1 + 4.193 = 5.193$ (to be compared with

$n^{\exp} = 5.34$), and $\kappa^{\min} = 1/5.19$ which is slightly smaller than the observed value of $\kappa = 1/4$. The b - parameter is however $0.165/0.042 \sim 4$ times smaller than the previous approach and that is due to approximation we used for deducing our data points in the first approach. The $\gamma_L^{\max}$ - parameters corresponding to two different analyses are more or less comparable.

(d) Analytical Approach (Two Dimensional Treatment)

Although Eqs. (4.28) and (4.29b) are giving appropriate results for the measured g_0 - values, but they suffer from some deficiencies. Firstly, when $l_{AMP} \to \infty$, g_0^{Theory} should approach to zero, or at least to a small quantity exhibiting the gain at slightly above the threshold, while it leads to m-parameter of 0.11 which is not small enough to be neglected. Secondly, our observations with the gain-values show that by increasing d_{AMP}, we have a drastic fall on the experimental gain-profile. This means that Eqs. (4.28) and (4.29b) are not completely explaining the geometrical dependency of the gain-profiles, or the d_{AMP} dependency of the gain-value must be included in m-parameter. To overcome this problem we propose, at least for the gain parameter, that a laser has to be viewed as a two dimensional apparatus, with the z and y directions, chosen parallel to the optical axis and electrodes direction, respectively, while the whole system as a traveling wave OSC must be faced with a single value for the photon lifetime given by $\tau_{ph} = l_{AMP}/c$, where it corresponds to the z-direction, i.e., to the direction of the laser output beam. Thus, with the assumption of z and y independency of the photon density, we can re-write two independent equations for the photon density, i.e.

$$\frac{\partial n_{ph,z}(z,t)}{\partial t} + c\frac{\partial n_{ph,z}(z,t)}{\partial z} = cg_{0z}n_{ph,z}(z,t) - \frac{n_{ph,z}(z,t)}{\tau_{ph}} \tag{4.31a}$$

and

$$\frac{\partial n_{ph,y}(y,t)}{\partial t} + c\frac{\partial n_{ph,y}(y,t)}{\partial y} = cg_{0y}n_{ph,y}(y,t), \tag{4.31b}$$

where in Eq. (4.31b), in contrast to Eq. (4.31a), we dropped the term corresponding to the photon-lifetime in the y-direction, otherwise its presence leads to an inconsistency with the measurements. Thus, up to this stage two independent gain-values corresponding to z and y-directions, i.e. g_{0z} and g_{0y} are introduced.

Now, if we multiply Eq. (4.31a) by $n_{ph,y}(y,t)$, and Eq. (4.31b) by $n_{ph,z}(z,t)$,then by adding two produced equations, we will have

$$\frac{\partial n_{ph}(z,y,t)}{\partial t} + c\frac{\partial n_{ph}(z,y,t)}{\partial z} + c\frac{\partial n_{ph}(z,y,t)}{\partial y} = c(g_{0z} + g_{0y})n_{ph}(z,y,t) - \frac{n_{ph}(z,y,t)}{\tau_{ph}}, \tag{4.32}$$

where we assumed that

$$n_{ph}(z,y,t) = n_{ph,z}(z,t)n_{ph,y}(y,t) \tag{4.33}$$

Eq. (4.32) shows that the special variations of the photon density can also occur in two independent directions of z and y. In addition, it shows that the overall gain-value can be realized that it consists of two independent gain-values corresponding to different directions, i.e., if we write Eq. (4.32) in the form of

$$\frac{\partial n_{ph}(z,y,t)}{\partial t} + c\nabla_t n_{ph}(z,y,t) = cg_0 n_{ph}(z,y,t) - \frac{n_{ph}(z,y,t)}{\tau_{ph}} \tag{4.34}$$

where ∇_t means $\frac{\partial}{\partial z} + \frac{\partial}{\partial y}$, a comparison of Eq. (4.34) with Eq. (4.32) gives

$$g_0 = g_{0z} + g_{0y} \tag{4.35}$$

Separation of g_0-value into two parts of g_{0z} and g_{0y} should not be considered as an unusual fact. For example, we already introduced two different measured values for the gain-parameter, where they are referring to polarizations of the laser beams in the directions of the z and y axes injected from the OSC-section into the AMP.

For evaluating the cavity effective lifetime, we can repeat our previous argument, or equivalently by introducing the overall photon density in an equivalent two-dimensional OSC; so we can write

$$n_{ph}(z,y,t) = n_{ph}(0,0,0)e^{-\frac{t}{\tau_{ph}^{eff}}} \exp\left[g_{0z}(z,\gamma_L^z)z + g_{0y}(z,\gamma_L^y)y\right], \tag{4.36}$$

where, we assumed that g_{0z} and g_{0y} are functions of z, γ_L^z; and y, γ_L^y, respectively. γ_L^z and γ_L^y are two independent power losses correspond to z and y-directions, respectively. The behavior of the overall γ_L with respect to z-direction (i.e. l_{AMP}) has been already been shown and also it is given in Figure 4.27, where in this figure $d_{AMP} = 7$mm has been kept fixed according to the relevant experiment. Analogously, for the power loss, γ_L^z, in the z-direction we can write

$$\gamma_L^z = \gamma_L^{z,\max} e^{-bz}, \ (z = l_{AMP}) \tag{4.37}$$

where $\gamma_L^{z,\max}$ and b are constants that can be determined.

For the y-direction, we can also write

$$\gamma_L^y = \gamma_L^{y,\max} e^{-ay}, \ (y = d_{AMP}) \tag{4.38}$$

where again $\gamma_L^{y,\max}$ and a are introduced as constants. The behavior of γ_L^y as given by Eq. (4.38) was found that is consistent with the experiment.

Upon substituting Eq. (4.36) into Eq. (4.32), we have

$$\frac{1}{\tau_{ph}^{eff}} = \frac{c}{z} + cz\left|\frac{\partial g_{0z}}{\partial z}\right| + cy\left|\frac{\partial g_{0y}}{\partial y}\right| \tag{4.39}$$

$\tau_{ph} = z/c$, where $z = l_{AMP}$, as appeared in Eq. (4.32), has been used.

At the threshold, we can introduce the following two equations, corresponding to two different directions

$$g_{0z}^{th} z = \gamma_L^z \quad , \quad \gamma_L^z > 0 \tag{4.40a}$$

$$g_{0y}^{th} y = \gamma_L^y \quad , \quad \gamma_L^y > 0 \tag{4.40b}$$

and by use of Eqs. (4.40a) and (4.40b), we can evaluate the last two terms appeared in Eq. (4.39), i.e.

$$cz\left|\frac{\partial g_{0z}^{th}}{\partial z}\right| = cz\left|\frac{z\,\partial\gamma_L^z/\partial z - \gamma_L^z}{z^2}\right| = c\left(\frac{\gamma_L^z + bz\gamma_L^z}{z}\right), \text{ where } z = l_{AMP} \tag{4.41a}$$

$$cy\left|\frac{\partial g_{0y}^{th}}{\partial y}\right| = cy\left|\frac{y\,\partial\gamma_L^y/\partial y - \gamma_L^y}{y^2}\right| = c\left(\frac{\gamma_L^y + ay\gamma_L^y}{y}\right), \text{ where } y = d_{AMP} \tag{4.41b}$$

Thus from Eq. (4.39), we have

$$\left(\tau_{ph}^{eff}\right)^{-1} = \frac{c}{l_{AMP}}\left(1 + \gamma_L^z + bl_{AMP}\gamma_L^z\right) + \frac{c}{d_{AMP}}\left(\gamma_L^y + ad_{AMP}\gamma_L^y\right) \tag{4.42}$$

Eq. (4.42), and also equivalently Eqs. (4.31) show that we can define two independent effective cavity lifetimes for the z and y directions, i.e.

$$\left(\tau_{ph}^{eff,z}\right)^{-1} = \frac{c}{l_{AMP}}\left(1+\gamma_L^z + bl_{AMP}\gamma_L^z\right) \tag{4.43a}$$

$$\left(\tau_{ph}^{eff,y}\right)^{-1} = \frac{c}{d_{AMP}}\left(\gamma_L^y + ad_{AMP}\gamma_L^y\right) \tag{4.43b}$$

and from Eq. (4.42), the overall effective cavity photon lifetime can be obtained

$$\left(\tau_{ph}^{eff}\right)^{-1} = \left(\tau_{ph}^{eff,z}\right)^{-1} + \left(\tau_{ph}^{eff,y}\right)^{-1} \tag{4.44}$$

Also for the g_0-value, we have

$$g_0 = m'' + \frac{1}{c\tau_{ph}^{eff}} = m'' + \frac{1+\gamma_L^z + bl_{AMP}\gamma_L^z}{l_{AMP}} + \frac{\gamma_L^y + ad_{AMP}\gamma_L^y}{d_{AMP}} \tag{4.45}$$

where m'' is a constant and is introduced to remove the inequality for $c\tau_{ph}^{eff} g_0 > 1$ condition.

By substituting Eqs. (4.37) and (4.38) into Eq. (4.45), we will get an exact analytical expression for the g_0-value in terms of the introduced parameters

$$g_0^{exact} = m'' + \frac{1}{l_{AMP}} + \gamma_L^{z,\max}\left(b + \frac{1}{l_{AMP}}\right)e^{-bl_{AMP}} + \gamma_L^{y,\max}\left(a + \frac{1}{d_{AMP}}\right)e^{-ad_{AMP}} \tag{4.46}$$

Or upon expansion of $\exp(-bl_{AMP})$ and $\exp(-ad_{AMP})$ terms, we have

$$g_0^{exact} = \left(m'' + \frac{1+\gamma_L^{z,\max}}{l_{AMP}} + \frac{\gamma_L^{y,\max}}{d_{AMP}}\right) + b\gamma_L^{z,\max}\sum_{i=1}^{\infty}\frac{i}{(i+1)!}\left(-bl_{AMP}\right)^i + a\gamma_L^{y,\max}\sum_{j=1}^{\infty}\frac{j}{(j+1)!}\left(-ad_{AMP}\right)^j \tag{4.47}$$

Eqs. (4.46) and (4.47) give a complete picture for the gain-behavior, either for d_{AMP} = fixed, and changing l_{AMP}; or l_{AMP} = fixed, and changing d_{AMP}, where the g_0-values corresponding to these two cases can be evaluated experimentally.

An approximate expression can also be obtained if we let two last terms in Eq. (4.47) to be negligible, i.e., we can have

$$g_0^{approx} = m'' + \frac{1+\gamma_L^{z,\max}}{l_{AMP}} + \frac{\gamma_L^{y,\max}}{d_{AMP}} \tag{4.48}$$

For a fixed d_{AMP}, and by changing l_{AMP}, we observed an empirical relation of the form $g_0 = m + n/l_{AMP}$, and for a fixed l_{AMP}, and by changing d_{AMP}, our observation in this respect is already introduced. Here we realized that the least squares fit of the measured gain – values to the function of type $g_0 = r + q/d_{AMP}$ with r and q some constants, can be most appropriately used as an empirical equation that satisfies the d_{AMP} dependency of the gain coefficient. This means that by observing Eq. (4.48) we have

$$m \cong m'' + \left.\frac{\gamma_L^{y,\max}}{d_{AMP}}\right|_{\text{for } d_{AMP} = fixed} \tag{4.49a}$$

$$r \cong m'' + \left.\frac{1+\gamma_L^{z,\max}}{l_{AMP}}\right|_{\text{for } l_{AMP} = fixed} \tag{4.49b}$$

These two cases have been proved numerically that they are consistent with the observations. Furthermore, Eq. (4.46) or (4.47) shows the l_{AMP}, and d_{AMP} dependency of the single dimensional gain – values g_{0z}^{SDG} and g_{0y}^{SDG}, i.e.

$$g_{0z}^{SDG} = \frac{1}{l_{AMP}} + \gamma_L^{z,\max}\left(b + \frac{1}{l_{AMP}}\right)e^{-bl_{AMP}} \tag{4.50a}$$

and

$$g_{0y}^{SDG} = \gamma_L^{y,\max}\left(a + \frac{1}{d_{AMP}}\right)e^{-ad_{AMP}} \tag{4.50b}$$

This means that by having two separate curves corresponding to g_{0z} and g_{0y} profiles, we are able to predict the overall gain - values either by using Eq. (4.45), or

$$g_0 = m'' + g_{0z}^{SDG} + g_{0y}^{SDG} \tag{4.51}$$

The profiles corresponding to g_{0z}^{SDG} and g_{0y}^{SDG} are given in Figure 4.30. It will be shown shortly that the parameters which have been used for drawing g_{0z}^{SDG} - and g_{0y}^{SDG} -profiles are based on experimental measurements.

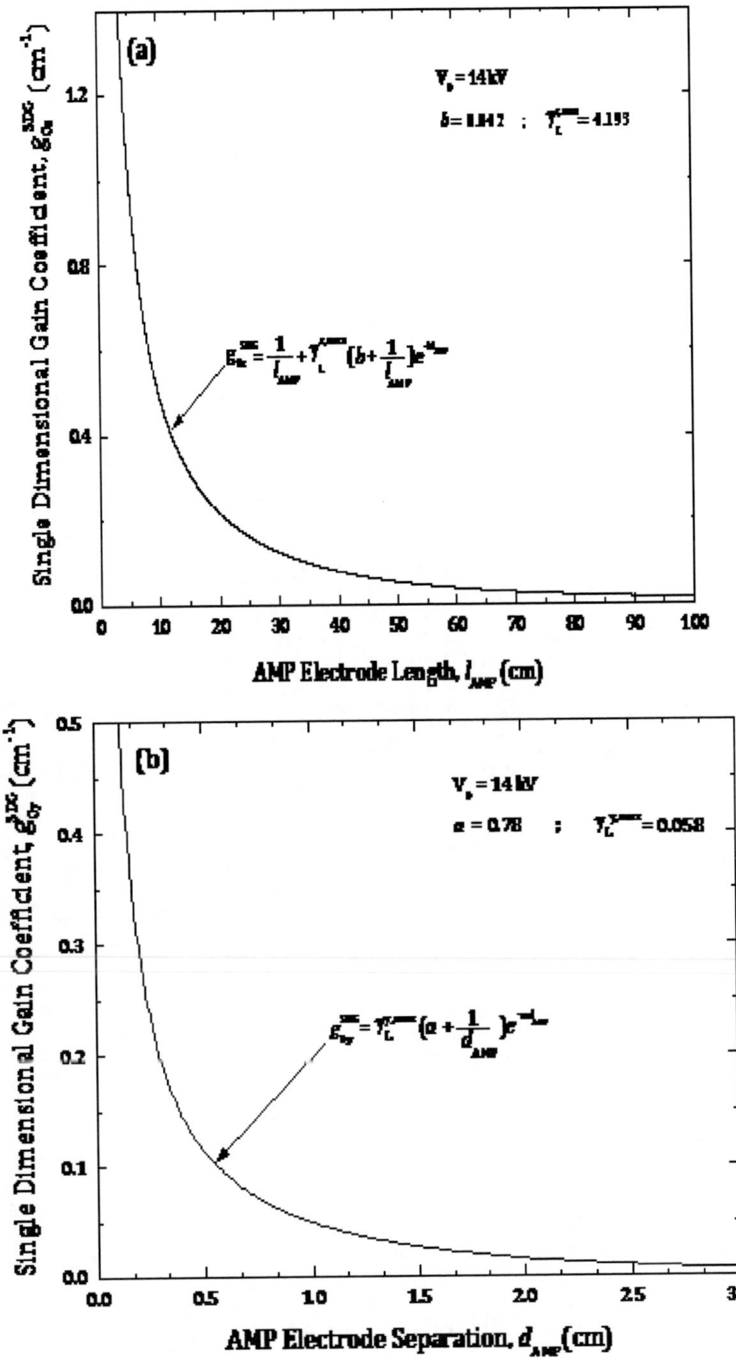

Figure 4.30. Plots, corresponding to single dimensional gain – values: (a) g_{0z}^{SDG} as given by Eq. (4.50a); and (b) g_{0y}^{SDG} as given by Eq. (4.50b) versus l_{AMP} and d_{AMP}, respectively. In both figures the parameters given to the corresponding equations deduced from the experiment according to Figures 4.29 and 4.30.

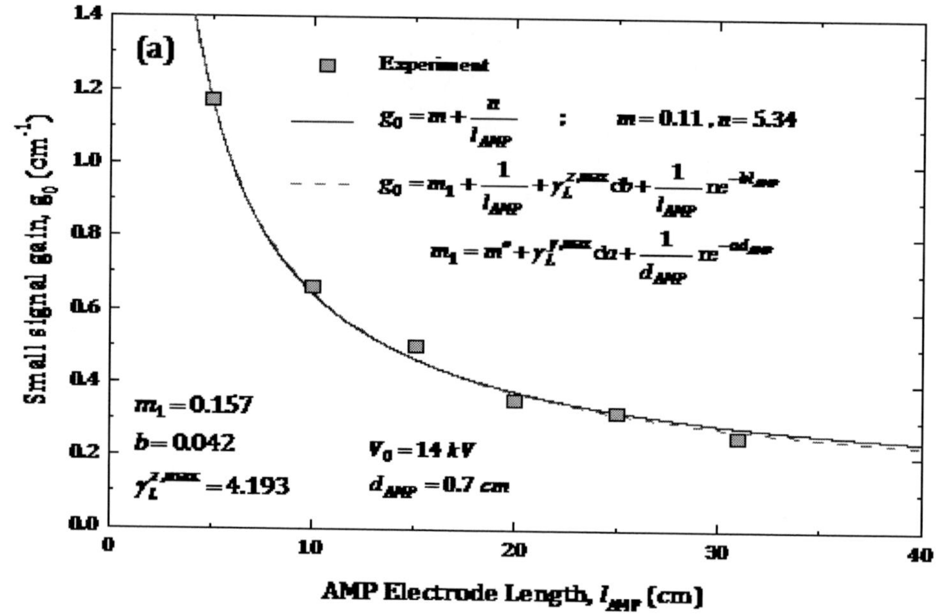

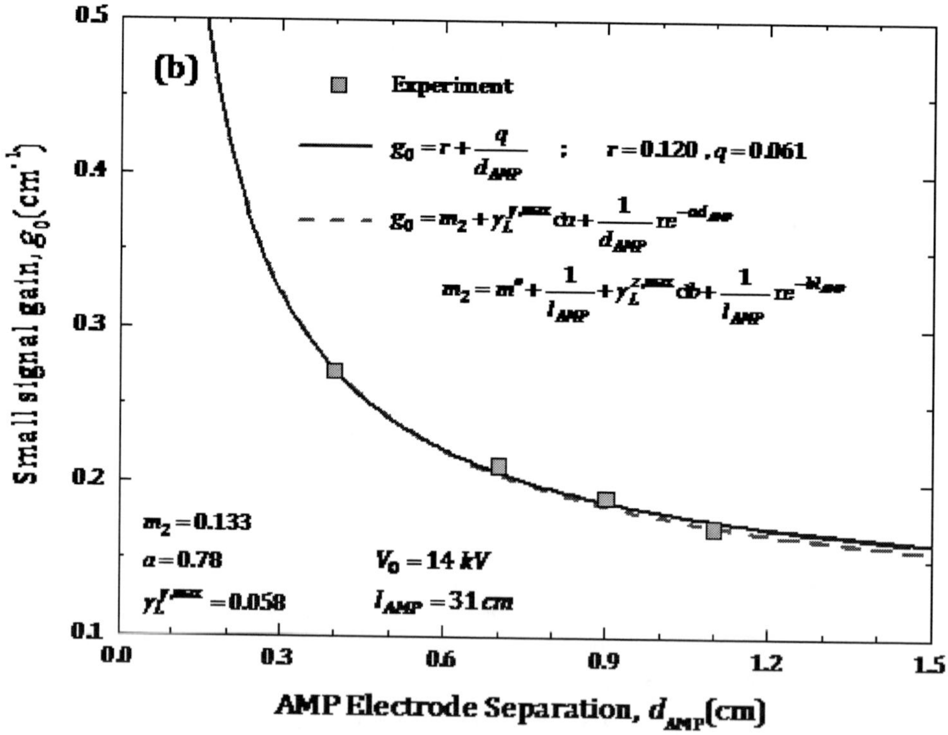

Figure 4.31. (a) Plots of g_0 vs. l_{AMP} for $d_{AMP} = 0.7\ cm$ to obtain numerical values for parameters introduced in Eqs. (4.46) and (4.47); (b) Plots of g_0 vs. d_{AMP} for $l_{AMP} = 31 cm$ to obtain numerical values for parameters introduced in Eqs. (4.46) and (4.47). Both figures give the required values for m_1, b, and $\gamma_L^{z,\max}$; m_2, a, and $\gamma_L^{y,\max}$. The parameters obtained by the least squares fit of the experimental measurements to the relevant functions given in the inset of the figures.

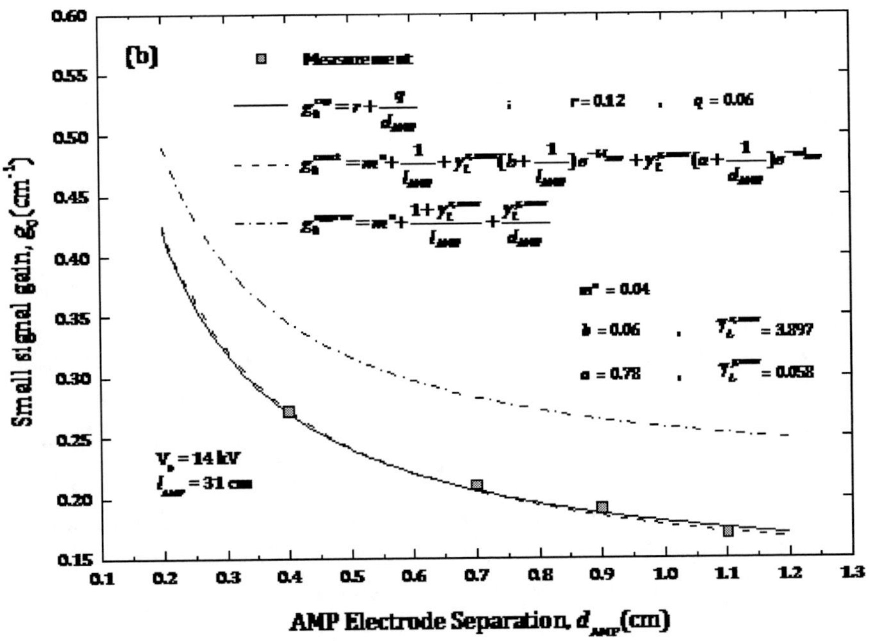

Figure 4.32. (a) Plots of g_0 vs. l_{AMP} ; and (b) g_0 vs. d_{AMP} with the measured gain – values, fitted to g_0^{exact} given in the inset of each figure to obtain the required values for the parameters introduced in the gain formulations. The profiles for the g_0^{approx} is also given for the comparison. The deduced parameters are compatible with those given in Figures 4.31a and 4.31b.

(e) Numerical Results

Experimentally, it is not possible to do measurement for determining two independent profiles of g_{0z} and g_{0y} with respect to l_{AMP} and d_{AMP}, respectively, so as to have plots of $g_{0z}(l_{AMP})$ and $g_{0y}(d_{AMP})$, since for g_{0z} vs. l_{AMP}, for example, as it is shown by Eqs. (4.46) and (4.47), we need to have a large value of d_{AMP} so that the fourth term in Eq. (27) can be ignored and this condition will never have the chance for the fulfillment. The reason is on the fact that for a laser operation some finite values of l_{AMP} and d_{AMP} at a specific input voltage, in addition of keeping a certain value for E/p, are required. Also, it is required to have a laser design with extremely large l_{AMP} to do g_{0y} vs. d_{AMP} measurements. Also, with the present conventional laser design performing any measurement across the electrodes gap separation is not possible. Thus, we are just allowed to use an OSC-AMP laser system with capability of using l_{AMP} of different lengths, or equivalently operating the AMP with the potential of using variable electrodes gap separation. The simplest approach to obtain numerical values for the parameters introduced in Eqs. (4.46) and (4.47) is to use the measured values of g_0 vs. l_{AMP} for a fixed d_{AMP}; and by the least squares fitting of these values to the relation given in one-dimensional form of the type $g_0 = m_1 + 1/l_{AMP} + \gamma_L^z(b + 1/l_{AMP})$; m_1, b, and $\gamma_L^{z,\max}$ will be obtained. In this case,

$m_1 = m'' + \gamma_L^{y,\max}(a + 1/d_{AMP})e^{-ad_{AMP}}$ is fixed throughout the experiment. This behavior is given in Figure 4.31a. To obtain m_2, a, and $\gamma_L^{y,\max}$ where $m_2 = m'' + 1/l_{AMP} + \gamma_L^{z,\max}(b + 1/l_{AMP})e^{-bl_{AMP}}$ for a fixed value of l_{AMP}, the profile of g_0 vs. d_{AMP} gives the corresponding required parameters, as it is shown in Figure 4.31b. m_1 and m_2 with the least squares fitting to the measured values are 0.157 and 0.133, as shown in figures 4.31a and 4.31b. If we use the expressions given in the insets of figures, as m_1 and m_2 with $m'' = 0.045$,they are 0.119 and 0.166, respectively. These values are very close to those introduced earlier, and are showing the consistency between two curves in this analysis. In another approach we can use the overall g_0 formulation given by Eq. (4.46), and then by use of experimental measurements of g_0 vs. l_{AMP}, and g_0 vs. d_{AMP} along with the least squares fittings of the data points to Eq. (4.46), all the necessary parameters as it is given in figures 4.32a and 4.32b are obtained. A comparison of the deduced parameters from figures 4.31 and 4.32 shows that between these parameters in both approaches the consistency is quite satisfactory. In addition, it is shown that as it was expected, $m'' = 0.04 - 0.05$ is a small quantity.

For the CVLs we used the collected data appeared in Ref. [12], where we draw a curve corresponding to g_0 vs. R_{AMP} (Figure 4.33). R_{AMP} is the radius of the CVL tube, and the gain measurements have been diameters reported by different autors. Our observation is indicating that the profile is following a curve given by Eq. (4.46), where R_{AMP} has been used instead of d_{AMP}. A slight deviation of $g_0^{exp} = r + q/R_{AMP}$ from the measured g_0-values has been observed, while equation of $g_0^{exact} = m_2 + \gamma_L^{R,\max}(a + 1/R_{AMP})e^{-aR_{AMP}}$ shows an excellent agreement with the experiment.

The formulations for the small-signal gain given by Eq. (4.46) or the approximate form of Eq. (4.48) with the parameters deduced from figures 4.30 and 4.31, and also given in Table 3, shown to be quite useful for the geometrical evaluation of the gain - values.

Still there are two other important parameters that have to be visualized, mainly the input operating voltage and current density. As the laser system is operating under the optimum operational condition, the effect of pressure has already been included in the formulations that are given. Since for the optimum operation, the $E/p = V/pd_{AMP}$ which is related to electron temperature, T_e, has to remain constant, so if the pressure, p, increases, for example, to atmospheric pressure, we have to reduce d_{AMP} - value, or increase the operating voltage of the laser system. Thus the effect of the gas pressure shows itself on d_{AMP} or input voltage. Also, we have already shown that g_0 increases almost linearly with the input voltage, and for the current density a rapid increasing and then saturation of the gain value with respect to current density is reported [41]. Therefore these two correcting effects must be taken into account for finalizing the given relationships.

Table 3. Values of calculated parameters appeared in the gain formulation. Empirical equations obtained from the measurements are:

$$g_0^{\exp}=m+n/l_{AMP} \text{ and } g_0^{\exp}=r+q/d_{AMP}.$$

One-dimensional model $d_{AMP}=7\ mm$		b	$\gamma_L^{\max}$	m'		
	From Figure 4.27	0.042	4.193	0.157		
	$g_0^{\exp}=m+\frac{n}{l_{AMP}}$, $g_0^{approx}=m'+\frac{1+\gamma_L^{\max}}{l_{AMP}}$ $\Rightarrow$				$m^{cal}=0.157$	, $m^{\exp}=0.11$
					$n^{cal}=1+\gamma_L^{\max}=5.193$	, $n^{\exp}=5.34$

Two-dimensional model		b	a	$\gamma_L^{z,\max}$	$\gamma_L^{y,\max}$	m''
	From Figure 4.30a	0.042	0.82	4.193	0.080	0.05
	From Figure 4.30b	0.060	0.78	3.897	0.058	0.04

$g_0^{\exp}=m+\frac{n}{l_{AMP}}$	$g_0^{approx}=m''+\frac{1+\gamma_L^{z,\max}}{l_{AMP}}+\frac{\gamma_L^{y,\max}}{d_{AMP}}$ $\Rightarrow$	for $d_{AMP}=7\ mm$	
		$m^{cal}=m''+\gamma_L^{y,\max}/d_{AMP}=0.164$	, $m^{\exp}=0.11$
		$n^{cal}=1+\gamma_L^{z,\max}=5.193$	, $n^{\exp}=5.34$
$g_0^{\exp}=r+\frac{q}{d_{AMP}}$		for $l_{AMP}=31\,cm$	
		$r^{cal}=m''+(1+\gamma_L^{z,\max})/l_{AMP}$ J 0.20	, $r^{\exp}=0.120$
		$q^{cal}=\gamma_L^{y,\max}=0.058$	, $q^{\exp}=0.061$

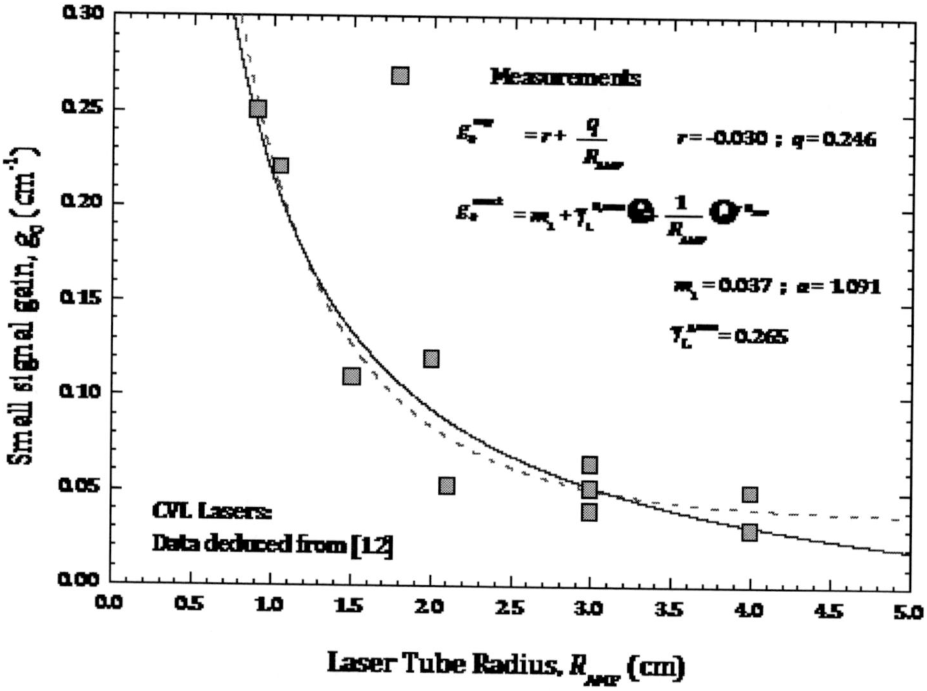

Figure 4.33. Plot of the measured values of CVLs gain-coefficient deduced from [12] versus CVL laser tube radius, R_{AMP}. Agreement of the g_0-values for the CVL lasers given by equation $g_0^{exact} = m_2 + \gamma_L^{R,max}(a + 1/R_{AMP})e^{-aR_{AMP}}$ is excellent.

At this stage it is worth mentioning some practical examples with the corresponding gain – values given in the literature. For $(l_{AMP}, d_{AMP}) = (94\,cm, 1.5\,cm)$ at 14 *kV*, we have a report of $g_0^{\exp} = 0.064\,cm^{-1}$ [19-33]. Our predictions are $g_{0z}^{SDG} = 0.014$, $g_{0y}^{SDG} = 0.027$, with $m'' = 0.045$, all in cm^{-1}, we have $g_0 = m'' + g_{0z}^{SDG} + g_{0y}^{SDG} = 0.086\,cm^{-1}$ which is very close to the experiment. For the work of Serafetinides et al [14], with $g_0^{\exp} = 0.22\,cm^{-1}$ at 760 *torr*, $l_{AMP} = 50\,cm$, and $d_{AMP} = 1.0\,cm$, we have $g_{0z}^{SDG} = 0.053\,cm^{-1}$, $g_{0y}^{SDG} = 0.049\,cm^{-1}$. Therefore we get $g_0 = 0.045 + 0.053 + 0.049 = 0.147\,cm^{-1}$. We have already shown that gain coefficient increases almost linearly when the input voltage increases. So, if we include the effect of the input voltage for their system operating at 20 *kV*, we have $g_0 = 0.147 \times \frac{20}{14} = 0.21\,cm^{-1}$, which is the same as their measurements of $0.22\,cm^{-1}$. This closeness of the experimental measurement for a laser operating at 760 torr, and our prediction with the measurements in the range 40 – 160 torr of N_2 gas pressure is indicating that, as it was mentioned earlier, the effect of pressure is included in the formulation through the E/p- value, or at least we can say that the operating pressure for deriving the present analytical expressions has a mild (or very week) contribution to the formulation of the gain – value. For the work of von Bergmann et al [41], for $l_{AMP} = 9\,mm$ and assumption of

$d_{AMP} = 2\,\text{mm}$, with the laser operating at 16 *kV*, we get $g_0 = 6.0 \times \frac{16}{14} = 6.9\,cm^{-1}$ which corresponds to their work for $\sim 10\ \text{kA/cm}^2$ current density. This means that the given relationships for g_0 – value can be valid up to a few kA/cm^2 of the current density. We can also estimate backward for obtaining the d_{AMP} - value if it is not given in the literature. For example in Ref. [13], we have $g_0^{\exp} = 0.277\,\text{cm}^{-1}$ at $V_0 = 16.5\,\text{kV}$, for $l_{AMP} = 45\,\text{cm}$. With the use of our formulation to find g_0 at 14 kV, we have $g_0 = 0.277 \times \frac{14}{16.5} = 0.235\,\text{cm}^{-1}$. Upon using Eq. (4.51), and the curve g_{0z}^{SDG} for $l_{AMP} = 45\,\text{cm}$ shown in Figure 4.50a, we get $g_{0y}^{SDG} = 0.132\,\text{cm}^{-1}$ where it corresponds to $d_{AMP} \sim 5$ mm.

Although the introduced approach given here to evaluate the correction factors due to effects of input voltage and current density on the gain coefficient is a rough estimation, the deduced results, however, are quite satisfactory. But, it seams that for more and better understanding the subject more elaborate mathematical works are needed in this respect.

ACKNOWLEDGMENT

I would like to acknowledge the contribution of S. Sarikhani for his work on the simulation studies on the gain parameter in N_2-lasers appeared in our latest publication, preparing graphs and typing the present manuscript.

REFERENCES

[1] Frantz, L. M. and Nodvik, J. S. (1963),"Theory of pulse propagation in a laser amplifier", *J. Appl. Phys.* vol.34, no.8, pp 2346-2349.

[2] Rigrod, W. W. (1963), "Gain saturation and output power of optical masers", *J. Appl. Phys.*, vol.34, no.9, pp 2602-2609.

[3] Siegman, A. E. (1986), *"Lasers"*, Mill Valley, CA. University Science Books.

[4] Hopf, F. A. (1974), *"High energy lasers and their applications"*, London, Addison-Wesley Publishing Company.

[5] Koechner, W. (1996), *"Solid-state laser engineering"*, (4th edition), New York, Berlin, Springr.

[6] Baranov, V. Yu., Kovalev, A. S., Persiantsev, I. G., Pis'mennyi, V. D., Rakhimov, A. T., and Starostin, A. N., editing by: Velikhov, E. P., Translated from the Russian by Kittell, S. (1981), *"Molecular gas lasers"*, Moscow, MIR Publications.

[7] Kriukov, P. G. and Letokhov, V. S. (1972), *in "Laser Handbook"*, edited by Arecchi, F. T. and Schulz-DuBois, E. O. North-Holland, Amsterdam, vol.1, p 561.

[8] Eggleston, J. M., Frantz, L. M., and Injeyan, H. (1989),"Derivation of the Frantz-Nodvik equation for zigzag optical path, slab geometry laser amplifiers", *IEEE J. Quant. Electron.,* vol.25, no.8, pp 1855-1862.

[9] Lowdermilk, W. H. and Murray, J. E. (1980), "The multipass amplifier: Theory and numerical analysis", J. Appl. Phys., vol.51, no.5, pp 2436-2444.

[10] Banic, J., Efthimiopoulos, T. and Stoicheff, B. P. (1980), *"Gain and saturation intensity measurements of a mode-locked KrF oscillator-amplifier"*, Appl. Phys. Lett., vol.37, no.8, pp 686-688.

[11] Hargrove, R. S., Grove, R. and Kan, T. (1979), "Copper vapor laser unstable resonator oscillator and oscillator-amplifier characteristics", *IEEE J. Quantum Electron.,* vol.QE-15, no.11, pp 1228-1233.

[12] Behrouzinia, S., Sadighi, R. and Parvin, P. (2003),"Pressure dependence of the small-signal gain and saturation intensity of a copper vapor laser", *Appl. Opt.,* vol.42, no.6, pp 1013-1018.

[13] Docchio, F., Magni, V. and Ramponi, R. (1984), "Thyratron-switched N_2 atmospheric-pressure oscillator, low-pressure amplifier laser system", *Rev. Sci. Instrum.,* vol.55, no.4, pp 477-481.

[14] Serafetinides, A. A. and Tsikrikas, G. N. (1990),"A semiconductively preionised TEA nitrogen oscillator-amplifier laser system", *Opt. Commun.,* vol.79, no.6, pp 448- 454.

[15] Farahbod, A. H., (1990), *"Active Q-switching of solid state-lasers",* MS thesis (advisor: A. Hariri).

[16] Farahbod, A. H. and Hariri, A., (1990) *Proceeding of Iranian Conference of phys*., p 19.

[17], (a) Farahbod, A. H., Barry, R., Morshedian, N., Tahsildaran, F., Tarkashvand, M. and Jaberi, M. (2009), *"Study of gain dependence on beam position in two-pass solid-state optical amplifiers",* Proceeding of 16th Iranian Conference of Optics and Photonics.

(b) Pearce, S., Ireland, C. L. M. and Dyer, P. E., (2005),"Simplified analysis of double-pass amplification with pulse overlap and application to Nd:YVO_4 laser", *Opt. Commun.* vol. 255, pp 297-303.

[18] Kagawa, K., Kobayashi, H. and Ishikane, M. (1979), "Oscillator-amplifier system in N_2-laser using a pressure dependent formative time-lag of discharge", *Jpn. J. Appl. Phys.,* vol.18, no.11, pp 2187-2188.

[19] Hariri, A., Tarkashvand, M. and Farahbod, A. H. (2003), "Experimental study of an oscillator-amplifier transversly excited molecular nitrogen laser system", *J. of Nuclear sci. and Tech.,* vol.29, pp17-24.

[20] Hariri, A., Tarkashvand, M. and Karami, A. (1990), "Corona preionized nitrogen laser with variable pulse width", *Rev. Sci. Instrum.,* vol.61, no.5, pp 1408-1412.

[21] Rahimian, K., Ghoreyshi, S. and Hariri, A. (2006), "Behavioral studies of gain and saturatin energy density in a N_2 laser with corona preionization", *Laser Phys.,* vol.16, no.3, pp 447-454.

[22] Armandillo, E. and Kearsley, A. J. (1982), "High-power nitrogen laser", *Appl. Phys. Lett.,* vol.41, no.7, pp 611-613.

[23] Brito Cruz, C. H., Loureiro, V., Tavares, A. D. and Scalabrin, A. (1984), "Characteristics of a wire preionized nitrogen laser with helium as buffer gas", *Appl. Phys.* B, vol,35, pp 131-133.

[24] Kunabenchi, R. S., Gorbal, M. R. and Savadatti, M. I. (1984), "Nitrogen lasers", *prog. Quant. Electr.*, vol.9, pp 259-329.

[25] Judd, O. (1976), "On the effect of addition of SF_6 to a N_2 electrical discharge laser", *IEEE, J. Quant. Electron.*, pp 78-80.

[26] Hall, D. W. and Weber, M. J. (1983), "Polarized Fluorescence line narrowing measurements of Nd laser glasses: Evidence of stimulated emission cross section anisotropy", *App. Phys. Lett.*, vol.42, no.2, pp 157-159.

[27] Sinha, B. K. and Kumbhare, S. R. (1989), "Evidence of polarization-dependent gain in a Nd:glass laser amplifier", *Opt. Commun.*, vol.73, no.3, pp 239-243.

[28] (a) Jelvani, S., Abolhosseini, S. and Mollabashi, M. (2010), "Polarization dependence of the small-signal gain and saturation energy of a dye laser", *Opt. Eng.*, vol.49, no.9, pp 091015 (1-4).

(b) Schäfer, F. P. (1977), *"Dye lasers"*, 2nd edition, edited by Schäfer, F. P., Berlin, Springer.

[29] Burns, D. J., Simpson, F. R. and McConkey, J. W. (1969), "Absolute cross sections for electron excitation of the second positive bands of nitrogen", *J. phys. B* (Atom. Molec. Phys.), Ser.2, vol.2, pp 52-64.

[30] Ghoreyshi, S., Rahimian, K. and Hariri, A. (2007), "Polarization-dependent gain and saturation energy density in a TE N_2-laser amplifier", *Opt. Commun.*, vol.272, pp 168-174.

[31] Martin, W. E. and Milam, D. (1982), "Gain saturation in Nd:doped laser materials", *IEEE J. Quant. Electron.*, vol.QE-18, no.7, pp 1155-1163.

[32] Strohwald, H. and Salzmann, H. (1976), "Picosecond uv laser pulses from gas discharges in pure nitrogen at pressures up to 6 atm", *Appl. Phys. Lett.*, vol.28, no.5, pp 272-274.

[33] Ghoreyshi, S., Rahimian, K. and Hariri, A. (2004), "Gain and saturation energy measurements in low pressure longitudinally excited N_2-lasers", *Opt. Commun.*, vol.238, pp 139-149.

[34] Theiss, F. J. (1973), "Dependence of peak power and pulse width for UV N_2 superradiant pulses on operating conditions in a coaxial laser device*", Opt. Commun.*, vol.9, no.1, pp 25-27.

[35] El-Osealy, M. A., Ido, T., Nakamura, K., Jitsuno, T. and Horiguchi, S. (2001), "Oscillation and gain characteristics of high power co-axially excited N_2 gas lasers", *Opt. Commun.*, vol.194, pp 191-199.

[36] Hariri, A. Ghoreyshi, S. and Rahimian, K. (2007), "Gain and saturation energy characteristics in transversely excited N_2 lasers", *J. Appl. Phys.*, vol.101, pp 033132 (1-7).

[37] Hariri, A., Jaberi, M. and Ghoreyshi, S. (2008), "Nitrogen lasers: Optical devices of variable gain coefficient", *Opt. Commun.*, vol.281, pp 3841-3852.

[38] Sarikhani, S and Hariri, A. (2010), "Nitrogen lasers, optical devices of variable gain coefficient: Theoretical considerations", *Opt. Commun.*, vol.283, pp 118-127.

[39] Hariri, A. and Sarikhani, S., *"A two dimensional theoretical approach for describing gain coefficient*" (unpublished)

[40] Svelto, O., (1998) "Principles of lasers", 4th ed., New York, Springer.

[41] von Bergmann, H. M. and Hasson, V., (1978) "Gain, fluorescence and energy extraction characteristics of photostaiblised atmospheric pressure UV nitrogen lasers", *J. Phys. D: Appl. Phys.*, vol.11, pp 2341 – 2351.

[42] Papakin, V. F. and Sonin, A. Yu., (1985) "Measurement of the gain of an ultraviolet nitrogen laser", *Sov. J. Quant. Electron.*, vol.15, no.4, pp 581 – 583.

[43] Godard, B., (1974) "A simple high-power large-efficiency N_2 ultraviolet laser", *IEEE, J. Quant. Electron.,* vol.QE-10, no.2, pp 147 – 153.

[44] Leonard, D. A., (1965) "Saturation of the molecular nitrogen second positive laser transition", *Appl. Phys. Lett.,* vol.7, no.1, pp 4 – 6.

[45] Fitzsimmons, W. A., Anderson, L. W., Riedhauser, C. E. and Vrtilek, J. M., (1976) "Experimental and theoretical investigation of the nitrogen laser", *IEEE, J. Quant. Electron.,* vol.QE-12, no.10, pp 624 – 633.

[46] Papadopoulos, A. D. and Serafetinides, A. A. (1990), "Characteristics of doubling circuits used in gas laser excitation: application to the N_2 laser", *IEEE, J. Quant. Electron.,* vol.26, no.1, pp 177 – 188.

[47] Crouch, J. H., Risk, W. S. (1972), "A compact high speed low impedance Blumlein line for high voltage pulse shaping", *Rev. Sci. Instrum.,* vol.43, no.4, pp 632 – 637.

[48] Tohyama, Y. and Nagata, T. (2005), "Electron Impact emission of N_2 2p (v', v") Bands studied under single-collision condition", *J. Phys. Soc. Japan,* vol.74, no.1, pp 326 – 332.

[49] Shemansky, D. E., Ajello, J. M. and Kanik, I. (1995), "Electron excitation function of N_2 second positive system", *Astrophys. J.,* vol.452, pp 472 – 479.

[50] Tabata, T., Shirai, T., Sataka, M. and Kubo, H. (2006), "Analitic cross sections for electron impact collisions with nitrogen molecules", *Atom. Data and Nucl. Data Tables,* vol.92, pp 375 – 406.

[51] Chang, S. H. and Teii, S. (1985), "Theoretical investigation of the traveling wave excitation in a TE N_2 laser", J. Appl. Phys., vol.57, no.5, pp 1478 – 1481.

[52] Chang, S. H. and Teii, S. (1985), "TE N_2 laser", *J. Appl. Phys.,* vol.58, no.7, pp 2467 – 2468.

In: Optical Amplifiers
Editor: Galina Nemova

ISBN 978-1-61209-835-7

Chapter 8

STABLE VORTICES IN THE MODEL OF A TWO-DIMENSIONAL AMPLIFYING MEDIUM

***D. Mihalache*[1], *D. Mazilu*[1], *V. Skarka*[2], *B. A. Malomed*[3], *H. Leblond*[2], *and N. B. Aleksić*[4]**
[1] Horia Hulubei National Institute for Physics and Nuclear Engineering (IFIN-HH), 407 Atomistilor, Magurele-Bucharest, Romania
[2] Laboratoire de Photonique d'Angers, EA 4464 Université d'Angers, 2 Bd. Lavoisier, 49045 Angers Cedex 01, France
[3] Department of Physical Electronics, Faculty of Engineering, Tel Aviv University, Tel Aviv, Israel
[4] Institute of Physics, Pregrevica 118, Belgrade, Serbia

Abstract

A ubiquitous model of bulk amplifying laser media is provided by two-dimensional complex Ginzburg-Landau (CGL) equations with the cubic-quintic (CQ) nonlinearity. Solutions to these equations of fundamental physical significance are solitons and solitary vortices. While the zero-vorticity solitons may easily be stable in this setting, a severe problem which impedes the stabilization of vortices and, thus, the observation of such modes in the experiment, is the lack of an effective diffusion term in the physically relevant CGL equations (light does not undergo diffusion in lasing cavities). The formal diffusion term is necessary for the stability of vortex solitons. Recently, it has been reported that the addition of a 2D periodic potential, which may be induced by a transverse grating in the laser cavity, readily stabilizes compound (four-peak) vortices. Nevertheless, the most fundamental axisymmetric "crater-shaped" vortices (CSVs), alias vortex rings, have not been found before in a stable form. In this chapter, we report families of stable compact CSVs with vorticity $S = 1$ in the CGL model with the CQ nonlinearity and an external potential of two different types: an axisymmetric harmonic-oscillator trap, and the periodic potential corresponding to a grating (in the latter case, the CSV must be essentially squeezed into a single cell of the grating). We identify stability regions for the CSVs, as well as for the fundamental solitons ($S = 0$).

If CSVs are unstable in the harmonic-oscillator potential, the vortex ring breaks up into robust dipoles. All the vortices with $S = 2$ are found to be unstable, splitting into *tripoles*. Stability regions for the dipoles and tripoles are reported too. The periodic potential does not stabilize CSVs with $S > 1$ either; instead, in this case unstable vortices with $S = 2$ evolve evolve into families of stable square-shaped quadrupoles.

1. Introduction

A broad class of pattern-formation models in nonlinear dissipative media is based on the complex Ginzburg-Landau (CGL) equations with the cubic-quintic (CQ) nonlinearity [1, 2]. These models find an important realization is lasing media, where the CQ terms account for the combination of nonlinear gain and loss (the linear loss is included into the CGL equation too) [3]. As models of laser systems, the CQ nonlinearity represents setups incorporating linear amplifiers and saturable nonlinear absorbers. In the one-dimensional (1D) setting, the CQ CGL equation is known to give rise to stable solitary pulses (dissipative solitons). These solutions and their physical implications have been studied in numerous works [4].

A challenging problem is looking for stable dissipative solitons in the two-dimensional (2D) version of CGL equations. In that case, the essential issues are the critical collapse induced by the cubic self-focusing term, and the aptitude of vortex solitons, which are shaped as vortex rings, to develop the instability against azimuthal perturbations which tend to split them [5, 6, 7]. Actually, the CGL equation with the CQ nonlinearity was originally introduced by Petviashvili and Sergeev [8] with the purpose to develop a model admitting stable localized 2D modes. Stable 2D solitary vortices (alias *spiral solitons*), with topological charge (vorticity) $S = 1$ and 2, were first reported in Ref. [9]. Stable vortex solitons were reported in the 3D version of the CQ CGL equation too [10]

In a uniform bulk medium with transverse coordinates (x, y), the general form of the CQ CGL equation for the amplitude of the electromagnetic field, $E(x, y, z)$, which propagates along axis z, is [10]

$$iE_z + \left(\frac{1}{2} - i\beta\right)(E_{xx} + E_{yy}) + i\delta E$$
$$+(1 - i\varepsilon)|E|^2 E + (\nu + i\mu)|E|^4 E = 0, \tag{1}$$

where δ is the linear-loss coefficient, the Laplacian with coefficient $1/2$ represents, as usual, the transverse diffraction in the paraxial approximation, β is an effective diffusion coefficient, ε is the cubic gain, the Kerr coefficient is normalized to be 1, and quintic coefficients $-\nu$ and μ account for the saturation of the cubic nonlinearity ($\nu > 0$ corresponds to the quintic self-focusing, which does not lead to the supercritical collapse, being balanced by the quintic loss [11]).

The realization of all terms in Eq. (2) is straightforward, except for the diffusion. This term is artificial in the application to optics, as light is not subject to diffusion. Nevertheless, $\beta > 0$ is a necessary condition for the stability of dissipative vortex solitons, unlike the fundamental ($S = 0$) solitons, which may be stable at $\beta = 0$ [9, 10]. Therefore, it is relevant to develop a physically relevant modification of the the 2D CGL model, without the diffusivity ($\beta = 0$), that can support *stable* localized vortices. Recently, it has been demonstrated that this problem can be resolved by adding a transverse periodic potential to Eq. (1), which casts the CGL equation into the following form [11]:

$$iE_z + \frac{1}{2}(E_{xx} + E_{yy}) + i\delta E + (1 - i\varepsilon)|E|^2 E$$
$$+(\nu + i\mu)|E|^4 E - V(x,y)E = 0. \tag{2}$$

The periodic potential can be induced by a grating, i.e., periodic modulation of the local refractive index in the plane of (x,y) :

$$V(x,y) = p[\cos(2x) + \cos(2y)], \; p > 0, \tag{3}$$

where p is proportional to the strength of the underlying grating, and the scaling invariance of of Eq. (2) was employed to fix the period of potential (3) to be π.

The laser-writing technology makes it possible to fabricate permanent gratings in bulk media [12]. In photorefractive crystals, *virtual* photonic lattices may be induced by pairs of laser beams illuminating the sample in the directions of x and y in the ordinary polarization, while the probe beam is launched along axis z in the extraordinary polarization [13].

It is relevant to notice that the equations of the CGL type describe laser cavities, where the mode-locked optical signal performs periodic circulations, as a result of averaging [14]. Therefore, the transverse grating (or a different structure inducing the effective transverse potential) is not required to fill the entire cavity; a layer localized within a certain segment, Δz, rather than uniformly distributed along z, may be sufficient to induce the effective potential in Eq. (2) [11].

Stationary solutions to Eq. (2) are sought for as $E(x,y,z) = e^{ikz}U(x,y)$, with real propagation constant k and complex function $U(x,y)$ satisfying the stationary equation,

$$[-k + i\delta - V(x,y)]U + \frac{1}{2}(U_{xx} + U_{yy})$$
$$+(1 - i\varepsilon)|U|^2 U + (\nu + i\mu)|U|^4 U = 0.$$

Stable vortices, supported by periodic potential (3), were constructed in Ref. [11] as compound objects, built of four separate peaks of the local power, which are set in four cells of the lattice. Two basic types of such vortices are “rhombuses”, alias *onsite vortices*, with a nearly empty cell surrounded by the four filled ones [15], and “squares”, alias *offsite vortices*, which feature a densely packed set of four filled cells [16]. The vorticity (topological charge) of these patterns is provided by phase shifts of $\pi/2$ between adjacent peaks, which corresponds to the total phase circulation of 2π around the pattern, as it should be in the case of vorticity $S = 1$. In the experiment, stable compound vortices with $S = 1$ were created in a conservative medium, *viz.*, the above-mentioned photorefractive crystals with the photoinduced lattice [17]. In Ref. [11], a stability region was identified for rhombus-shaped compound vortices, with $S = 1$, in the framework of the CGL equation (2) with potential (3), and examples of their stable square-shaped counterparts (which are essentially less stable than the rhombuses) were produced too. In addition, Ref. [11] reported examples of stable rhombic quadrupoles, i.e., four-peak patterns with alternating signs of the peaks (and zero vorticity).

A challenging issue remains to find conditions providing for the stability of compact "crater-shaped" vortices (CSVs, alias vortex rings) which, unlike the compound vortical structures, are squeezed into a *single cell* of the periodic potential (typical examples of stable "craters" supported by 2D periodic potentials can be seen below in Fig. 13). These nearly axisymmetric vortices are most similar to their counterparts found in the free space [9]. As mentioned above, in the absence of the potential the vortices may only be stabilized by the diffusion term in Eq. (1), with $\beta > 0$; otherwise, azimuthal perturbations break them into sets of fragments. A natural expectation is that the trapping potential may stabilize "craters" in the model with $\beta = 0$. Nevertheless, no examples of stable CSVs were reported in Ref. [11].

The search for stable CSVs is also a challenging problem in the studies of 2D conservative models with lattice potentials. In particular, only unstable vortices of this type were reported in the 2D nonlinear Schrödinger (NLS) equation with the CQ nonlinearity and a checkerboard potential [18, 19] (see also Ref. [20]). On the other hand, stable *supervortices*, i.e., chains formed by compact craters with $S = +1$ and an independent global vorticity, $S' = \pm 1$, imprinted onto the chain, were found as stable objects in 2D NLS equations with periodic potentials and the cubic or CQ nonlinearities [21, 19]. Eventually, a stability region for CSVs was recently identified in the cubic NLS equation, provided that the periodic potential is strong enough [22].

The main objective of the present work is to demonstrate that crater-shaped dissipative vortex solitons may be *stabilized*, in the framework of the CQ CGL equation (2), by external potentials. To this end, we consider two potentials: the periodic one, taken as per Eq. (3), and also the axisymmetric trapping potential,

$$V(x,y) = \left(\Omega^2/2\right) r^2, \tag{4}$$

where $r^2 \equiv x^2 + y^2$. The consideration of potential (4) is suggested by known results for the 2D NLS equation (in that context, it is introduced as the Gross-Pitaevskii equation for the Bose-Einstein condensate), which demonstrate that potential (4) can stabilize localized vortices with $S = 1$ against the splitting [23]. Actually, potential (4) can be realized in the laser cavity merely by inserting a lens with focal length $f' = k/(L\Omega^2)$, where k is the wavenumber and L the cavity length. Then, averaging over the cyclic optical path yields potential (4), within the framework of the paraxial approximation. It is possible to check that the generic situation for vortices and other types of dissipative solitons generated by Eq. (2) may be adequately represented by fixing $\delta = 1/2$, $\mu = 1$, and $\nu = -0.1$, which is assumed below. Two remaining parameters, that will be varied in this work, play a crucially important role in the model: cubic gain ε and the strengths, Ω^2 or p, of the trapping potentials.

The rest of the paper is organized as follows. In Sections 2 and 3, we consider the stabilization of the CSVs in axisymmetric potential (4). First, we apply the generalized variational approximation (VA), which was developed in Ref. [24] for a class of CGL equations, as an extension of the well-known VA for conservative nonlinear-wave systems [25]. In Section 3 we continue the consideration of the CSVs in the same potential by means of numerical methods. Both the VA and direct simulations reveal the existence of a broad stability region for these vortices with $S = 1$. Unstable CSVs split into stable patterns in the form of *dipoles*. Vortices with $S = 2$ can also be constructed, but they all are unstable (similar to the situation in the NLS equation [23]), splitting into *tripole* patterns. Dipoles and tripoles are studied in Section 4, where their stability regions are identified.

A stability region for CSVs with $S = 1$ in periodic potential (3) is reported in Section 5. This is a novel result, as stable CSVs have never before been reported in CGL models without the diffusion term. Stable CSVs with vorticities $S > 1$ are not found; instead, families of robust compact square-shaped quadrupoles are found to exist at different values of the strength of the periodic potential. The paper is concluded by Section 6.

2. The Variational Approximation for Vortices in the Axisymmetric Potential

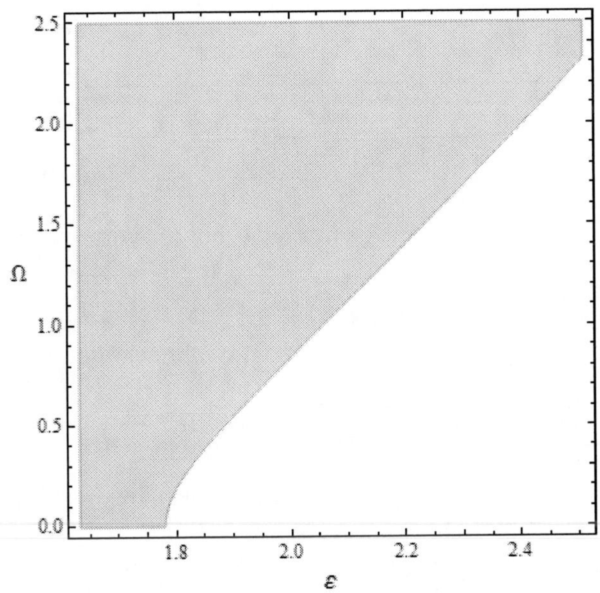

Figure 1. (Color online) The stability domain for fundamental solitons ($S = 0$), which is situated on the left-hand side of the plotted curve (the shaded area), in the parameter plane of (ε, Ω), as predicted by the variational approximation, which pertains to the CGL equation with the axisymmetric trapping potential (4). Other parameters are fixed as said above, i.e., $\delta = 1/2, \mu = 1$, and $\nu = -0.1$.

The VA for dissipative systems, elaborated in Ref. [24], is applied here to look for axisymmetric vortex solutions to Eq. (2) with potential (4), using the following ansatz, written in polar coordinates r and θ:

$$E = A_0 A \left(\frac{r}{R_0 R}\right)^S \exp\left[R_0^{-2}\left(-\frac{r^2}{2R^2} + iCr^2\right) + iS\theta + i\psi\right]. \quad (5)$$

Here, S is the integer vorticity, and real variational parameters are amplitude A, width R, wave-front curvature (spatial chirp) C, and phase ψ, that all may be functions of z.

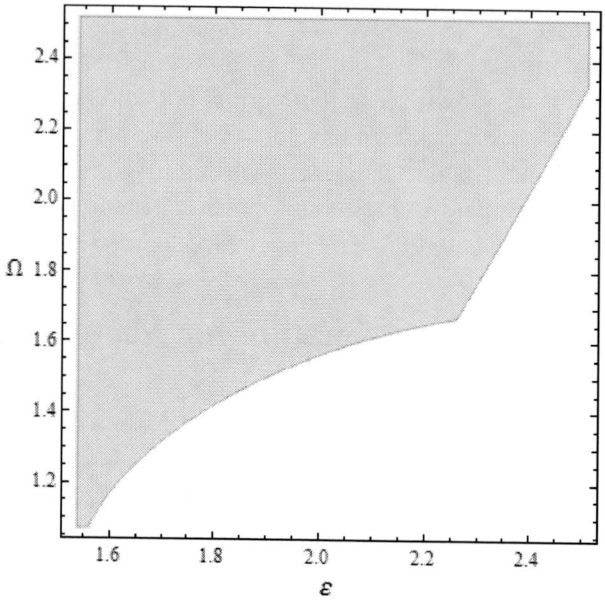

Figure 2. (Color online) The same as in Fig. 1, but for vortex solitons with $S = 1$.

The ansatz includes normalization factors, $A_0 = 3 \cdot 2^{-(S+1)} \sqrt{[3^{3S}(2S)!]/[2(3S)!]}$ and $R_0 = 2^{S+1/2} A_0^{-1} \sqrt{(S+1)!/(2S)!}$. A natural characteristic of the soliton is its total power,

$$P = 2\pi \int_0^\infty |E(r)|^2 \, r dr, \tag{6}$$

which takes value $P = A^2 R^2$ for ansatz (5) (in fact, normalization factors A_0 and R_0 were introduced so as to secure this simple expression for P).

Skipping technical details, the application of the generalized VA technique, along the lines of Ref. [24], leads to the following system of the first-order evolution equations for the parameters of ansatz (5):

$$\frac{dA}{dz} = \frac{A}{R_0^2} \left[\frac{3+2S}{2} \varepsilon A^2 - \frac{5+3S}{4} \mu A^4 - R_0^2 \delta - 2C \right], \tag{7}$$

$$\frac{dR}{dz} = \frac{R}{2R_0^2} \left(4C - \varepsilon A^2 + \mu A^4 \right), \tag{8}$$

$$\frac{dC}{dz} = \frac{1}{2R_0^2} \left(\frac{1}{R^4} - \frac{A^2}{R^2} - \nu \frac{A^4}{R^2} - 4C^2 - \Omega^2 R_0^4 \right), \tag{9}$$

$$\frac{d\psi}{dz} = \frac{(S+1)}{R_0^2} \left(\frac{3}{2} A^2 + \frac{5}{4} \nu A^4 - \frac{1}{R^2} \right). \tag{10}$$

The VA predicts steady states as fixed-point solutions to Eqs. (7)-(9). A straightforward analysis yields two such solutions:

$$A^2 = \frac{2}{3\mu} \left[\varepsilon \pm \sqrt{\varepsilon^2 - 3\mu R_0^2 \delta (S+1)^{-1}} \right] \equiv \left(A^\pm \right)^2, \tag{11}$$

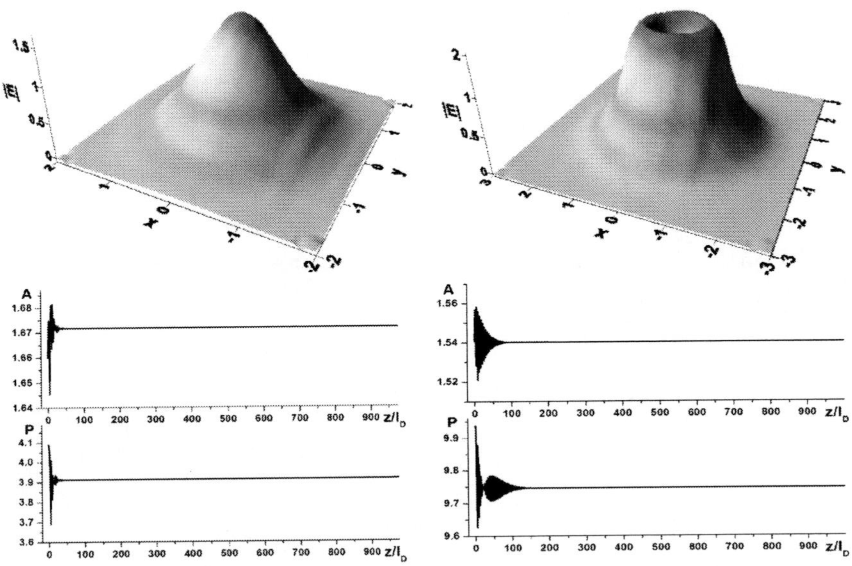

Figure 3. (Color online) Left and right panels display the self-trapping of stable fundamental ($S = 0$) and vortical ($S = 1$) solitons, respectively, from inputs predicted by the variational approximation, at parameter values $\Omega = 1.5$, $\varepsilon = 2.2$ for $S = 0$, and $\Omega = 1.7$, $\varepsilon = 2.22$ for $S = 1$. The 3D images are the established shapes of the solitons, while plots $A(z/l_D)$ and $P(z/l_D)$ show the evolution of the amplitude and total power, see Eq. (6), with z measured in units of the respective diffraction lengths, l_D.

$$R^2 = 2\left[A^2(1+\nu A^2)+ \sqrt{A^4\left[(1+\nu A^2)^2+(\varepsilon-\mu A^2)^2\right]+4\Omega^2 R_0^4}\right]^{-1},$$

$$C = \left(A^2/4\right)\left(\varepsilon-\mu A^2\right).$$

In particular, the nonzero value of C (the wave's front curvature) in the stationary solution is an essential difference from stationary solitary vortices in conservative models described by the NLS equations.

Further, the calculation of eigenvalues of small perturbations around the fixed points demonstrates that solution A^+ is stable, while A^- is not, cf. Ref. [24]. Finally, the VA predicts stability domains for the fundamental ($S = 0$) solitons and vortices with $S = 1$ in the plane of the free parameters, ε and Ω. The domains are displayed, respectively, in Figs. 1 and 2, cf. Ref. [26]. In these plots, the vertical borders of the stability regions on the left-hand side correspond to the existence condition of solution (11), i.e., $\varepsilon > \sqrt{3\mu R_0^2 \delta (S+1)^{-1}}$. In particular, for $S = 0$ it is $\varepsilon > 2\sqrt{2/3} \approx 1.63$, and for $S = 1$, the existence region is $\varepsilon > 8/\left(3\sqrt{3}\right) \approx 1.54$. These existence limits are found to be in excellent agreement with the corresponding values obtained from direct numerical simulations reported below in Fig. 4.

The accuracy of the solutions for dissipative solitons and vortices predicted by the VA was checked by running direct simulations of the full CGL equation (2), using the respective wave forms, given by Eqs. (5) and (11), as initial conditions. Typical results of such simulations over 1000 diffraction lengths are displayed in Fig. 3, for the solitons with $S = 0$ and $S = 1$. It is seen that the input wave forms predicted by the VA quickly relax into the finally established soliton shapes, which are shown in Fig. 3.

3. Numerical Results for Vortices in the Axisymmetric Trapping Potential

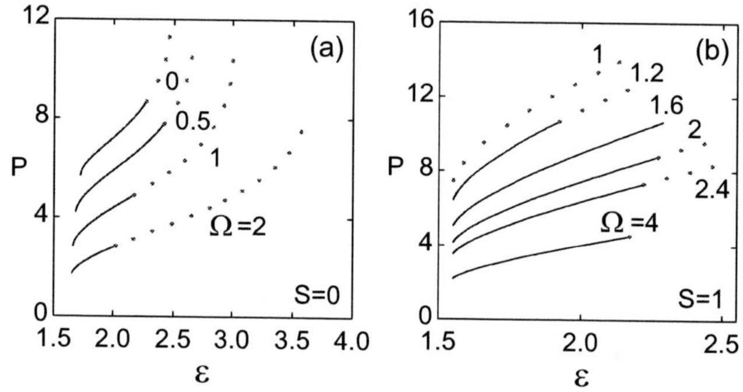

Figure 4. (Color online) The total power, P, versus the cubic gain, ε, for families of (a) fundamental solitons ($S = 0$) and (b) vortices with $S = 1$ at different values of the trapping frequency in potential (4). Solid lines: stable solutions; dotted lines: unstable ones.

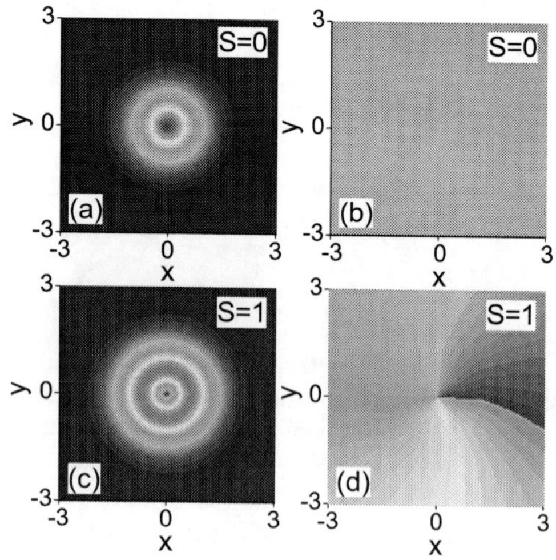

Figure 5. (Color online) Examples of stable dissipative solitons with vorticities $S = 0$ and $S = 1$, for $\varepsilon = 1.8$ and potential (4) with $\Omega = 2$. Panels (a,c) and (b,d) display the amplitude and phase distributions, respectively.

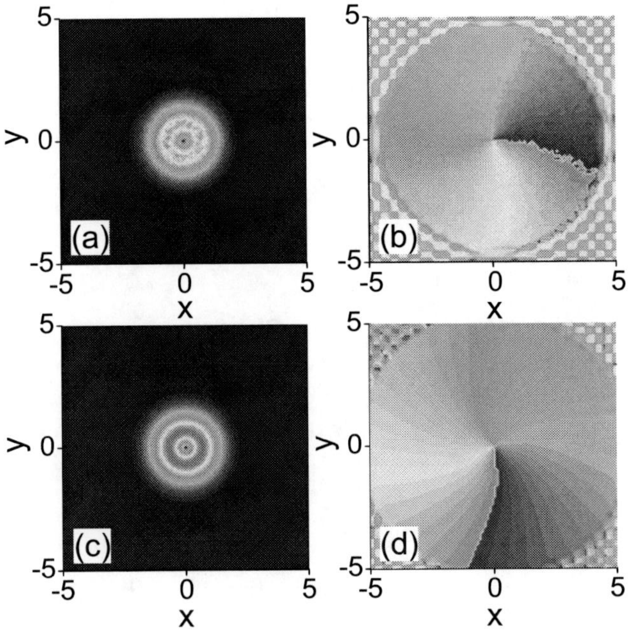

Figure 6. (Color online) The recovery of a perturbed stable vortex soliton with $S = 1$ at $\varepsilon = 1.8$ and $\Omega = 2$, in the case of the axisymmetric potential (4): (a) and (b) initially perturbed amplitude and phase distributions; (c) and (d) self-cleaned amplitude and phase distributions, at $z = 200$.

Looking for axisymmetric solutions to Eq. (2) with potential (4) in the numerical form, we substitute $E(z,x,y) = U(z,r)\exp(iS\theta)$, which yields the evolution equation for complex amplitude $U(z,r)$:

$$iU_z + \frac{1}{2}\left(U_{rr} + \frac{1}{r}U_r - \frac{S^2}{r^2}U\right) - \frac{1}{2}\Omega^2 r^2 U$$
$$+(1 - i\varepsilon)|U|^2U + (\nu + i\mu)|U|^4U + i\delta U = 0. \quad (12)$$

We note that stationary solutions to Eq. (12) must decay exponentially at $r \to \infty$, and as $r^{|S|}$ at $r \to 0$.

Stationary dissipative solitons, both fundamental ($S = 0$) and vortical ones, were generated as *attractors* by direct simulations of Eq. (12). To this end, we simulated Eq. (12), starting with the input field in the form of the Gaussian corresponding to vorticity S,

$$U_0(r) = A_0 r^S \exp\left[-\left(r^2/w_0^2\right)\right], \quad (13)$$

with real constants A_0 and w_0, until the solution would self-trap into a stable dissipative soliton. The so found established solutions can be eventually represented in the form of $U(z,r) = u(r)\exp(ikz)$, where propagation constant k is, as a matter of fact, an eigenvalue determined by parameters of Eq. (12).

The simulations of Eq. (12) were run using a 2D Crank-Nicolson finite-difference scheme, with transverse and longitudinal stepsizes $\Delta r = 0.05$ and $\Delta z = 0.002$. The resulting

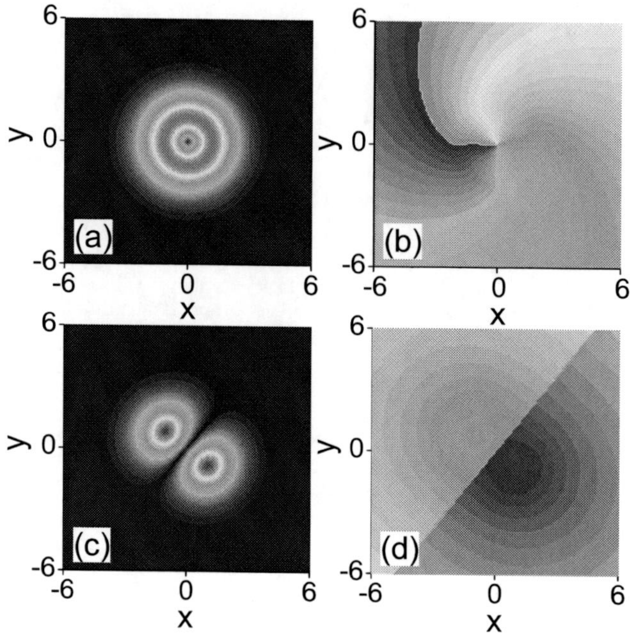

Figure 7. (Color online) The spontaneous breakup of an unstable vortex with $S = 1$, which is shown in panels (a) and (b), into a stable dipole soliton, displayed in (c) and (d) at $z = 400$. The parameters are $\varepsilon = 1.8$ and $\Omega = 0.5$, for the axisymmetric potential (4).

nonlinear finite-difference equations were solved using the Picard iteration method, and the ensuing linear system was then dealt with using the Gauss-Seidel elimination procedure. To achieve reliable convergence, eight Picard and six Gauss-Seidel iterations were sufficient. The wave number, k, was found as the value of the z-derivative of the phase of $U(z,r)$. The solution was considered as the established one if k ceased to depend on z and r, up to five significant digits. After a particular stationary solution was found by the direct integration of Eq. (12), it was then used as the initial configuration for a new run of simulations, with slightly modified values of the parameters, aiming to generate the solution corresponding to the new values.

When localized states could not self-trap in the course of the evolution, or existed temporarily but eventually turned out to be unstable, $U(z,r)$ would eventually decay to zero, or evolve into an apparently random pattern filling the entire integration domain. Naturally, the decay to zero was observed when the cubic-gain coefficient, ε, was too small. In the opposite case, with ε too large, the random pattern was generated.

If the simulations of Eq. (12) converged to stationary localized modes, their full stability was then tested by adding white-noise perturbations at the amplitude level of up to 10%, and running direct simulations (in the Cartesian coordinates) of the underlying equation (2). In the course of the stability tests, the evolution of both the total norm, $P(z)$, and the amplitude of the solution was monitored. The solution was identified as a stable one if its amplitude and shape had relaxed back to the unperturbed configuration.

Results of the numerical analysis are summarized in Fig. 4, which represents both stable

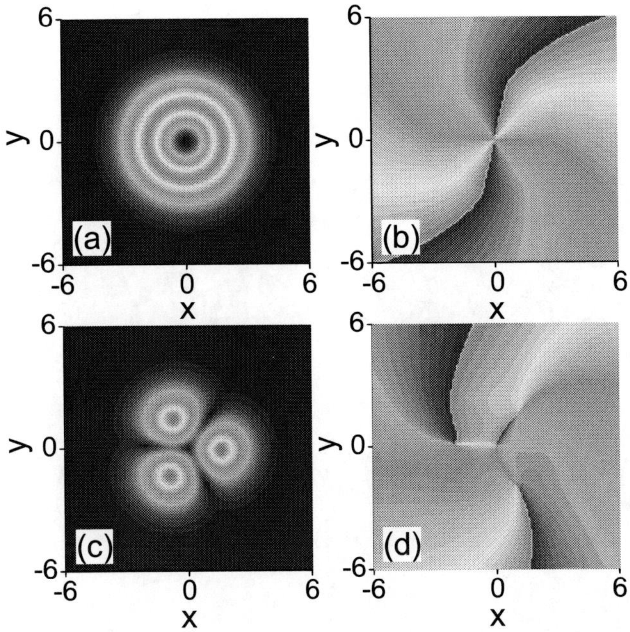

Figure 8. (Color online) The breakup of an unstable vortex with $S = 2$, which is shown in panels (a) and (b), into a stable tripole, displayed in (c) and (d) at $z = 340$. The parameters are $\varepsilon = 1.7$ and $\Omega = 0.5$.

(solid lines) and unstable (dotted lines) soliton families with $S = 0$ and $S = 1$, in terms of the dependence of total power P on nonlinear gain ε. The stability of each family is limited to a particular interval, $\varepsilon_0 < \varepsilon < \varepsilon_{cr}$ (as said above, at $\varepsilon > \varepsilon_{cr}$ the solitons evolve into a random pattern filling the entire transverse domain).

Both the VA and direct simulations predict that the stability of the vortices requires relatively large values of trapping frequency Ω. Note that the fundamental solitons ($S = 0$) have a stability domain at $\Omega = 0$ [see Fig. 4(a)], in accordance with Ref. [9]. For some values of Ω, the stability intervals predicted by the VA are in good agreement with those produced by the direct simulations: compare, for example, the intervals for the fundamental solitons at $\Omega = 1$ in Figs. 1 and 4(a), and for the vortices with $S = 1$ at $\Omega = 2$ in Figs. 2 and 4(b). However, the agreement is worse in some other cases. Indeed, the VA gives only an approximate prediction for the stability of the zero-vorticity solitons, because ansatz (5) does not accommodate all possible modes of the instability.

Higher-order vortex solitons, with $S \geq 2$, are found to be completely unstable. If vortices with $S = 1$ are unstable, they spontaneously split into stable *dipoles*, whereas those with $S = 2$ split into *tripoles*, see below. Typical examples of the amplitude and phase structure of stable dissipative solitons with vorticities $S = 0$ and $S = 1$ are displayed in Fig. 5. For the same case, the recovery of the vortex soliton perturbed by the random noise at the 10% amplitude level is displayed in Fig. 6.

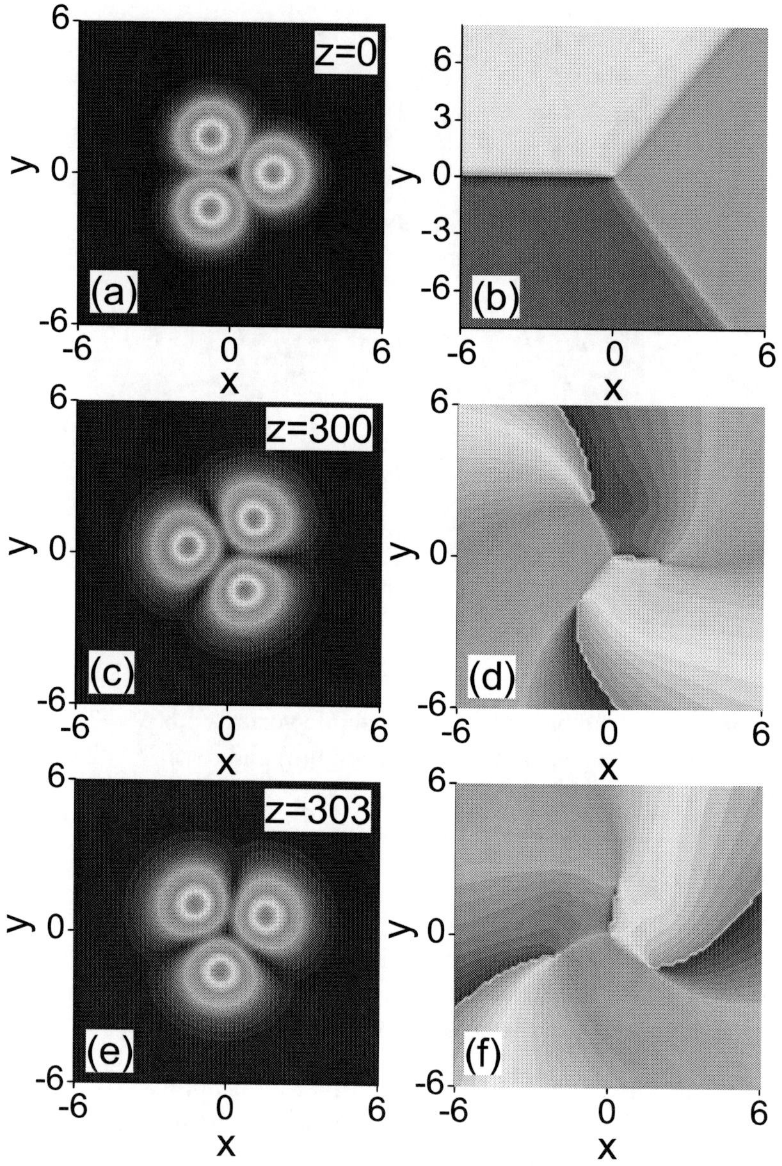

Figure 9. (Color online) The generation of a robust *rotating* tripole in potential (4) from an input cluster formed by three Gaussians with phase differences $2\pi/3$ between them. Left panels: the input field (a), and the established field amplitude $|A(x,y)|$ at $z = 300$ (c) and at $z = 303$ (e). Right panels: the phase of the input field (b), and the phases of the established pattern at $z = 300$ (d) and at $z = 303$ (f). The parameters are $\varepsilon = 1.7$ and $\Omega = 0.5$.

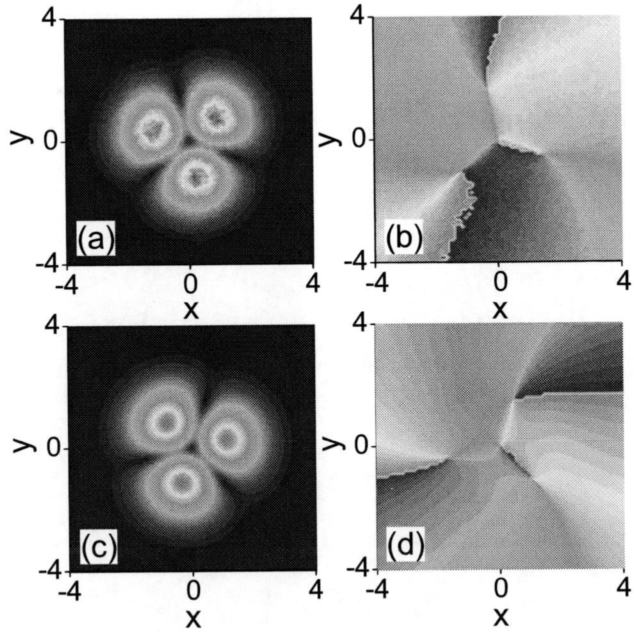

Figure 10. (Color online) The recovery of a perturbed stable tripole, at $\varepsilon = 1.7$ and $\Omega = 1$, in the axisymmetric trapping potential (4): (a) and (b) perturbed initial distributions of the amplitude and phase; (c) and (d) self-cleaned amplitude and phase distributions at $z = 200$.

4. Dipoles and Tripoles in the Axisymmetric Trap

As said above, the evolution of those vortices with $S = 1$ which are unstable, and of the vortices with $S = 2$ (recall they all are unstable), leads to their breakup into other types of robust modes, *viz.*, *dipoles* and *tripoles*, which feature phase shifts π and $2\pi/3$, respectively, between their components. Typical examples of the breakup are displayed in Figs. 7 and 8.

Both the dipole and tripole modes can also be readily generated from initial clusters, formed, respectively, by two Gaussians with the phase shift of π between them, or by three Gaussians with phase differences $2\pi/3$. An example of the formation of a stable tripole from the cluster is shown in Fig. 9. Notice the fast rotation of the tripole, which is clearly seen from comparison of panels 9(c) and 9(e). The rotation is possible thanks to the absence of the diffusion, as there is no effective friction that would brake the motion of solitons, cf. Ref. [27]. Nevertheless, the dipoles generated by the direct numerical simulations do not feature the rotation.

The stability of the dipoles and tripoles was verified by means of systematic direct simulations of initially perturbed patterns, similar to how it was done above for the fundamental solitons and vortices with $S = 1$. The random perturbations were imposed at the amplitude level of 10%. A typical example of the relaxation of a perturbed stable tripole is displayed in Fig. 10 (the self-cleaning of stable dipoles is quite similar).

Results of the systematic analysis of the stability of the dipole and tripole modes are summarized in terms of the respective $P = P(\varepsilon)$ curves in Fig. 11, cf. Fig. 4 for the

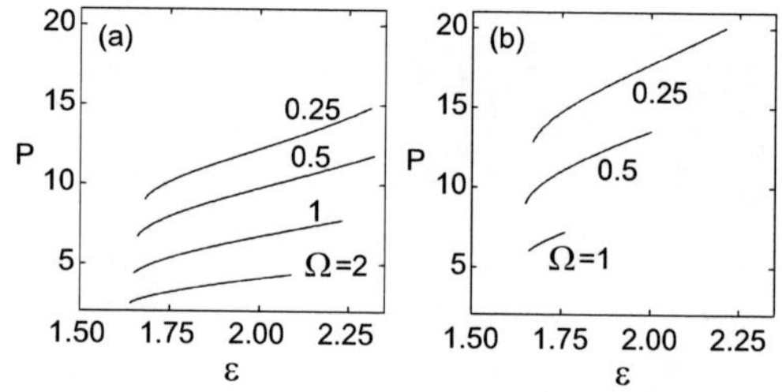

Figure 11. Power P versus cubic gain ε for stable dipole solitons (a) and stable tripole solitons (b) trapped in the axisymmetric potential (4), at different values of frequency Ω.

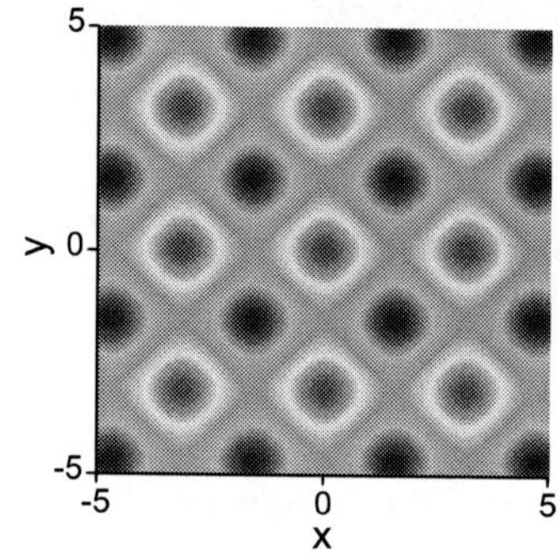

Figure 12. (Color online) The amplitude distribution of the periodic potential, $V(x,y) = p[\cos(2x) + \cos(2y)]$, for $p = 1$.

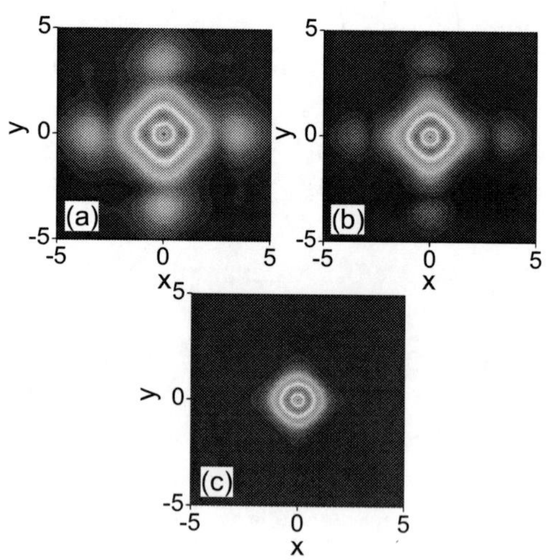

Figure 13. (Color online) The shapes of stable compact ("crater-shaped") vortices with $S = 1$ for $\beta = 0$ and $\varepsilon = 1.8$. The strength of the periodic potential (3) is $p = 1$ (a), $p = 2$ (b), and $p = 5$ (c).

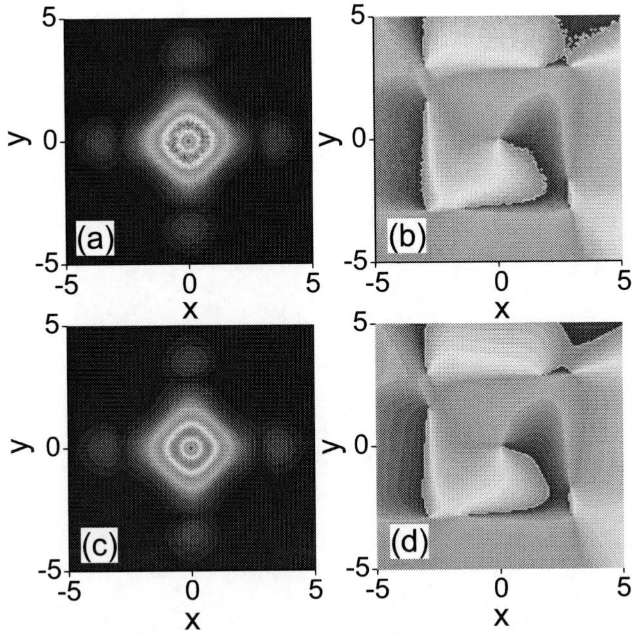

Figure 14. (Color online) Top: the amplitude (a) and phase (b) of a perturbed compact vortex ("crater") with $S = 1$, for $\beta = 0$, $\varepsilon = 2$, and $p = 2$, in the axisymmetric potential (4). Bottom: the amplitude (c) and phase (d) of the self-cleaned vortex soliton at $z = 200$.

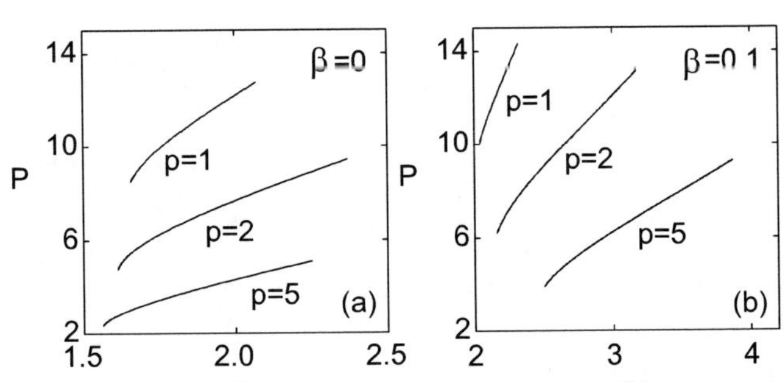

Figure 15. Power P versus cubic gain ε for families of stable compact vortices ("craters") with $S = 1$, at several values of strength p of the periodic potential (3), for $\beta = 0$ (a) and $\beta = 0.1$ (b).

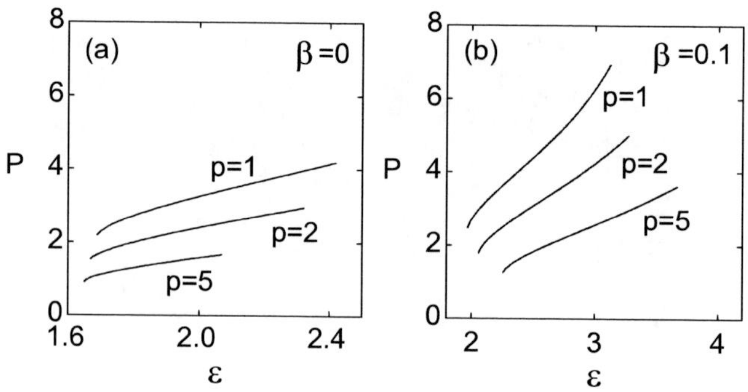

Figure 16. The same as in Fig. 15, but for families of stable fundamental ($S = 0$) solitons.

fundamental and $S = 1$ solitons. The dipole and tripole modes are stable in the intervals of ε in which curves $P = P(\varepsilon)$ are plotted.

5. Stability of Crater-Shaped Vortices and Square-Shaped Quadrupoles in the Periodic Grating

In this section we consider the model based on the CGL equation (2) with the periodic potential taken as per Eq. (3). Our first objective is to construct the CSVs with $S = 1$, which are squeezed, essentially, into a single cell of the grating potential, and identify their stability regions (if any). Note that choosing $p > 0$ in Eq. (3) implies the presence of a potential maximum at the center of the grating cell, $x = y = 0$. This choice complies with the expected minimum of the local power (the "hole") at the center of the compact vortex.

Families of relevant solutions were generated by simulating Eq. (2) with potential (3), starting with a Gaussian input corresponding to vorticity S, in the form of

$$E_0(x,y) = a_0 \exp\left[-(r-r_0)^2/w_0^2\right]\exp(iS\theta) \tag{14}$$

[cf. input (13) which created vortices in the axisymmetric parabolic trap (4)], with real constants a_0, r_0, and w_0, in anticipation of a self-trapping of the input field distribution into a ring-shaped pattern with a radius close to r_0. The so found established dissipative solitons can be eventually represented, as before, in the form of $E(x,y,z) = u(x,y)\exp(ikz)$, with some propagation constant k. This propagation constant was found as the value of the z-derivative of the phase of $E(z,x,y)$, at the eventual stage of the evolution, when k ceased to depend on x, y and z, up to five significant digits. The stability of the solitons was then tested, as before, against random perturbations with the relative amplitude of up to 10%.

As in the previous section, the Crank-Nicolson algorithm was used for the numerical simulations, with transverse and longitudinal stepsizes $\Delta x = \Delta y = 0.1$ and $\Delta z = 0.005$, for the grating strength $p = 1$. For larger values of p, it was necessary to use smaller stepsizes: $\Delta x = \Delta y = 0.08$, $\Delta z = 0.004$ for $p = 2$, and $\Delta x = \Delta y = 0.06$, $\Delta z = 0.003$ for $p = 5$. Using the same algorithm as mentioned above, the nonlinear finite-difference equations were solved using the Picard iteration method, and the resulting linear system was handled by means

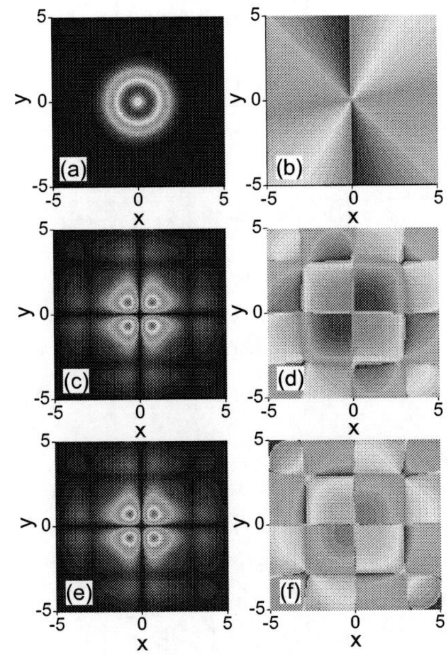

Figure 17. (Color online) A typical example of the generation of a stable quadrupole from an input ring-like field distribution with vorticity $S = 2$ (see Eq. 14). Here $p = 2$, $\varepsilon = 1.8$, and $\beta = 0$. (a) input, $z = 0$; (b) $z = 20$; (c) $z = 112$.

of the Gauss-Seidel iterative procedure. To achieve good convergence, ten Picard and five Gauss-Seidel iterations were needed.

In Fig. 12 we show an illustrative plot of the 2D periodic potential (3) with strength $p = 1$. The numerical simulations demonstrate that *fully stable* CSVs may be indeed supported by the periodic potential (3), see Figs. 13 and 14. This result is significant, as no example of stable compact vortices, squeezed into a single cell of the supporting lattice, was earlier reported in 2D CGL models. A set of typical examples of stable craters is displayed in Fig. 13, and the stability of such vortices (in the form of the self-cleaning against random perturbations) is illustrated by Fig. 14. Further analysis (not shown here) demonstrate that the shape of the craters and their self-cleaning after the addition of random perturbations seem essentially the same if the diffusion term with a small coefficient β is added (which means that the grating's potential remains a stronger stabilizing factor than the weak diffusion, if any).

In Fig. 15, the CSV families with $S = 1$ are represented, as before, by the corresponding $P(\varepsilon)$ curves, which are plotted in intervals of values of the cubic gain ε where the CSVs are stable. For the sake of comparison, in Fig. 16 we display similar diagrams for the fundamental solitons ($S = 0$) in the same model. In Figs. 15 and 16, we additionally display the stability domains found at a small nonzero value of the diffusion parameter, $\beta = 0.1$. The comparison demonstrates that the stability regions for both the fundamental solitons and compact vortices shift to larger values of ε at $\beta > 0$, which is natural, as a larger value of the cubic gain is needed to compensate the loss incurred by the diffusion term.

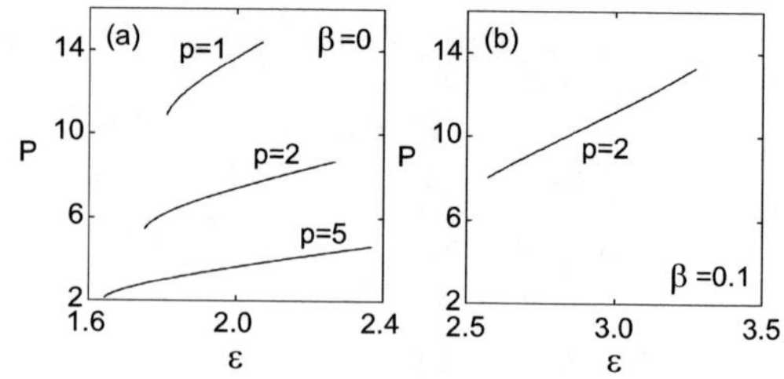

Figure 18. Power P versus cubic gain ε for families of stable quadrupoles at different values of the periodic-potential's strength, p, for $\beta = 0$ (a) and $\beta = 0.1$ (b).

Stable CSVs with vorticities $S = 2$ have not been found in direct simulations of Eq. (2) with the periodic potential; instead, families of robust compact square-shaped *quadrupoles*, into which unstable vortices with $S = 2$ are spontaneously transformed, were found at different values of the periodic-potential's strength, p. A typical example of the transformation, for $a_0 = 1.2$, $r_0 = 0.7$, and $w_0 = 1$, is displayed in Fig. 17. The stability of the quadrupoles against random perturbations was tested in the same way as done above for the vortices with $S = 1$. The results are again summarized by means of the respective $P(\varepsilon)$ curves, which are displayed, both for $\beta = 0$ and $\beta = 0.1$, in Fig. 18. Finally, in Fig. 19 we show typical examples of the amplitude and phase structure of compact square-shaped quadrupoles for $\beta = 0$, $\varepsilon = 2$, and three different values of the strength of the periodic potential, $p = 1, 2$, and 5.

6. Conclusions

The objective of this work is to find stable compact "crater-shaped" vortices, with topological charge $S = 1$, as solutions to the complex Ginzburg-Landau model of laser cavities, which does not include the artificial diffusion term. The stabilization of the compact vortices in this system is provided by external potentials, which were taken in two different forms: as the axisymmetric parabolic trap (4), and the periodic potential of the grating (3). In the experiment, the effective axisymmetric potential can be realized by means of a focusing lens inserted into the cavity. In both cases, stability regions for the crater-shaped vortices have been identified. Parallel to that, the stability regions of the fundamental solitons ($S = 0$) were indetified too, for the sake of the comparison. In the case of the axisymmetric potential, those crater-shaped vortices which are unstable split into robust *dipoles*. All the vortices with $S = 2$ are unstable, splitting into stable *tripoles*, that may freely rotate. The stability regions for the dipole and tripole modes were also identified. As concerns the periodic potential, it cannot stabilize crater-shaped vortices with $S > 1$. Instead, families of stable compact square-shaped quadrupoles were found at different values of the strength of the periodic potential.

A challenging extension suggested by the present work is to find stable compact solitons with embedded vorticity in the three-dimensional (spatiotemporal) version of the complex Ginzburg-Landau equation with the periodic potential. This possibility is especially inter-

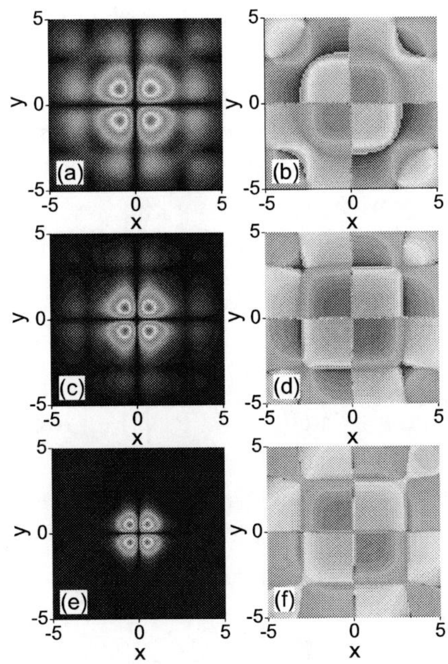

Figure 19. (Color online) The amplitude and phase structure of stable square-shaped quadrupoles for $\beta = 0$ and $\varepsilon = 2$. The strength of the periodic potential (3) is $p = 1$ (a)-(b), $p = 2$ (c)-(d), and $p = 5$ (e)-(f).

esting because this model does not support stable compound vortices [28].

Acknowledgements

This work was supported, in a part, by a grant on the topic of "Nonlinear spatiotemporal photonics in bundled arrays of waveguides" from the High Council for Scientific and Technological Cooperation between France and Israel, and by the Romanian Ministry of Education and Research through Grant No. IDEI-497/2009, as well as by the Ministry of Science of the Republic of Serbia under the Project No. OI 141031. Also acknowledged is a partial support provided by Deutsche Forschungsgemeinschaft (DFG), Bonn.

References

[1] I. S. Aranson and L. Kramer, Rev. Mod. Phys. **74**, 99 (2002); B. A. Malomed, in *Encyclopedia of Nonlinear Science*, p. 157, A. Scott (Ed.), Routledge, New York, 2005.

[2] N. Akhmediev and A. Ankiewicz (Eds.), *Dissipative Solitons: From Optics to Biology and Medicine,* Lect. Notes Phys. **751** (Springer, Berlin, 2008).

[3] N. N. Rosanov, *Spatial Hysteresis and Optical Patterns* (Springer, Berlin, 2002); S. Barland *et al.*, Nature (London) **419**, 699 (2002); Z. Bakonyi, D. Michaelis, U. Peschel, G. Onishchukov, and F. Lederer, J. Opt. Soc. Am. B **19**, 487 (2002); E. A. Ultanir, G. I. Stegeman, D. Michaelis, C. H. Lange, and F. Lederer, Phys. Rev. Lett. **90**, 253903 (2003); P. Mandel and M. Tlidi, J. Opt. B: Quantum Semiclass. Opt. **6**, R60 (2004); N. N. Rosanov, S. V. Fedorov, and A. N. Shatsev, Appl. Phys. B **81**, 937 (2005); C. O. Weiss and Ye. Larionova, Rep. Progr. Phys. **70**, 255 (2007); N. Veretenov and M. Tlidi, Phys. Rev. A **80**, 023822 (2009); P. Genevet, S. Barland, M. Giudici, and J. R. Tredicce, Phys. Rev. Lett. **104**, 223902 (2010).

[4] B. A. Malomed, Physica D **29**, 155 (1987); O. Thual and S. Fauve, J. Phys. (Paris) **49**, 1829 (1988); S. Fauve and O. Thual, Phys. Rev. Lett. **64**, 282 (1990); W. van Saarloos and P. C. Hohenberg, Phys. Rev. Lett. **64**, 749 (1990); V. Hakim, P. Jakobsen, and Y. Pomeau, Europhys. Lett. **11**, 19 (1990); B. A. Malomed and A. A. Nepomnyashchy, Phys. Rev. A **42**, 6009 (1990); P. Marcq, H. Chaté, R. Conte, Physica D **73**, 305 (1994); N. Akhmediev and V. V. Afanasjev, Phys. Rev. Lett. **75**, 2320 (1995); H. R. Brand and R. J. Deissler, Phys. Rev. Lett. **63**, 2801 (1989); R. J. Deissler and H. R. Brand, *ibid.* **72**, 478 (1994); V. V. Afanasjev, N. Akhmediev, and J. M. Soto-Crespo, Phys. Rev. E **53**, 1931 (1996); J. M. Soto-Crespo, N. Akhmediev, and A. Ankiewicz, Phys. Rev. Lett. **85**, 2937 (2000); H. Leblond, A. Komarov, M. Salhi, A. Haboucha, and F. Sanchez, J. Opt. A: Pure Appl. Opt. **8**, 319 (2006); W. H. Renninger, A. Chong, and F. W. Wise, Phys. Rev. A **77**, 023814 (2008); J. M. Soto-Crespo, N. Akhmediev, C. Mejia-Cortes, and N. Devine, Opt. Express **17**, 4236 (2009); D. Mihalache and D. Mazilu, Rom. Rep. Phys. **61**, 175 (2009); C.-K. Lam, B. A. Malomed, K. W. Chow, and P. K. A. Wai, Eur. Phys. J. Special Topics **173**, 233 (2009); C. H. Tsang, B. A. Malomed, C.-K. Lam, and K. W. Chow, Eur. Phys. J. D **59**, 81 (2010); D. Mihalache, J. Optoelectron. Adv. Mat. **12**, 12 (2010); Y. J. He, B. A. Malomed, D. Mihalache, F. W. Ye, and B. B. Hu, J. Opt. Soc. Am. B **27**, 1266 (2010).

[5] B. A. Malomed, D. Mihalache, F. Wise, and L. Torner, J. Opt. B: Quantum Semiclass. Opt. **7**, R53 (2005).

[6] A. S. Desyatnikov, Y. S. Kivshar, and L. Torner, Progr. Opt. **47**, 291 (2005).

[7] W. J. Firth and D. V. Skryabin, Phys. Rev. Lett. **79**, 2450 (1997); L. Torner and D. V. Petrov, Electron. Lett. **33**, 608 (1997); D. V. Petrov, L. Torner, J. Martorell, R. Vilaseca, J. P. Torres, and C. Cojocaru, Opt. Lett. **23**, 1787 (1998).

[8] V. I. Petviashvili and A. M. Sergeev, Dokl. AN SSSR **276**, 1380 (1984) [Sov. Phys. Doklady **29**, 493 (1984)].

[9] L.-C. Crasovan, B. A. Malomed, and D. Mihalache, Phys. Rev. E **63**, 016605 (2001); Phys. Lett. A **289**, 59 (2001).

[10] D. Mihalache, D. Mazilu, F. Lederer, Y. V. Kartashov, L.-C. Crasovan, L. Torner, and B. A. Malomed, Phys. Rev. Lett. **97**, 073904 (2006); D. Mihalache, D. Mazilu, F. Lederer, H. Leblond, and B. A. Malomed, Phys. Rev. A **76**, 045803 (2007); *ibid.* **75**, 033811 (2007).

[11] H. Leblond, B. A. Malomed, and D. Mihalache, Phys. Rev. A **80**, 033835 (2009); D. Mihalache, D. Mazilu, V. Skarka, B. A. Malomed, H. Leblond, N. B. Aleksić, and F. Lederer, *ibid*. 82, 023813 (2010).

[12] A. Szameit, J. Burghoff, T. Pertsch, S. Nolte, and A. Tünnermann, Opt. Exp. **14**, 6055 (2006).

[13] J. W. Fleischer, M. Segev, N. K. Efremidis, and D. N. Christodoulides, Nature **422**, 147 (2003).

[14] H. Leblond, M. Salhi, A. Hideur, T. Chartier, M. Brunel, and F. Sanchez, Phys. Rev. A **65**, 063811 (2002); A. Komarov, H. Leblond, and F. Sanchez, Phys. Rev. E **72**, 025604(R) (2005); M. Salhi, A. Haboucha, H. Leblond, F. Sanchez, Phys. Rev. A **77**, 033828 (2008).

[15] B. B. Baizakov, B. A. Malomed, and M. Salerno, Europhys. Lett. **63**, 642 (2003); T. Mayteevarunyoo, B. A. Malomed, B. B. Baizakov, and M. Salerno, Physica D **238**, 1439 (2009).

[16] J. Yang and Z. H. Musslimani, Opt. Lett. **28**, 2094 (2003); J. Wang and J. Yang, Phys. Rev. A **77**, 033834 (2008).

[17] D. N. Neshev, T. J. Alexander, E. A. Ostrovskaya, Y. S. Kivshar, H. Martin, I. Makasyuk, and Z. Chen, Phys. Rev. Lett. **92**, 123903 (2004); J. W. Fleischer, G. Bartal, O. Cohen, O. Manela, M. Segev, J. Hudock, and D. N. Christodoulides, *ibid*. **92**, 123904 (2004).

[18] R. Driben, B. A. Malomed, A. Gubeskys, and J. Zyss, Phys. Rev. E **76**, 066604 (2007).

[19] R. Driben and B. A. Malomed, Eur. Phys. J. D **50**, 317 (2008).

[20] A. Gubeskys and B. A. Malomed, Phys. Rev. A **76**, 043623 (2007).

[21] H. Sakaguchi and B. A. Malomed, Europhys. Lett. **72**, 698 (2005).

[22] H. Sakaguchi and B. A. Malomed, Phys. Rev. A **79**, 043606 (2009).

[23] T. J. Alexander and L. Bergé, Phys. Rev. E **65**, 026611 (2002); D. Mihalache, D. Mazilu, B. A. Malomed, and F. Lederer, Phys. Rev. A **73**, 043615 (2006); L. D. Carr and C. W. Clark, Phys. Rev. Lett. **97**, 010403 (2006).

[24] V. Skarka and N. B. Aleksić, Phys. Rev. Lett. **96**, 013903 (2006); N. B. Aleksić, V. Skarka, D. V. Timotijević, and D. Gauthier, Phys. Rev. A **75**, 061802 (2007); V. Skarka, D. V. Timotijević, and N. B. Aleksić, J. Opt. A: Pure Appl. Opt. **10**, 075102 (2008).

[25] B. A. Malomed, in: Progr. Optics **43**, 71 (E. Wolf, editor: North Holland, Amsterdam, 2002).

[26] V. Skarka, N. B. Aleksić, M. Derbazi, and V. I. Berezhiani, Phys. Rev. B **81**, 035202 (2010).

[27] H. Sakaguchi, Physica D **210**, 138 (2005); G. Wainblat and B. A. Malomed, *ibid*. **238**, 1143 (2009).

[28] D. Mihalache, D. Mazilu, F. Lederer, H. Leblond, and B. A. Malomed, Phys. Rev. A **81**, 025801 (2010).

In: Optical Amplifiers
Editor: Galina Nemova

ISBN 978-1-61209-835-7

Chapter 9

HIGH-POWER CONTINUOUS-WAVE FIBER OPTICAL PARAMETRIC OSCILLATORS

R. Malik, [*] ***and M. E. Marhic***
School of Engineering, Swansea University,
Swansea, UK

ABSTRACT

Fiber optical parametric amplifiers (OPAs) are based on a highly-efficient four-wave mixing process. Their capability to give very high gain and large bandwidths has made them an attractive candidate for numerous applications. One of them is continuous-wave (CW) fiber optical parametric oscillators (OPOs) using them as a gain medium.

In this chapter we present a novel architecture for CW fiber OPOs, which has allowed us to significantly extend the performance of these devices. To do so we used: (i) a short highly-nonlinear fiber (HNLF) as the parametric gain medium; (ii) a narrowband tunable filter in the fiber feedback loop; (iii) a high coupling fraction from the HNLF into the feedback loop (up to 3 dB). With these new features, we have been able to obtain excellent performance, even with a reduced pump power. With only about 2 W of pump power, we have obtained the following record performance: (i) tuning range of 254 nm; (ii) output power in excess of 1 W at some wavelengths; (iii) external conversion efficiency in excess of 60% at some wavelengths; (iv) linewidth as low as 8 GHz.

This approach can be used for providing narrowband tunable high-power CW radiation over hundreds of nanometers. Such sources could find applications in remote sensing, optical communication, nonlinear optics, etc.

Keywords: Fiber optical parametric amplifier, four-wave mixing.

---

[*] E-mail: R.Malik.394771@swansea.ac.uk

1. INTRODUCTION

The demand for high-power continuous-wave (CW) fiber optic sources has always been high, because they have potential applications in telecommunications, medicine, spectroscopy, etc. Conventional lasers can produce coherent light over different spectral ranges. However, there are wavelength regions where conventional lasers do not work, and the alternative is generation of coherent sources using nonlinear effects with high optical intensities. Optical parametric oscillation is one such nonlinear effect.

Using optical feedback, in principle any optical gain medium can be converted into an oscillator. When the gain is provided by a parametric process (superfluorescence), the result is known as an optical parametric oscillator (OPO). If a fiber optical parametric amplifier (OPA) is used as a gain medium, the oscillator is known as a fiber OPO. Much work has already been done on $\chi^{(2)}$-based OPOs [1-4]. They have already been in the market for a long time now. They can deliver hundreds of milliwats of average power, high conversion efficiency and enormous tuning ranges (from 500 nm to 2000 nm). On the other hand, there has not been much work done on CW fiber OPOs in the past. The reasons are high pump power requirements, and the existence of commercial $\chi^{(2)}$-based OPOs. But fiber OPOs have some added advantages such as all-fiber configuration, which is useful in telecommunication applications, pumping fiber Raman amplifiers, etc.

Fiber OPOs have been investigated in recent years, as they have the potential for providing tunable radiation in regions of the optical spectrum not well covered by the main laser systems. Most of the work has been performed with pulsed pumps, which can provide high peak powers, and hence lead to short OPOs that can be made from relatively lossy fibers, such as microstructured fibers. Powers obtainable with CW pumps are lower, and so larger lengths of low-loss fibers are required. This limits performance compared to pulsed pumps, and as a result few CW fiber OPOs have been investigated [5-10].

Fiber OPAs can provide large gain over very wide range [11]. With the availability of highly-nonlinear fibers (HNLFs), CW fibers OPAs with very high gain (70 dB) have been reported [12]. Also, ultrawide tuning range fiber OPOs have been reported. Using a similarly configured fiber OPA as a gain medium for fiber OPO, very large tuning range and large output powers are possible.

The layout of this chapter is as follows: In Section 2 we introduce fiber OPAs, the gain medium for fiber OPOs. In Section 3 the threshold equations for fiber OPOs are given. In Section 4 we show the architecture for fiber OPOs, and in Section 5 we present and discuss the results obtained. In Section 6 fiber OPO centred at 1593 nm is described. In Section 7 we conclude the chapter.

2. FIBER OPA AS GAIN MEDIUM

Optical parametric amplification (OPA) in fibers is based on the four-wave mixing (FWM) process. Availability of new high output power light sources, and commercialization of optical fibers with nonlinearity coefficient almost 10 times higher than for conventional fibers, has increased the interest in fiber optical parametric amplifiers. Unlike other types of optical amplifiers, they can provide amplification over large bandwidths. For instance,

Erbium-doped fiber amplifiers (EDFAs) have gain bandwidth of only about 30 nm, while fiber OPAs can give amplification over hundreds of nanometres.

2.1. Four Wave Mixing and Gain Equations for One-pump Fiber OPA

For the single-pump fiber OPA, we need to consider the degenerate four-wave mixing process. If ω_p, ω_s, and ω_i are respectively the radian frequencies of pump, signal and idler, we have the equation between frequencies

$$2\omega_p = \omega_s + \omega_{i.} \tag{1}$$

The linear phase matching requirement for FWM is

$$\Delta\beta = 0;\ \text{where}\ \Delta\beta = \beta_s + \beta_i - 2\beta_p, \tag{2}$$

where β is the propagation constant, which is a function of frequency.

Taking the nonlinear phase shift into account, the total phase mismatch term becomes

$$\kappa = \Delta\beta + 2\gamma P_0 \tag{3}$$

where: γ is the nonlinearity coefficient of the fiber, given by the expression $\gamma = \frac{2\pi n_2}{\lambda_p A_{eff}}$ $n2$ is the change in the refractive index due to a high-intensity wave propagating in the medium; λ_p is the is the pump wavelength; A_{eff} is the effective core area of the fiber; P_0 is the incident pump power. The small-signal parametric gain expression can be found as in Ref. [13]

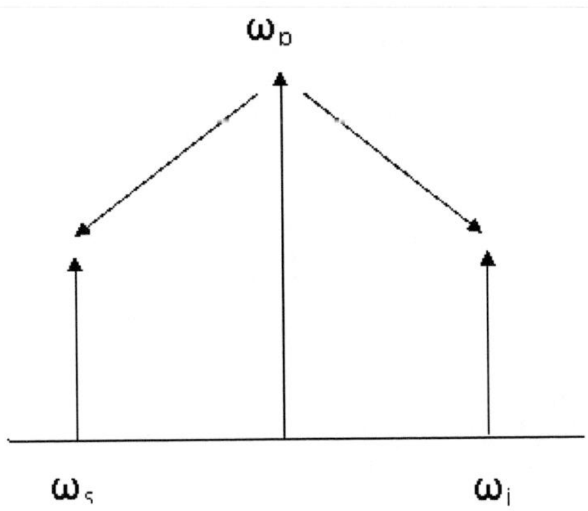

Figure 1. Frequency assignments for FWM in fibers.

$$G_s(L) = \frac{|E_s(L)|^2}{|E_s(0)|^2} = 1 + \left[\frac{\gamma P_0}{g}\sinh(gL)\right]^2 \tag{4}$$

where, $E_s(z)$ is the amplitude of the electric field envelope at the distance z from the input,

$$g = \sqrt{(\gamma P)^2 - (\kappa/2)^2}\,, \tag{5}$$

and L is the fiber length.

Using a Taylor series expansion around the pump frequency, Eq. (2) can be written as:

$$\Delta\beta = 2\sum_{m=1}^{\infty} \frac{\beta_{2m}}{2m!}(\omega_s - \omega_p)^{2m} \tag{6}$$

where β_{2m} denotes the $2m^{th}$ derivative of β at ω_p.

It is clear that by tailoring the fiber dispersion properties, and choosing the right pump wavelength, fiber OPA gain bandwidth and spectrum shape can be altered. Also, by looking at Eq. (4), it is clear that $\Delta\beta$ is only a function of even powers of (ω_s-ω_p). By combining Eq.(4) with Eq. (3), we see that the gain spectrum is independent of the odd orders of the dispersion: only even orders of dispersion contribute in the gain spectrum of fiber OPAs.

Considering the perfect phase matching condition ($\kappa = 0$), the gain for the fiber OPA can be written as:

$$G = \cosh^2(\gamma PL) \tag{7}$$

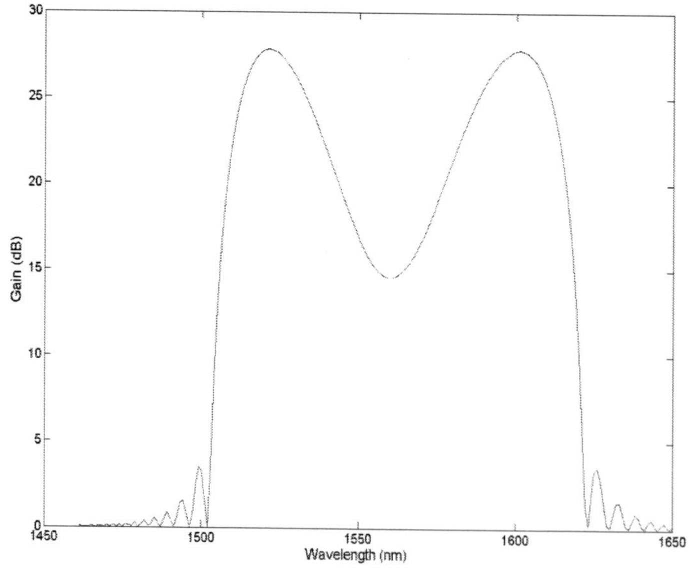

Figure 2. Simulated OPA spectrum.

Figure 2 depicts a simulated fiber OPA gain spectrum. The parameters considered for the simulations were as follows: L = 350 m; P0 = 1 W: fiber ZDW 1560 nm; λp= 1561 nm; □ = 15 W-1km-1.

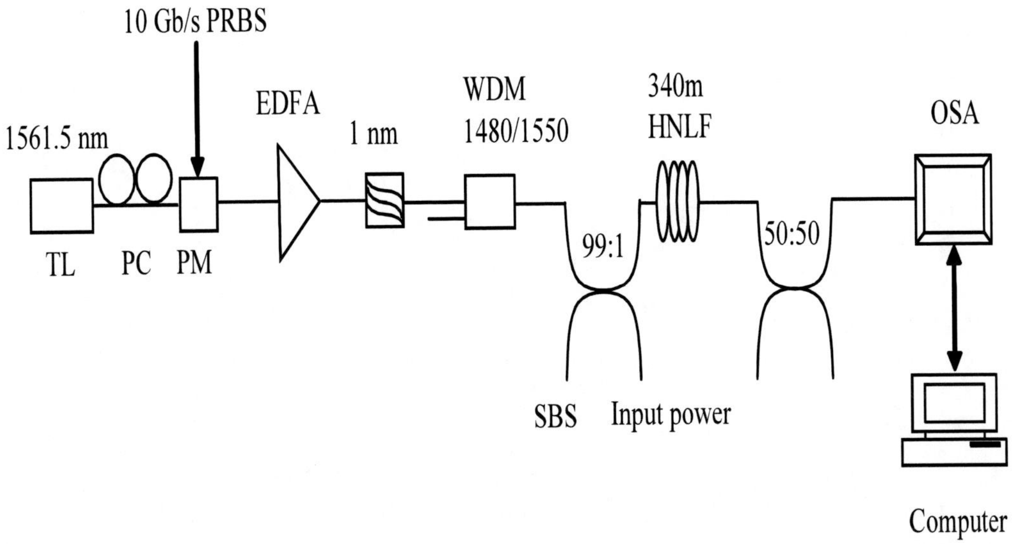

Figure 3. Experimental set-up for the fiber OPA.

2.2. Experimental Set-up for Fiber OPA and Gain Spectra

Figure 3 depicts the experimental set-up for the fiber OPA. A tunable laser (TL) serves as the pump at 1561.5 nm. It passes through a polarization controller (PC) to align the state of polarization (SOP) with the phase modulator (PM). PM is driven by a 10 Gb/s pseudo-random bits sequence (PRBS) to suppress the stimulated Brillouin scattering (SBS). Then it passes through a 3-W Erbium-doped fiber amplifier to boost its power to a high level. It then passes through the 1-nm bandwidth tunable bandpass filter (TBPF), to suppress the amplified spontaneous emission (ASE). This is then followed by a 1480/1550 nm wavelength division multiplexer (WDM), which has very low insertion loss (less than 1 dB) at the pump wavelength; the other port (1480 nm) is used to launch the signal into the system. A 20-dB coupler was used for the purpose of keeping track of input power and back-reflected light. Then the pump passes through an HNLF, with L = 340 m. □ is about 15 $W^{-1}km^{-1}$, and the ZDW is 1560 nm. Then the output is attenuated, and passes to the optical spectrum analyzer (OSA).

Figure 4 shows the fiber OPA spectrum obtained experimentally P_0 = 2 W. It shows a gain bandwidth in excess of 200 nm. The idler asymmetry is due to the Raman gain contribution. The latter is due to both the real and imaginary parts of the third-order susceptibility; it is less pronounced at lower pump power.

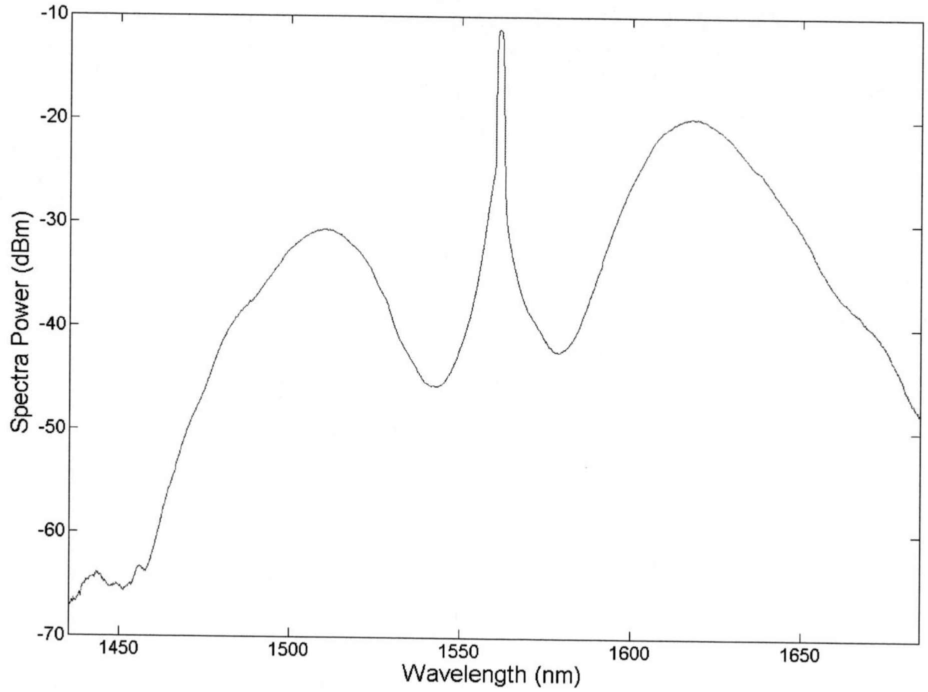

Figure 4. Experimental fiber OPA spectrum observed with 1561.5 nm pump wavelength.

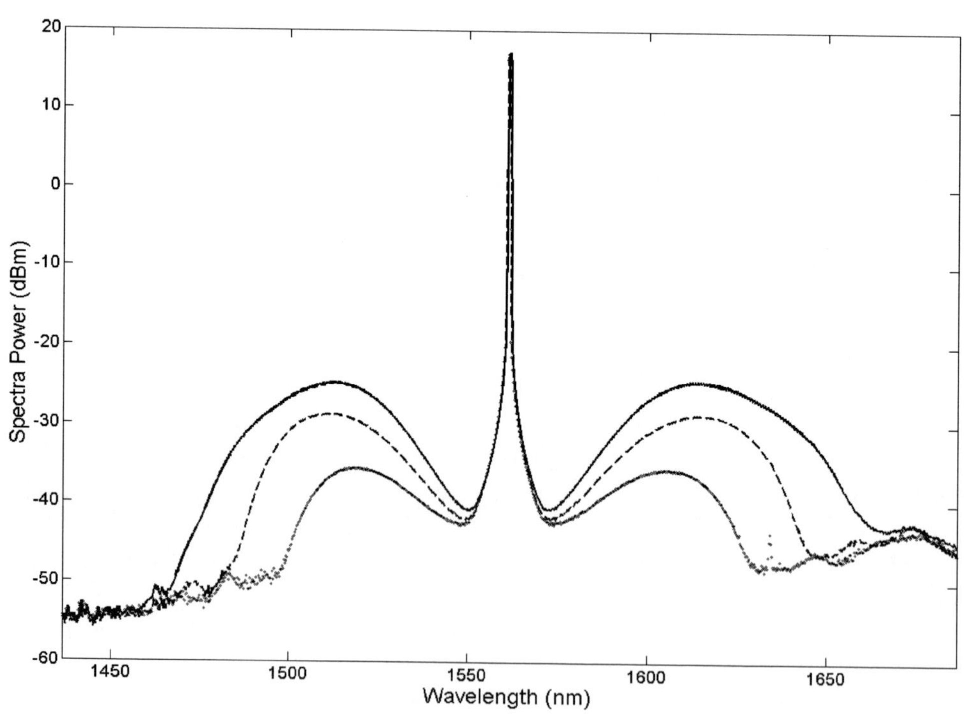

Figure 5. Fiber OPA output spectra corresponding to different pump wavelengths.

2.3. Fiber OPA Bandwidth vs Pump Wavelength

Figure 5 shows the experimental fiber OPA output spectra P_0 = 30 dBm, for different pump wavelengths around the ZDW. Slight changes lead to prominent changes in the shape and bandwidth of the fiber OPA spectra. Corresponding to the pump wavelength of 1561.5 nm, the bandwidth and the gain are optimum. This spectrum is shown in solid line.

3. Threshold Pump Power for Fiber OPO

The combined effects of Raman and parametric gain were presented in Ref. [14]. The combined gain can be written as:

$$G = \left| \cosh(\gamma R P_0 L) \pm \frac{i(K-q)}{R} \sinh(\gamma P_0 R L) \right|^2 \quad (8)$$

where: $R = \sqrt{K(2q-K)}$; K is the phase mismatch wave vector; from Eq (3), relation between K and κ can be written as $K = \kappa / \gamma P$; $q(\Omega) \equiv 1 - f + f\chi_R^{(3)}(-\Omega)$; f is the fractional strength of the Raman contribution; $\chi_R^{(3)}$ is the Raman susceptibility, which includes both real and imaginary parts; Ω is the radian frequency detuning from the pump. The Raman contribution gives asymmetry to the gain curve. For optimal phase matching, K = 1, and using a binomial expansion, $\sqrt{2q-1} = q$, leading to $R = q$. In the high gain limit, $\gamma P_0 L >> 1$, and Eq. (8) reduces to:

$$G = \frac{1}{4}\left| 1 \pm \frac{i(1-q)}{q} \right|^2 \exp(2\gamma P_0 \operatorname{Re}(q) L) . \quad (9)$$

Introducing the transmissivity α of the cavity, the threshold condition becomes,

$$\alpha G = 1 . \quad (10)$$

For our cavity α = 0.1 near 1480 nm, and so more than 10 dB of gain is required for oscillation to take place. According to Ref. [9], the thresholds become similar for anti-Stokes and Stokes waves if the transmissivity (and Q-factor) of the cavity is small. Because of the Raman contribution the gain was higher on the Stokes side, but since our cavity has a low Q-factor, thresholds for both Stokes and anti-Stokes sides can be assumed to be similar, and so we resonated the cavity on the anti-Stokes side.

4. OPO ARCHITECTURE

The experimental setup is shown in Figure 6. It is similar to that of Figure 3, with the addition of components to introduce the feedback loop. A wideband WDM coupler (1480/1550) was used to couple with low loss pump and intracavity signal into a 340-m long HNLF.

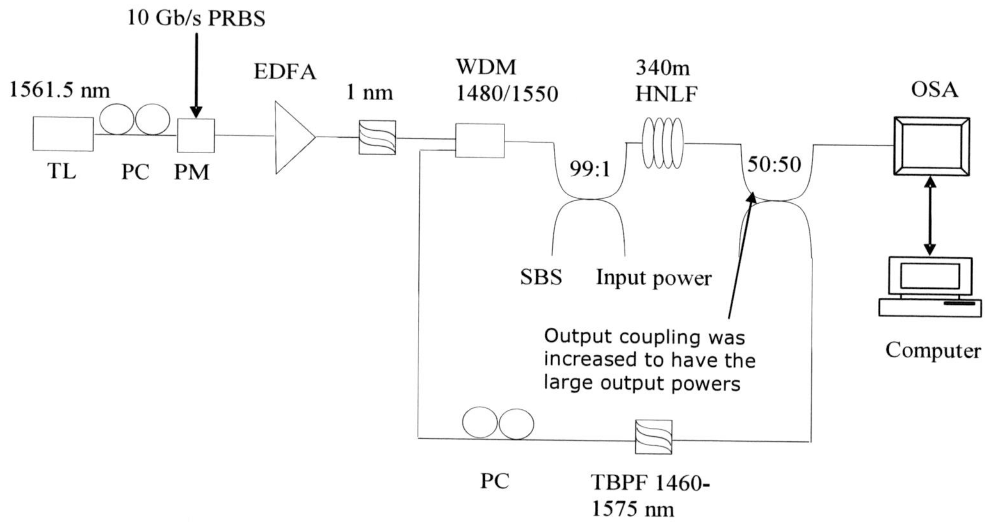

Figure 6. Experimental set-up for the fiber OPO.

After the WDM coupler a 20-dB coupler was used to measure the SBS power, and the input power to the HNLF, which was 33 dBm. After the HNLF a 3-dB coupler was used to couple half of the power out of the resonator, and to couple the other half back into it. Then in the resonator a narrow TBPF was used, which had an insertion loss of 5 dB, and a tuning range extending from 1460 to 1575 nm, i.e. primarily to the short-wavelength side of the pump. After TBPF, we used a PC in the cavity to maximize the output powers on the idler side. Total cavity loss was measured to be 9.8 dB.

5. OUTPUT SPECTRUM AND OUTPUT POWERS

To get the OPO to oscillate, the OPA open-loop gain for the signal must be higher than the intracavity losses. With the pump power injected into the OPA the maximum parametric gain was of the order of 50 dB, and so oscillation could readily be obtained upon closing the loop.

Figure 7 shows OPO output spectra obtained for various settings of the intracavity tunable filter.

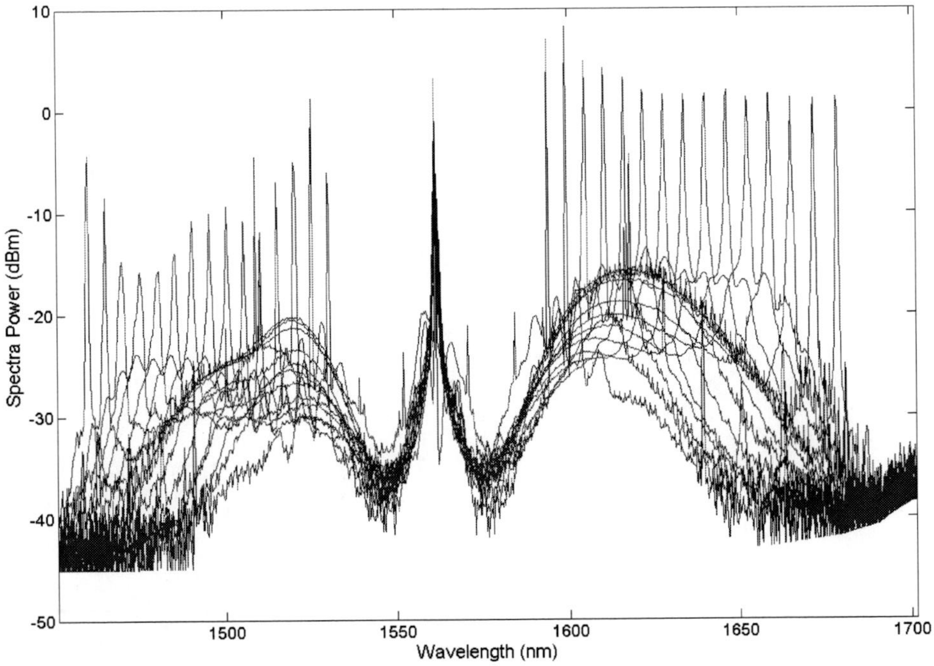

Figure 7. Fiber OPO spectra obtained by selecting different wavelengths by TBPF.

This shows that OPO lasing could be obtained from 1463 to 1674 nm, which corresponds to a 211-nm wide tuning range. The peaks on the anti-Stokes side of the pump correspond to light passing through the filter, which we call the signal; the peaks on the Stokes side are the corresponding idlers. The peak output power variation with respect to signal wavelength is shown in Figure 9. This shows an output power in the vicinity of 1 W from 1600-1670 nm. Peak powers on the Stokes side are as much as 7.1 dB higher than the pump output power; this indicates strong pump depletion, and hence high external conversion efficiency (ECE).

5.1. OPO Spectra with Variable Coupler to Level the Idler Spectra

As we saw the Raman contribution at high pump powers renders the parametric gain spectrum asymmetric. For this reason, the peak OPO outputs on both sides are unequal. This is a potentially undesirable feature of this OPO. To counter this problem, we used variable output coupling. Figure 8 shows the OPO spectrum achieved with variable output coupling.

It can be seen that the amplitudes of the idler peaks are much more even this way. But the only variable output coupler that we had in our laboratory had an insertion loss varying in a nonlinear manner in terms of wavelength; it also had some insertion loss (1.5 dB). So we did not pursue this technique further. But this data shows that if a variable coupler is used, a flatter output spectrum can be achieved.

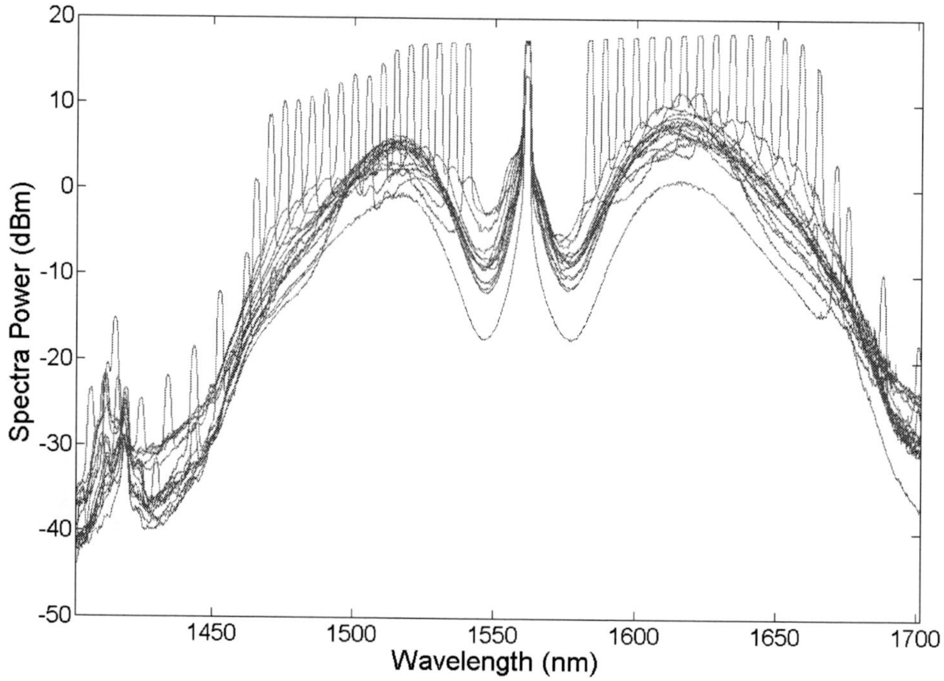

Figure 8. Fiber OPO spectra obtained using variable output coupling.

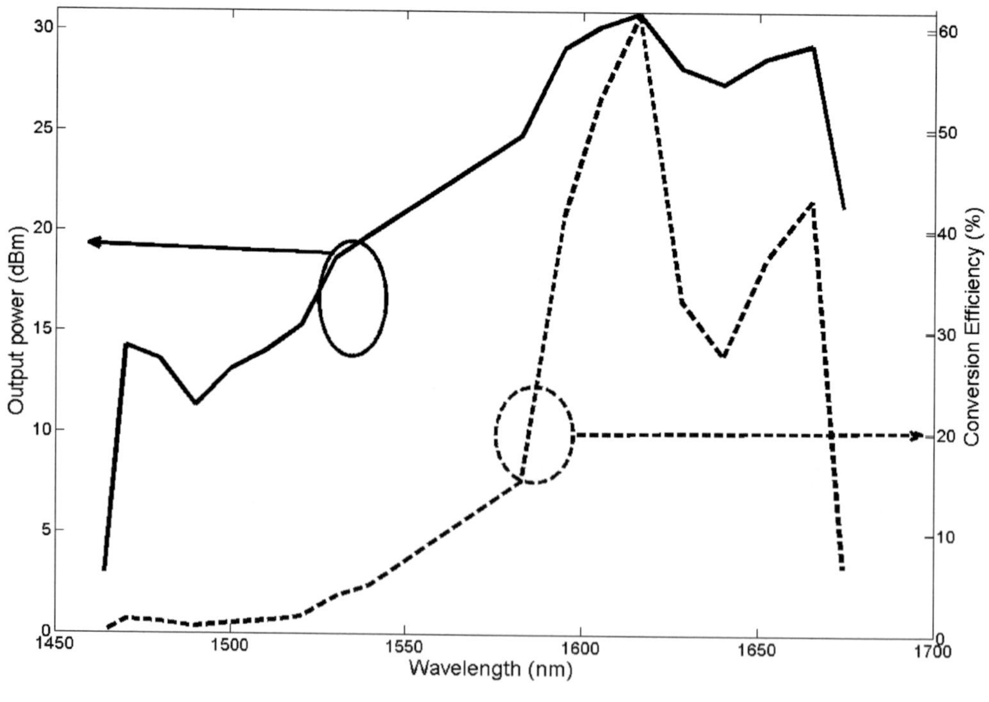

Figure 9. Output power and external conversion efficiency vs wavelength.

5.2. MEASUREMENT OF OUTPUT POWER AND EXTERNAL CONVERSION EFFICIENCY

We define the external conversion efficiency (ECE) as the ratio of output power and pump power launched into the HNLF. Figure 9 shows the ECE variation with wavelength, which reaches as high as 61 %. This is much higher than achieved in previous work [9]: the output ECE in the most recent work, Ref. [6], was only about 2.5 %. This is to our knowledge the best ECE achieved to date over such a large tuning range. So, we achieved a tuning range comparable to the maximum achieved so far [9], but with higher ECE and output power.

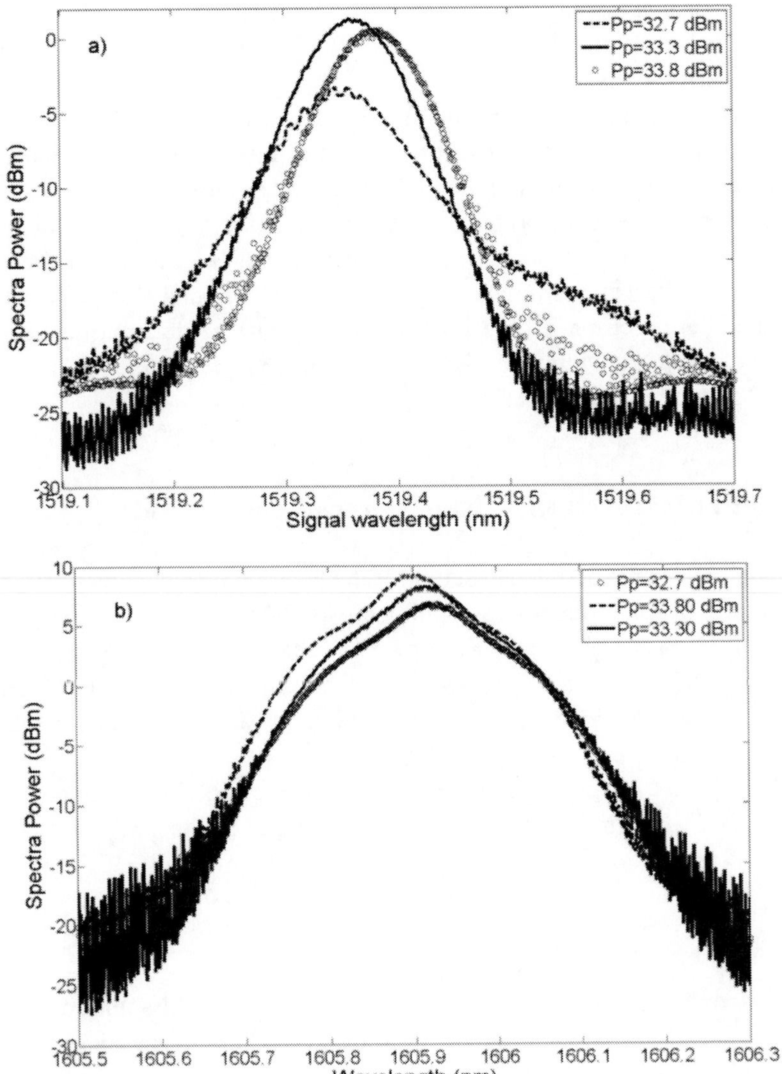

Figure 10. (a) Signal output spectra at different pump wavelengths; (b) Idler output spectra at different pump wavelengths.

5.3. Linewidth Measurements with a High Resolution OSA

Our OPO output spectra are narrower than in [9], as we used a narrower intracavity filter (0.25 nm FWHM). Figure 10 shows typical signal and idler output spectra, obtained for several values of the pumpower near its maximum.

We note that the idler peak power increases monotonically with the pump power, but that the signal power is maximum at an intermediate pump power. This may be due to the fact that the Raman influence becomes even more prominent at higher power. This introduces further losses on the anti-Stokes side.

Another difference is that signal and idler have rather different spectral shapes. Looking at the 3-dB bandwidths, we see that the signal (idler) has about a 0.1 (0.2) nm linewidth. For these measurements, the resolution of the OSA was kept at its highest value (0.06 nm).

Actually, considering the fact that the OSA has 0.06 nm resolution, we see that the signal is considerably narrower than the idler.

In fact we have also measured the signal spectrum by heterodyne with a narrow-linewidth laser, and we found it to be about 10 GHz (0.08 nm), much narrower than the tunable filter itself (the electrical spectrum analyzer had a 250 ms sweep time).

By contrast, even after deconvolving the OSA resolution, the accurate idler linewidth was found to be of the order 0.15 nm. The difference in the linewidths can be understood as follows. The signal spectrum is forced to fit within the filter bandwidth, and so is necessarily narrow; but the idler spectrum is broadened during the parametric amplification, and acquires about twice the pump linewidth. The 20-dB linewidths are considerably larger, being of the order of 0.3 (0.5) nm for the signal (idler).

5.4. Linewidth Measurements Using Heterodyne Method

Figure 11 shows the experimental setup for the heterodyne technique to measure the OPO output linewidth. The OPO ouput was combined with the output of a narrow-linewidth (a few MHz) tunable laser (TL). The sum of the two was then detected by a photodiode (PD). TL was scanned to find the beat signal between signal and idler. Its output then went to an electrical spectrum analyser (ESA). We used the fastest available sweep time of the ESA (250 ms).

Figure 12 depicts the signal spectrum observed on the ESA. Signal linewidth is measured to be about 8 GHz, which matches well with that measured on the OSA, using deconvolution.

5.5. Discussion

We have shown that by using a 3-dB output coupler for extracting power from a fiber OPO, we can btain over 1 watt of output power and 61 % external conversion efficiency at long wavelengths (1600 - 1670 nm), together with a tuning range over 200 nm. The 3-dB signal linewidth is as small as 0.08 nm. This type of light source is uncommon at long wavelengths, and it could find many applications, for instance: pumps for Raman amplifiers; large seed signals for Raman amplification; optical communication; gas detection, etc.

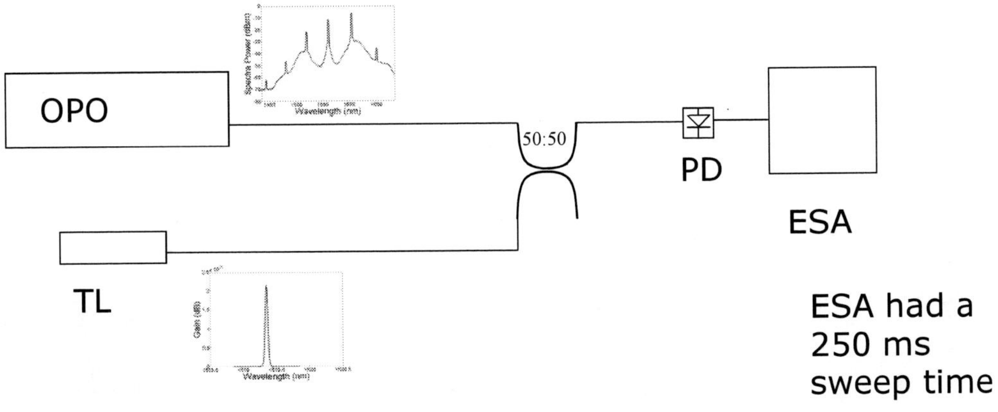

Figure 11. Heterodyne setup to measure the linewidth of the OPO ouput.

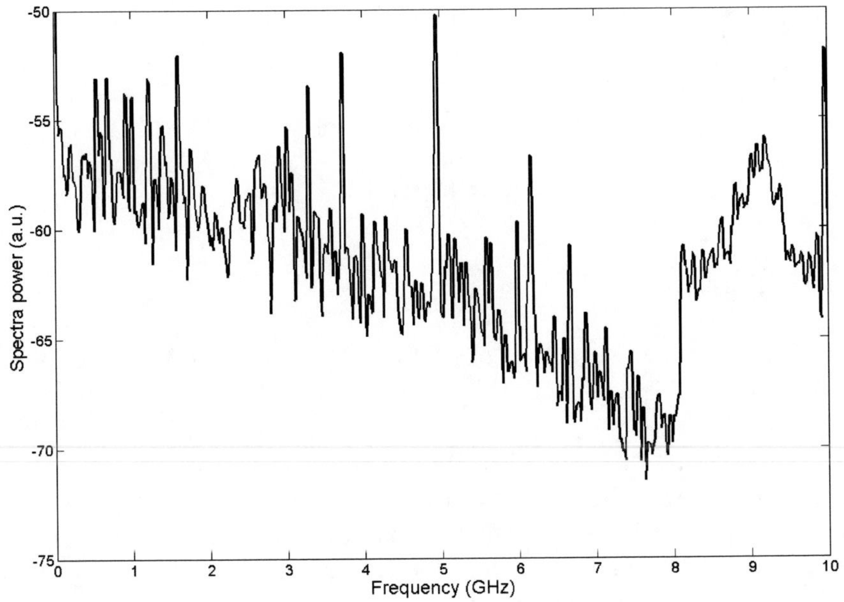

Figure 12. Signal spectrum observed at the ESA.

6. CW Fiber OPO Centred at 1593 nm

To exploit the fact that the OPA bandwidth depends on the dispersion parameters and ZDW of the fiber used, we then used a fiber with a ZDW near 1600 nm to build a second fiber OPO. ZDWL closer to 1600 nm gave us gain beyond 1700 nm. When we reduced various losses, such as that due to output coupling, record tuning range (254 nm) was achieved. Also, when the output coupling was changed to 50%, we obtained very large output powers, and very high ECEs.

The summary of the chapter is as follows: Section 6.1 states the gain medium for this work. Section 6.2 describes the experimental setup. Section 6.3 contains all the results.

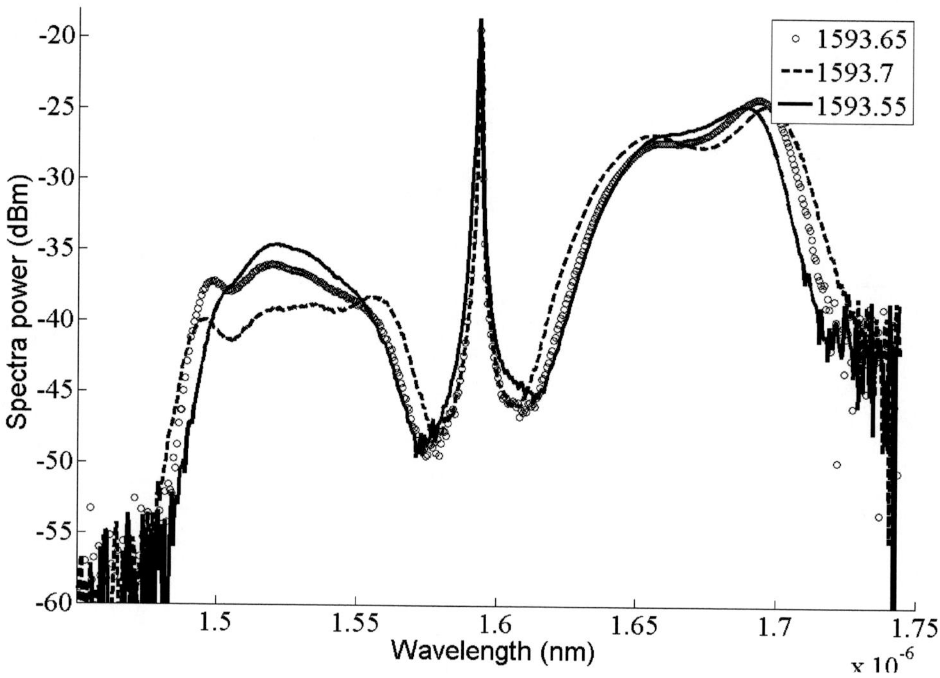

Figure 13. OPA spectra at different pump wavelengths.

6.1. Gain Medium

For this fiber OPO, we used 520 m of HNLF, with zero dispersion wavelength at 1592.9 nm, and nonlinear coefficient of about 15 $W^{-1}km^{-1}$. We used 3 watts of CW pump power,

Figure 13 shows the different fiber OPA ASE spectra obtained with different pump wavelengths (recorded with the feedback loop open). The circles correspond to the 1593.65 nm pump wavelength, which gives optimum bandwidth and gain to obtain tunability over more than 250 nm. The asymmetry in the OPA ASE spectra can be attributed to the Raman gain contribution. In order to make the OPO oscillate the gain should be higher than the losses in the ring. The lowest wavelength we could lase on was 1476 nm which gave us an idler at 1730 nm, giving us a total tuning range of 254 nm. (The OPA gain was measured by amplifying a tunable narrowband signal laser.)

6.2. OPO Architecture

The experimental setup is shown in Figure 14. A tunable laser source at 1593.65 nm was used as pump. The pump light was then passed through a PC, and PM to be phase-modulated by a 10 Gb/s PRBS, to suppress SBS. After this a 10-W L-band EDFA was used in order to boost the pump power. It was followed by a 1-nm bandwidth filter to remove the EDFA ASE.

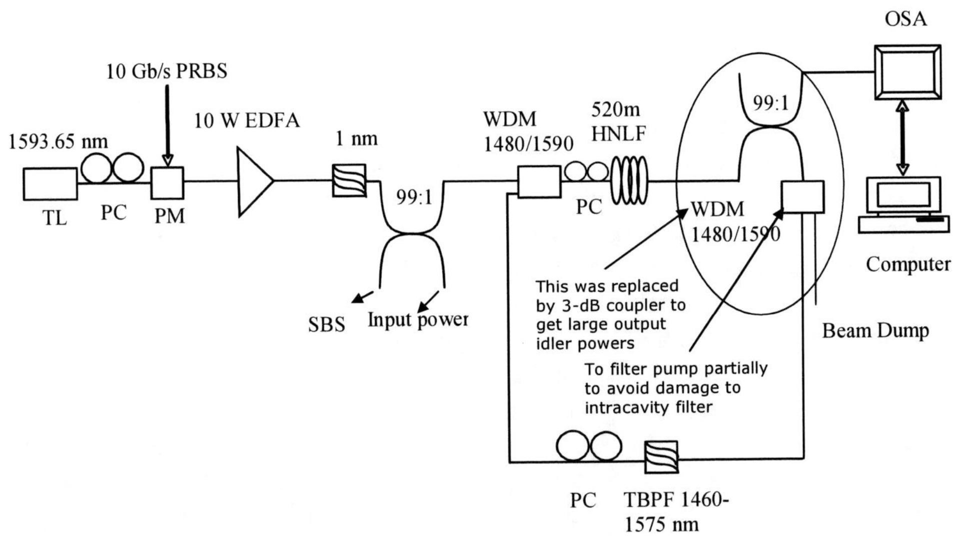

Figure 14. Experimental set-up for fiber OPO.

After this a 20-dB coupler was used to measure the SBS power and the input power to the HNLF, which was about 3 W (35 dBm). The SBS-reflected pump light level was up to 18 dBm. Then a wideband WDM coupler (1480 nm/1590 nm) was used to couple with low loss pump and intracavity signal into a 520-m HNLF. Before the HNLF, we used another PC to obtain different shapes for the OPA gain spectra, in order to maximize the gain at the desired wavelengths. The HNLF had the following parameters: nonlinearity coefficient γ = 15 W^{-1} km^{-1}; ZDWL = 1592.9 nm; fourth-order dispersion coefficient β_4 = 3×10^{-5} ps^4 km^{-1}; dispersion slope $D_\lambda = 0.025$ ps nm^{-2} km^{-1}.

The OPO output coupler was placed after the HNLF. We initially used a 20-dB coupler in order to minimize the loop loss, so as to maximize the OPA gain bandwidth, and therefore the tuning range. (Later on we replaced the output coupler by a 3-dB coupler in order to increase the output power, at the expense of a slightly reduced tuning range). The 20-dB coupler coupled 1% of the power out of the resonator, and the other 99% was coupled back into it. We then inserted a 1480 nm/1590 nm WDM coupler to partially remove the pump power and prevent damage to the intracavity filter. In the resonator a narrowband TBPF was used, which had an insertion loss of 5 dB, and a tuning range extending from 1460 to 1575 nm. The TBPF had a 3-dB bandwidth of 0.3 nm, and a 20-dB bandwidth of 0.8 nm. It was followed by another PC to match the SOPs of the signal and the pump.

Total cavity loss was measured to be 9.8 dB in the 1480-nm signal wavelength region. The loss in the cavity was kept lowest near 1480 nm by using the WDM couplers, as the OPA gain was lowest near 1480 nm. But the loss in the cavity increased as the signal approached the pump, due to the WDM couplers. The main purpose of using a 1480 nm/1590 nm WDM coupler after the 20 dB coupler was to partially filter out the pump, so the power incident on the intracavity filter was below 1 W (damage threshold power). This increased the intracavity losses as the signal approached the pump, but the OPO still oscillated, as the gain in this region was quite high. The 1% power coupled out of the ring was fed into an OSA, which was connected to the computer by GPIB.

6.3. Results

Figure 15 shows the fiber OPO spectra obtained for different settings of the intracavity filter. For each filter setting, there are two peaks: at the filter wavelength (signal), and at its symmetric with respect to the pump (idler). The right-hand side (idler) peaks have larger amplitudes, and this can be explained by the asymmetry of the fiber OPA ASE spectra due to the Raman contribution. The largest signal (filter) wavelength for which oscillation was obtained is 1575 nm. Oscillation could in principle be obtained closer to the pump, but our intracavity filter could not go beyond 1575 nm.

To obtain large output powers, we then replaced the combination of 20 dB and 1480 nm/1590 nm WDM couplers by a 3 dB coupler. This increased the output powers, but it reduced the tuning range by a margin of 12 nm, as this slightly increased the loop losses.

Figure 16 shows the resulting OPO spectra with a fixed input pump power of 34 dBm. The amplitudes of the peaks are saturated from 1670 nm to 1710 nm. This is because the gain is very high and intracavity loss is comparatively quite low. The idler peaks between 1670 nm and 1593 nm are not saturated because the cavity loss has increased due the wavelength-dependent loss of the WDM couplers. Saturation effects in fiber OPAs have been observed [15]. The saturated gain can be written as [15]

$$G_{Ps} = \frac{G_0(\omega_s)}{1 + \frac{P_s}{P_{sat}}} \tag{11}$$

where P_{sat} is the saturation power defined as the input signal power required to reduce the gain by 3 dB. G_0 is the low-power gain.

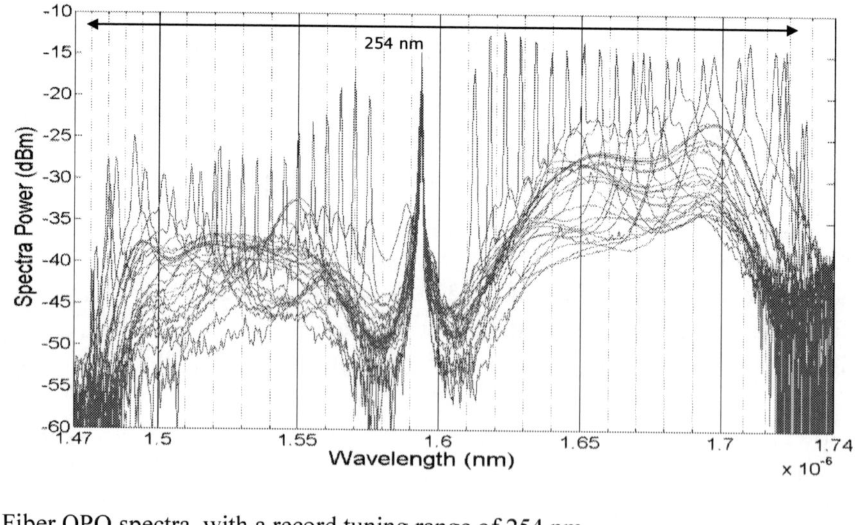

Figure 15. Fiber OPO spectra, with a record tuning range of 254 nm.

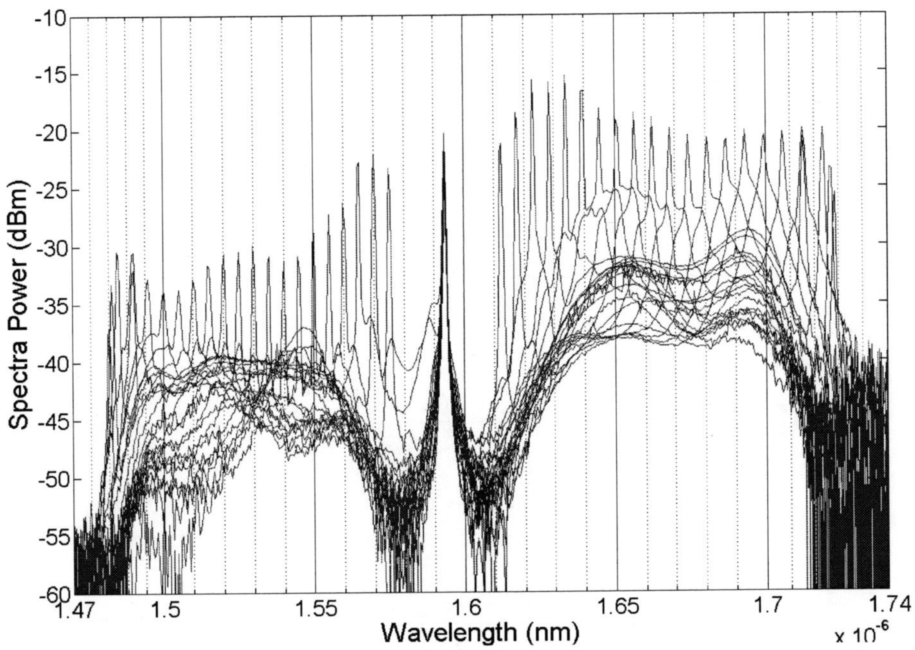

Figure 16. Fiber OPO spectra with 50% output coupling.

To avoid saturation, pump power should be adjusted. To maximize the output idler peaks, pump power was adjusted at every wavelength. This optimized pump power follows according to the losses in the cavity.

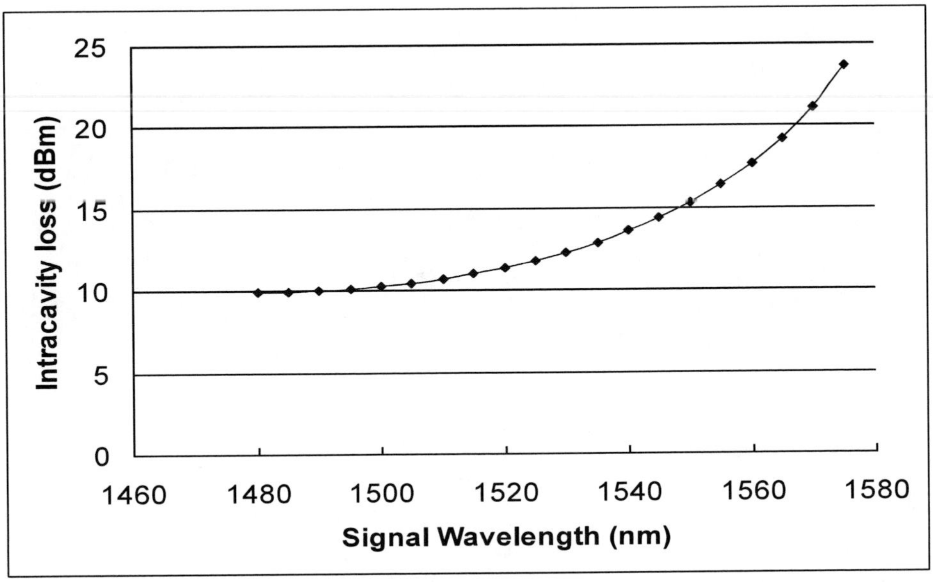

Figure 17. Intracavity losses versus the signal wavelength.

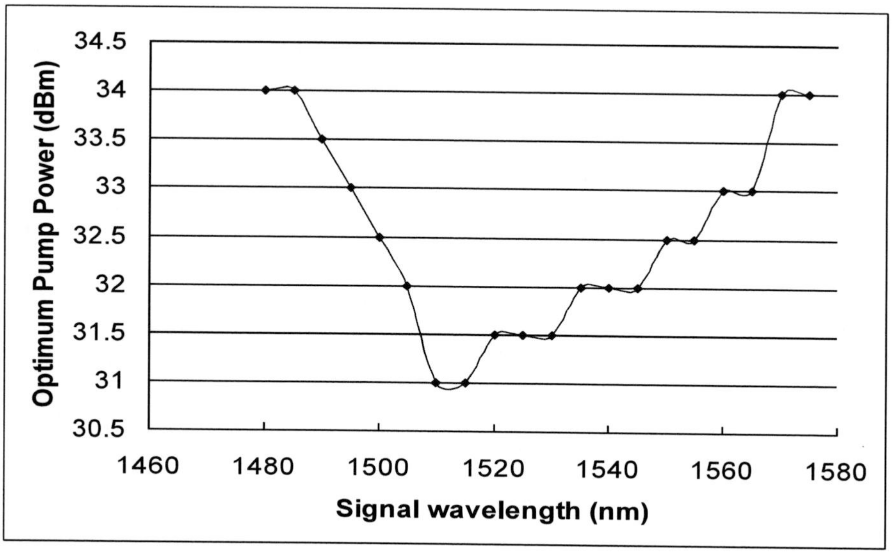

Figure 18. Optimized pump power used versus the signal wavelength.

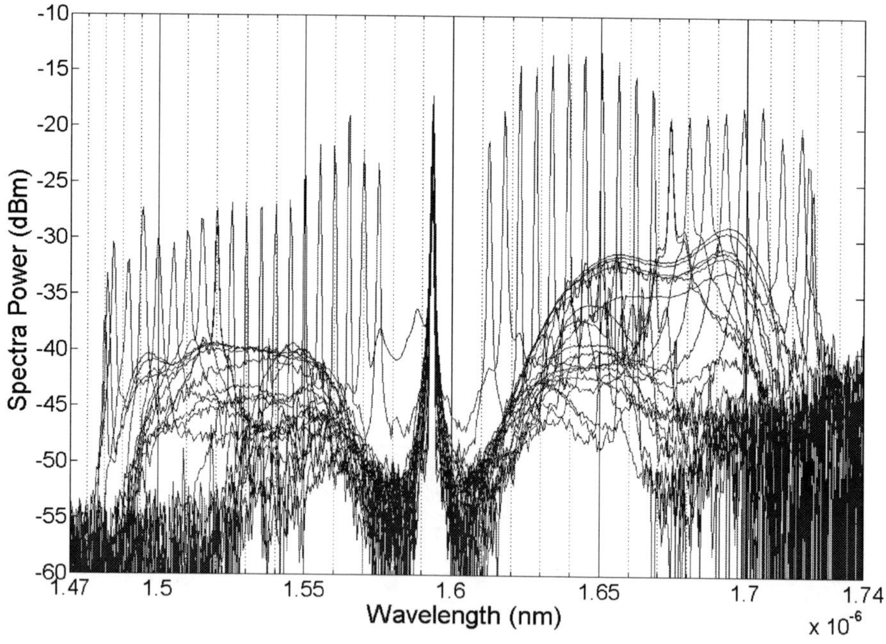

Figure 19. Fiber OPO spectra obtained versus the variable pump power.

Figure 17 plots the intracavity losses versus signal wavelength selected by the intracavity filter. To mitigate the saturation effects, pump power was optimized for each signal wavelength to obtain high signal and idler amplitudes. Figure 18 shows the corresponding plot of optimized pump power versus signal wavelength. The step for pump power optimization was taken as 0.5 dB.

Figure 19 shows the OPO spectra when the pump power was optimized for each setting of the intracavity filter. As can be seen from the comparison between figure16 and Figure 19, peaks in Figure 4 and 19 from 1670 nm to 1710 nm are higher by 2-3 dB, which shows the benefit of the pump power optimization.

Figure 20 shows the plot of output power versus wavelength. Output powers are higher than 15 dBm from 1550 nm to 1720 nm, and higher than 20 dBm from 1615 nm to 1720 nm. (There was 40 dB attenuation between the OPO output and the OSA, in order to avoid damage to the latter.)

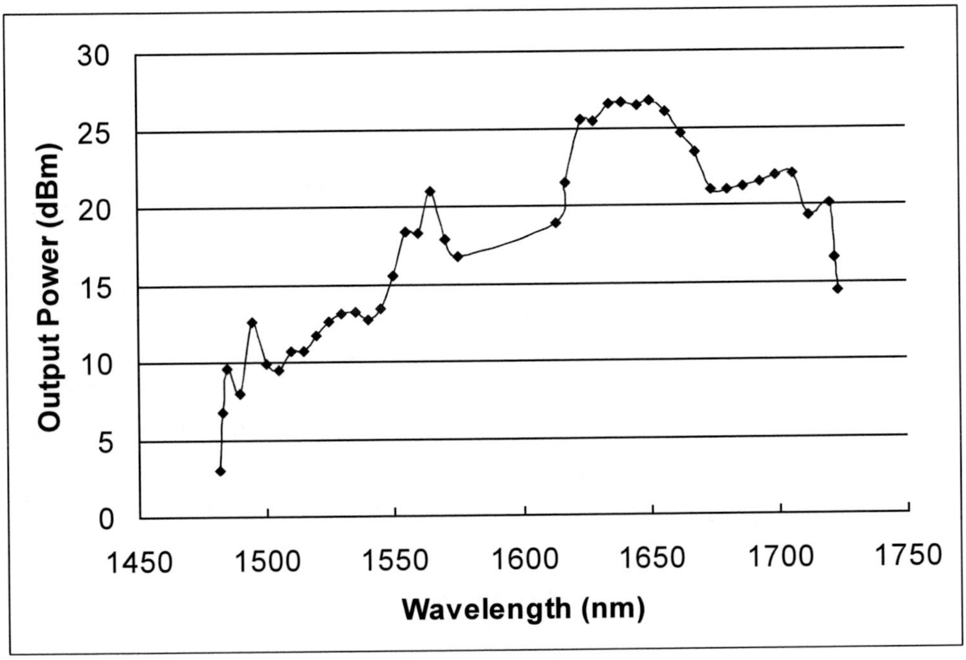

Figure 20. Output powers obtained versus the signal wavelength selected by the TBPF, and its corresponding generated idler wavelength.

Conclusion

We have investigated fiber OPOs as potential high power CW sources mainly in the L- and U-band. We exploited the large bandwidth and gain of fiber OPAs to make OPOs with high output power. First we used a fiber OPO with pump wavelength around 1561 nm. This gave us 211 nm of tuning range, and watt-level output power from 1600 nm to 1670 nm. We then measured the OPO output linewidth, with a highest-resolution OSA, and deconvolving its resolution. We also measured the linewidth by heterodyne detection. By both methods, signal and idler linewidths were measured to be 0.08 nm and 0.15 nm respectively.

We then switched to a different HNLF with a longer ZDWL. With it we obtained the largest tuning range for a CW fiber OPO that has been achieved so far, to the best of our knowledge. Also, when the output coupling fraction was increased, we obtained large output powers (20-25 dBm) from 1610 nm to 1720 nm. Laser sources with this much output power and in this wavelength region are uncommon. Such sources could potentially find applications in remote sensing, Raman pumping or seeding, and other nonlinear applications.

ACKNOWLEDGMNENT

We thank EPSRC for providing research grant and Sumitomo Electric Industries for providing the HNLF.

REFERENCES

[1] J. A. Giordmaine and R. C. Miller, "Tunable coherent parametric oscillation in $LiNbO_3$ at optical frequencies", *Phys. Rev. Lett.* vol. 14, p. 973, 1965.

[2] L. E. Myers, R. C. Eckardt, M. M. Fejer, R. L. Byer, W. R. Bosenberg, and J. W. Pierce, "Quasi-phase-matched optical parametric oscillators in bulk periodically poled $LiNbO_3$," *J. Opt. Soc. Am. B* vol. 12, p. 2102, 1995.

[3] P. E. Powers, T. J. Kulp, and S. E. Bisson, "Continuous tuning of a continuous-wave periodically poled lithium niobate optical parametric oscillator by use of a fan-out grating design," *Opt. Lett.*, vol. 23, p. 159, 1998.

[4] R. C. Eckardt, C. D. Nabors, W.J. Kozlovsky, and R.L. Byer, "Optical parametric oscillator frequency tuning and control," *J. Opt. Soc. Am.B,* vol. 8, p. 646, 1991.

[5] J. E. Sharping, "Microstructure Fiber Based Optical Parametric Oscillators," *J. Lightwave Technol.*,vol. 26, p. 2184, 2008.

[6] M. E. Marhic, K. K.-Y. Wong, L. G. Kazovsky, and T.-E. Tsai, "Continuous-wave fiber optical parametric oscillator," *Opt. Lett.,* vol. 27, p. 1439, 2002.

[7] C. J. S. de Matos, J. R. Taylor, and K. P. Hansen, "Continuous-wave, totally fiber integrated optical parametric oscillator using holey fiber," *Opt. Lett.*, vol. 29, p. 983, 2004.

[8] Z. Luo, W.D. Zhong, M.Tang, Z. Cai, C. Ye, and X. Xiao, "Fiber-optic parametric amplifier and oscillator based on intracavity parametric pump technique," *Opt. Lett.*, vol. 34, p. 214, 2009.

[9] Y. Q. Xu, S. G. Murdoch, R. Leonhardt, and J. D. Harvey, "Raman-assisted continuous-wave tunable all-fiber optical parametric oscillator," *J. Opt. Soc. Am. B* vol. 26, p. 1351, 2009.

[10] R. Malik and M. E.Marhic, "Tunable Continuous-Wave Fiber Optical Parametric Oscillator with 1 W Output Power," *In Optical Fiber Communication (OFC)/OSA' 2010*, paper JWA18 (March 21-25, San Diego, CA, USA).

[11] M. E. Marhic, "Fiber Optical Parametric Amplifiers, Oscillators and Related Devices", Cambridge, 2008.

[12] T. Torounidis, P. A. Andrekson, and B.-E. Olsson, "Fiber optical parametric amplifier with 70-dB gain," *IEEE Photon. Technol. Lett.*, vol.18, p. 1194, 2006.

[13] G. P Agrawal, "Nonlinear Fiber Optics", fourth edition, Academic Press, San Diego, USA, 2007.

[14] A. S. Y. Hsieh, G. K. L. Wong, S. G. Murdoch, S. Coen, F. Vanholsbeeck, R. Leonhardt, and J. D. Harvey, "Combined effect of Raman and parametric gain on single-pump parametric amplifiers," *Opt. Express*, vol. 15, p. 8104, 2007.

[15] P. Kylemark, H. Sunnerud, M. Karlsson, and P. A. Andrekson, "Semi-Analytic saturation theory of fiber optical parametric amplifiers," *J. Lightwave Technol.* vol. 24, p. 3471, 2006.

INDEX

D

G

H

I

J

K

L

M

Q

R

S

T

U

V

W

X

Y

Z